Nanoparticle–Protein Corona
Biophysics to Biology

Issues in Toxicology

Editor-in-chief:
Diana Anderson, *University of Bradford, UK*

Series editors:
Michael D. Waters, *Michael Waters Consulting, USA*
Timothy C. Marrs, *Edentox Associates, UK*

Editorial advisor:
Alok Dhawan, *CSIR-Indian Institute of Toxicology Research, Lucknow, India*

Titles in the Series:
1: Hair in Toxicology: An Important Bio-monitor
2: Male-mediated Developmental Toxicity
3: Cytochrome P450: Role in the Metabolism and Toxicity of Drugs and other Xenobiotics
4: Bile Acids: Toxicology and Bioactivity
5: The Comet Assay in Toxicology
6: Silver in Healthcare
7: *In Silico* Toxicology: Principles and Applications
8: Environmental Cardiology
9: Biomarkers and Human Biomonitoring, Volume 1: Ongoing Programs and Exposures
10: Biomarkers and Human Biomonitoring, Volume 2: Selected Biomarkers of Current Interest
11: Hormone-disruptive Chemical Contaminants in Food
12: Mammalian Toxicology of Insecticides
13: The Cellular Response to the Genotoxic Insult: The Question of Threshold for Genotoxic Carcinogens
14: Toxicological Effects of Veterinary Medicinal Products in Humans: Volume 1
15: Toxicological Effects of Veterinary Medicinal Products in Humans: Volume 2

16: Aging and Vulnerability to Environmental Chemicals: Age-related Disorders and their Origins in Environmental Exposures
17: Chemical Toxicity Prediction: Category Formation and Read-across
18: The Carcinogenicity of Metals: Human Risk Through Occupational and Environmental Exposure
19: Reducing, Refining and Replacing the Use of Animals in Toxicity Testing
20: Advances in Dermatological Sciences
21: Metabolic Profiling: Disease and Xenobiotics
22: Manganese in Health and Disease
23: Toxicology, Survival and Health Hazards of Combustion Products
24: Masked Mycotoxins in Food: Formation, Occurrence and Toxicological Relevance
25: Aerobiology: The Toxicology of Airborne Pathogens and Toxins
26: Chemical Warfare Toxicology, Volume 1: Fundamental Aspects
27: Chemical Warfare Toxicology, Volume 2: Management of Poisoning
28: Toxicogenomics in Predictive Carcinogenicity
29: Human Stem Cell Toxicology
30: The Comet Assay in Toxicology, 2nd edition
31: Computational Systems Pharmacology and Toxicology
32: Ecotoxicology and Genotoxicology: Non-traditional Terrestrial Models
33: Ecotoxicology and Genotoxicology: Non-traditional Aquatic Models
34: Oxidative Stress and Redox Signalling in Parkinson's Disease
35: Nanotoxicology: Experimental and Computational Perspectives
36: Regulatory Toxicology in the European Union
37: Formaldehyde: Exposure, Toxicity and Health Effects
38: Chemical Health Threats: Assessing and Alerting
39: The Micronucleus Assay in Toxicology
40: Nanoparticle–Protein Corona: Biophysics to Biology

How to obtain future titles on publication:
A standing order plan is available for this series. A standing order will bring delivery of each new volume immediately on publication.

For further information please contact:
Book Sales Department, Royal Society of Chemistry, Thomas Graham House, Science Park, Milton Road, Cambridge, CB4 0WF, UK
Telephone: +44 (0)1223 420066, Fax: +44 (0)1223 420247
Email: booksales@rsc.org
Visit our website at www.rsc.org/books

Nanoparticle–Protein Corona
Biophysics to Biology

Edited by

Ashutosh Kumar
Ahmedabad Univeristy, India
Email: ashutosh.kumar@ahduni.edu.in

and

Alok Dhawan
CSIR-Indian Institute of Toxicology Research, India
Email: alokdhawan@iitr.res.in

ROYAL SOCIETY OF CHEMISTRY

Issues in Toxicology No. 40

Print ISBN: 978-1-78801-391-8
PDF ISBN: 978-1-78801-630-8
EPUB ISBN: 978-1-78801-853-1
Print ISSN: 1757-7179
Electronic ISSN: 1757-7187

A catalogue record for this book is available from the British Library

The Royal Society of Chemistry is a charity, registered in England and Wales, Number 207890, and a company incorporated in England by Royal Charter (Registered No. RC000524), registered office: Burlington House, Piccadilly, London W1J 0BA, UK, Telephone: +44 (0) 20 7437 8656.

For further information see our web site at www.rsc.org

Printed in the United Kingdom by CPI Group (UK) Ltd, Croydon, CR0 4YY, UK

Preface

Nanoparticles have numerous biomedical applications such as in drug delivery, bone implants and imaging. It is now well understood that a material is always covered by proteins immediately upon contact with a physiological environment. This phenomenon is key to understanding the bio–nano interface. The reactivity of nanomaterials with biological fluids is predominantly affected by various interactions between the surface of nanoparticles and the medium used for their dispersion. These interactions evolve over time and have led to the emergence of the field of protein–nanoparticle complexes.

The formation of a nanoparticle–protein corona changes the behavior of the nanoparticle, resulting in new biological characteristics and influencing the circulation lifetime, accumulation, toxicity, cellular uptake and agglomeration. The formation of the nanoparticle–protein corona has fueled the issues related to their transport and fate in the environment, animals and humans.

This book, *Nanoparticle–Protein Corona: Biophysics to Biology*, covers various aspects of the biological significance of the nanoparticle–protein corona. This phenomenon opens new avenues to study the protein–protein interaction at a molecular level. This book delves into the effect of the nanoparticle–protein corona in cellular assays, methods to characterize these coronas, computational approaches to understand these interactions, and their applications and challenges. The book will provide great insights for biomedical and material scientists, toxicologists, pharmaceutical chemists, and interdisciplinary scientists.

This book is a culmination of the long-term scientific association and collaboration between the editors and authors.

Ashutosh Kumar
Alok Dhawan

Issues in Toxicology No. 40
Nanoparticle–Protein Corona: Biophysics to Biology
Edited by Ashutosh Kumar and Alok Dhawan

Published by the Royal Society of Chemistry, www.rsc.org

Contents

Issues in Toxicology No. 40
Nanoparticle–Protein Corona: Biophysics to Biology
Edited by Ashutosh Kumar and Alok Dhawan

Published by the Royal Society of Chemistry, www.rsc.org

CHAPTER 1

Nanoparticle–Protein Corona Complex: Composition, Kinetics, Physico–Chemical Characterization, and Impact on Biomedical Applications

JUHI SHAH AND SANJAY SINGH*

Division of Biological and Life Sciences, School of Arts and Sciences, Ahmedabad University, Central Campus, Navrangpura, Ahmedabad 380009, Gujarat, India
*E-mail: sanjay.singh@ahduni.edu.in

1.1 Introduction

Due to recent advancements in the area of materials science, it is possible to prepare nanomaterials with precise shapes, sizes, compositions and surface modifications with desired ligands. The use of such fascinating materials has offered significant advances in the development of nanomedicines such as therapeutics and diagnostics.[1] To achieve the biomedical applications possible, these nanomaterials (NMs) are required to be administered to the human/animal body, which leads to the exposure of these materials to the complex composition of biological fluids including a variety of salts, nutrients, and proteins. Interactions between the constituents of biological fluids

Issues in Toxicology No. 40
Nanoparticle–Protein Corona: Biophysics to Biology
Edited by Ashutosh Kumar and Alok Dhawan

Published by the Royal Society of Chemistry, www.rsc.org

and NMs may alter the physiochemical characteristics of the NMs. Therefore, it is essential to understand the interactions and predict the fate of NMs intended for the use of drug delivery and therapeutic applications for alteration in their systemic circulation, bio-distribution in different organs, and bio-availability at the targeted site. Although there could be several constituents affecting the properties of NMs, in this chapter we will only focus on the interactions of NMs with proteins dispersed in the complex biological fluids and their effects on the potential applications. It is well documented that dispersion of NMs in biological fluids facilitates the formation of a NM–protein complex, termed as a "protein corona". The protein corona (PC) consists of various proteins adsorbed on the surface of NMs. The composition of the PC is governed by the size, shape, surface charge, composition and surface capping molecules, therefore, potentially each nanoparticle with a unique set of PC constituents represents a new biological identity.

The PC established at a nanoparticle (NP)–protein boundary can be broadly divided into two parts: the hard corona and soft corona. These forms of PC can be differentiated by considering the binding strength and rate of exchange of proteins from the surface of NMs.[2] Proteins constituting a hard corona are reported to have a high binding affinity with the NM's surface and, therefore, their rate of exchange is extremely slow but faster than the time needed for the internalization of NPs.[2] Such an adsorption behaviour of protein has been ascribed as irreversible binding.[3] Contrary to this, the soft corona consists of proteins that are loosely attached to the NP surface or sometimes associated with an already-formed hard PC by a weak protein–protein interaction.[4,5] Proteins in a soft corona display a rapid exchange of molecules from the surface of the NP and are, therefore, substituted easily in a biological environment of dispersed proteins. It has been reported that in a typical PC, the hard corona proteins are arranged in the inner layer of the corona whereas the soft corona proteins distribute them in the upper layer of the corona.[6] In a soft corona, outer layer proteins exhibit a high exchange rate due to the abundance of proteins in a biological fluid as well as due to their direct contact.[7] The binding affinity between proteins and the NP surface dictates the stability of the PC, where the soft corona is considered less stable than the extremely stable hard corona.[6] Formation of a hard or soft PC depends on the size, surface charge, morphology, and association/dissociation rate of proteins from the NPs.[8,9]

Adsorption and desorption of a protein/biomolecule to the surface of NPs is a dynamic process.[9–12] The association and dissociation rates of proteins can be represented by the factors k_{on} and k_{off}. The balance between k_{on} and k_{off} provides some important clues about the affinity of a protein/biomolecule for a NP, which is commonly referred to as "dissociation constant" (K_d).[9–12] The value of k_{on} is dependent on the contact frequency of the protein/biomolecule with NPs.[10–13] It has been observed that high concentrations of protein/biomolecules will diffuse quickly and thus display high k_{on} values. However, the value of k_{off} could be influenced by the binding energy of the protein–NP complex. Proteins with higher binding energies show lower k_{off} values.

In complex situations, proteins compete for binding sites available on NP surfaces through adsorption, which depends on their distinctive individual K_d values.[9–12] In general, K_d values calculated for the adsorption of certain proteins in isolation to NPs are reported to be in the range of 10^{-4} and 10^{-9} M.[10,14] This interaction could correlate well with antibody–antigen interactions. Recently, Tenzer *et al.* reported that the composition of the corona is mainly altered quantitatively but not qualitatively over time.[15] Older models suggest that the highly dynamic PC composition changes over time owing to constant protein association and dissociation events and is controlled by the Vroman effect.[8,16,17] However, new models assume that pristine NPs can exist only for a short time in a complex environment; however, the developed PC is arranged in the form of multiple core–shell structures leading to a 'Christmas tree-like' structure.[18]

1.2 Physico–Chemical Parameters of NPs Controlling Protein Corona Formation

Although it is well known that as soon as any NP is dispersed in a protein-rich solution, it undergoes spontaneous protein corona formation; however, certain physico–chemical aspects of NPs control the composition of the PC.

1.2.1 Protein Corona Composition Varies with NP Type

Considering the affinity of certain materials towards the particular functional group present in proteins, the composition of PC can be different for different NMs. Using commercially available NMs such as ZnO, SiO_2, and TiO_2, Deng *et al.* studied the binding efficacy of proteins present in human blood plasma. Considering the same surface charge, they found that SiO_2 NPs adsorbed a distinct group of proteins (*e.g.*, alpha-2-acid glycoprotein, apolipoprotein D, clusterin), which are significantly different from those adsorbed by ZnO NPs (haptoglobin-alpha, Ig heavy chain alpha, transferrin).[14] It has also been suggested that the library of proteins making PCs is not only controlled by protein–NP interactions but also protein–protein interactions. An experiment was performed using SiO_2 NPs interacting with the top 10 proteins adsorbed on the SiO_2 surface as well as the most abundant protein in human blood plasma, albumin. The results revealed that apart from these 10 proteins, other proteins, such as complement proteins, fibrinogen, and immunoglobulins (Ig), are also found in the PC, which significantly affects the opsonisation, circulation, and targeting efficiency of NMs.[4,19]

1.2.2 Effect of NP Size Over Protein Corona Formation

It has been reported that the dimensions of NPs significantly control the PC composition.[20] Among the different dimensions, the surface curvature of NPs is considered as an essential parameter, which facilitates the adsorption and change in conformation of proteins and thus the composition of PCs.[4,21] It has been reported that since the surface curvature of NMs is higher than their bulk counterparts, the binding affinities of protein molecules remain different for materials of a similar composition.[22] Further, protein–protein interactions are decreased at highly curved surfaces resulting in a diverse composition of PCs. It has been shown that NMs with high surface curvatures induce a lower alteration of corona composition than proteins binding onto the flat surfaces of materials with a similar composition.[23] Tenzer *et al.* studied PC formation on SiO_2 NPs of different sizes when dispersed in blood plasma.[15] They observed that the size of the SiO_2 NPs significantly affects the binding of ~37% of the identified proteins. It was found that a variation of 10 nm in particle size leads to a significant change in the composition of PCs. Similarly, clustsacaerin lipoproteins were found to have a better affinity towards smaller SiO_2 NPs whereas prothrombin or the actin regulatory protein (gelsolin) showed absorbance on larger SiO_2 NPs. Authors also report that the adsorption of some proteins such as IgG or actin was not associated with NP size. Dobrovolskaia *et al.* observed that gold NPs (AuNPs) of 30 and 50 nm incubated in blood plasma resulted in the adsorption of more proteins on the smaller sized particles than larger AuNPs.[24] Additionally, Piella *et al.* reported that the thickness and density of PCs is predominantly dependent on the size of NPs. The diverse temporal patterns for the development of protein coatings are generally faster for smaller particles than larger particles.[25]

1.2.3 Effect of NP Shape on Protein Corona Formation

NP shape is another important parameter, which controls composition and overall PC formation. In this concern, García-Álvarez *et al.* have reported that 70 nm gold nanostars (AuNSs) showed a high concentration of absorbance of proteins due to the greater surface area of AuNSs than that of gold nanorods (AuNRs).[26] The shape of AuNPs has an immense influence on their interactions with cell layers; in particular, spherical AuNPs have a better adsorption tendency with mammalian cells than rod-shaped AuNPs.[27] Deng *et al.* studied the influence of titanium dioxide NPs' shape on protein binding and found that clusterin and apolipoprotein D have a binding affinity for spherical NPs but not for nanorods or nanotubes.[14] Additionally, Ma *et al.* have revealed the adsorption kinetics of human serum albumin (HSA), γ-globulin, fibrinogen (HSF) proteins, and RBC hemolysis when exposed to two different spherical and rod-shaped mesoporous silica NPs (MSNPs). They observed that the initial adsorption rate for HSF was high for rod-shaped MSNPs.[28]

1.2.4 Effect of Hydrophobicity and Hydrophilicity on Protein Corona Formation

Hydrophobic interactions and hydrophilic repulsion are the two major phenomena of aqueous media and they, therefore, significantly influence the adsorption and desorption of proteins as well as attenuate these processes over a NP surface. Roach *et al.* reported that NPs bearing hydrophobic surfaces demonstrate better protein adsorption than NPs with hydrophilic surfaces. They observed that hydrophobic surface NPs are frequently involved in the denaturation of surface-adsorbed proteins and also cause a conformational change in the native structures of proteins.[29] Similarly, Cedervall *et al.* and Gessener *et al.* have shown that during the development of the PC, hydrophobic NPs display a higher binding affinity for apolipoproteins; however, hydrophilic NPs favour the adsorption of fibrinogen, IgG and albumin.[12,30] Cedervall *et al.* have further demonstrated that there is a direct correlation between an increase in protein stoichiometry and an increase in hydrophobicity. The residence time of albumin on hydrophobic particles was found to be shorter than on hydrophilic particles.[7] Additionally, Dominguez-Medina *et al.* have demonstrated that citrate-capped AuNPs dispersed in aqueous solutions undergo aggregation followed by precipitation due to the interaction of NPs with oppositely charged ions in solutions of similar pH and ionic strength to human blood plasma.[31] Colloidal stability was improved by the association of BSA molecules on the surface of negatively charged (citrate coated) AuNPs. Hydrophobic interactions between BSA molecules present on the AuNP surface could impart colloidal stability after corona formation. Hulander *et al.* studied the capability of pre-adsorbed IgG on AuNPs to elicit the immune complement (IC) after modification in surface hydrophobicity.[32] They found that surface-bound hydrophilic AuNPs diminished the ability of IgG to activate the IC. In contrast, an increase in surface hydrophobicity favours the activation of IC by IgG.

1.2.5 Effect of a NP's Surface Charge on Protein Corona Formation

The charge present on the surface of a NP is an essential feature for the determination of PC composition and formation. The interaction between the charges of the NP and protein molecules determines the adsorption kinetics and composition of a PC. For example, opsonins can promptly identify the NPs with a positive surface charge, which leads to recognition of the NPs by the reticuloendothelial system (RES) and subsequent deposition in organs such as the liver and spleen.[23,33] Considering this, it has been suggested that opsonization can be avoided by coating the NP's surface with negatively charged groups, which exhibits a negative zeta potential varying from −30 to −50 mV in a biological environment. Ehrenberg *et al.* have shown that NPs dispersed in a physiological environment lead to the adsorption of proteins, which can induce a remarkable drop in surface charge. This observation suggests that there is a direct correlation between the nature of the PC and the colloidal stability of the developed NP–protein complex.[34] In a further study, Lynch *et al.* have reported that AuNPs bearing positive, negative and neutral charges could induce protein denaturation to a different extent. Charged AuNPs (positive or negative) were found to cause a higher protein denaturation than a neutral charge thus allowing the proteins to retain their native structure.[35] Similarly, Gessner *et al.* studied the surface charge density effect of polymeric NPs bearing a negative charge and observed that as the surface charge density increases, the plasma protein adsorption also increases.[36] Moreover, it has been revealed that for polystyrene NPs, proteins with an isoelectric point of <5.5 are likely to bind on the positively charged surface, whereas proteins with an isoelectric point >5.5 prefer negatively charged NPs.

1.2.6 Effect of Temperature on Protein Corona Formation Over NPs

Although the average internal temperature of the human body is considered as 37.0 °C (98.6 °F), it differs from person to person, based on various factors such as body type, gender, and physical activities. Additionally, a person cannot have a constant body temperature throughout the day. However, following a person's circadian rhythm, the cycle of the temperature of the body varies throughout the day and night. It is reported that exposure of parts of the body to cold surroundings could lead to a drop in body temperature to 28 °C.[37] It has also been reported that the temperature of the intracellular parts of living cells is not similar. The temperature of the female body is little higher compared to the male body, which varies further according to the female hormone cycle.[38] Physical activity can also lead to an enhanced body temperature by up to ~2 °C and the body temperature decreases during sleep. In general, the body temperature varies from 35 to 39 °C; however, during a fever, the body temperature can increase to 41 °C.[39] Therefore, it is also expected that PC formation may also be greatly influenced by varying the temperature of the surrounding environment. In this context, Mahmoudi *et al.* have studied the influence of alteration in temperature on the PCs formation and its concomitant composition.[40] Here, dextran-coated FeOx NPs with different surface charges were incubated with FBS at various temperatures and the authors found that it was not only the composition of the PC that was significantly affected but also that the pattern of cellular uptake was altered. Mahmoudi *et al.* reported that the composition of the PC on AuNRs can be influenced by the plasmonic heat induction.[41] Here, cetyltrimethylammonium bromide (CTAB)-coated AuNRs were incubated with different FBS concentrations and a change in PC formation was monitored prior to and after plasmonic heating by constant exposure to a laser. The alteration of the PCs composition on AuNRs was found to be dependent upon the type of heating applied such as plasmonic heating or conventional heating.

1.2.7 Effect of pH on Protein Corona Formation

Altering the pH can also be regarded as one of the key factors in assessing the formation of a NP–protein complex. Variation in environmental pH may vary protein binding affinity,[42] which can induce significant modification of the adsorbed protein pattern. During cellular uptake, NPs are exposed to biological fluids with varying pH, such as blood (pH 7.4), dispersion media (pH 6.9–7.4), intracellular fluid (pH 6.8), and lysosomes (pH 4.5–5.0). Once NPs are administered, they tend to enter the blood circulation and are eventually deposited in different organs.[43] It is well known that cancer cells/tissues generate an acidic microenvironment, thus represent specific types of groups of proteins that can alter the PCs around NPs. These events could lead to the altered property of NPs, which is necessary for the therapeutic application and biological availability of NPs. Understanding the effect of pH on the formation of a NP-PC complex decides the fate of NPs in a biological environment present in the body.[43] When isoelectric points and pH values are very close to each other, the NPs' stability is decreased due to the significant effect of pH on PC formation and its correlation with the surface charges of NPs. Considering the anionic nature of natural organic matter (NOM) and prokaryotic or eukaryotic cells,[44] it was observed that their interaction with positively charged NPs is predominantly electrostatic in nature. Conversely, Chen *et al.* have reported that the binding of bacterial exopolysaccharides on a AgNP surface or humic acid on TiO_2 is significantly reduced when the pH exceeds the IEP.[45] Such examples indicate that the electrostatic interaction between NPs and other biomolecules during PC formation is not the sole mechanism; rather, ligand exchange or hydrophobic interactions are also essential for NPs, for, *e.g.*, fullerenes adsorbing NOMs by hydrophobic interactions.[46,47]

1.2.8 Effect of Surface Modification of NPs and Protein Corona Formation

Strategies are required to prevent undesirable proteins being absorbed on NP surfaces that results in PC formation. Such events may negatively affect the targeting moiety of NMs that are surface modified for selective binding to tissues or organs. A method to modify the NP surface by incorporation with certain functional groups prevents NPs being caught by immune cells. Coating NP surfaces with polymers, such as polyethylene glycol (PEG), reduces the binding of proteins from biological fluids and also prevents recognition by the RES. By preventing the formation of the PC, significantly quicker transport of NPs across the endothelium can be obtained. Further, PEGylation with a controlled density of PEG on the NMs' surface makes them suitable for longer circulation in the bloodstream. Gref *et al.*, Perry *et al.* and Engin *et al.* showed that silicon can be used as a coating material exhibiting a similar-favourable impact on PC formation.[48–50] Polystyrene coated with poloxam ine 908 indicated a drop in fibronectin adsorption compared to uncoated nanospheres.[51] Ehrenberg *et al.* demonstrated the interaction of polystyrene NPs with various functional groups on the endothelium cells. They found that the capability of the NPs' surface to adsorb protein is an indicator of their tendency to interact with cells and the bound proteins do not affect cell–NP association.[34] Aggarwal *et al.* have shown that NPs coated with different molecules, such as PEG, poloxamer, poloxamine, dextran, pluronic F127, polysorbate and poly(oxyethylene) *etc.*, exhibited improved NP–protein interactions, protein conjugation and NP distribution in a biological system.[52] Dutta *et al.* reported that amorphous SiO_2 NPs and single-walled carbon nanotubes coated with pluronic F127 result in a better dispersion of NPs and also a considerable decrease in the adsorption of serum proteins.[53] Contrary to the aforementioned parameters, Muller *et al.* have reported the interactions of polymeric NPs and other inorganic NPs with plasma proteins obtained from different sources including human, rabbit, sheep and mouse.[54] The data obtained from liquid chromatography–mass spectroscopy (LC-MS) (Figure 1.1) shows that the composition of the PC differed when the NPs were dispersed in plasma proteins from the different sources.[54] Thus, it can be concluded that it is imperative to investigate the fate of NPs dispersed under *in vitro* and *in vivo* experimental systems by monitoring their interactions with plasma proteins from the corresponding cell culture or animal model. Additionally, reports suggest that the nature of the protein ultimately decides the fate of NPs in a particular organism.

1.3 Composition of the Protein Corona

Human blood plasma is the most abundant biological reservoir for PC formation. A layer of protein that covers a NPs surface influences its cellular uptake as well as biodistribution. In this context, Monopoli *et al.* have shown that fibrinogen, IgG, and complement factors lead to stimulation of phagocytosis and eventually elimination of NPs from the bloodstream; however, HSA and apolipoproteins extend the NPs circulation time in the blood.[55] Plasma proteins, such as fibrinogen, IgG, albumin, and apolipoproteins have a tendency for rapid adsorption onto metal-based NPs during the early phase of PC formation.[4,56] These proteins are observed in the hard coronas of all the studied NPs and are substituted by apolipoproteins and coagulation factors during the slow phase of corona formation. Walkey *et al.* introduced the term "adsorbome" to represent the group of 125 of the most prevailing proteins present in plasma.[17] Adsorbome is involved in a variety of physiological processes including lipid and ion transportation, activation of the complement system, recognition of pathogens, and blood clotting. Metallic NPs favour the adsorption of albumin, the most abundant plasma protein, during PC formation. At pH 7.4, albumin has a net negative charge due to ~60 positively charged amino acid lysine molecules, which facilitates the interaction of albumin with NPs (positive and negatively charged NPs). Irrespective of the net charge on the NP surface, albumins generally produce anionic PC complexes with NPs.[57] Additionally, Fleischer *et al.* have reported that the configuration of albumin protein could be altered by cationic NPs; however, anionic NPs did not show this effect.[57] An alteration in protein structure affects the interactions of the albumin–NP complex at a cellular level and, therefore, cell receptors bind to the PC on anionic NPs; however, albumin–NP complexes formed from

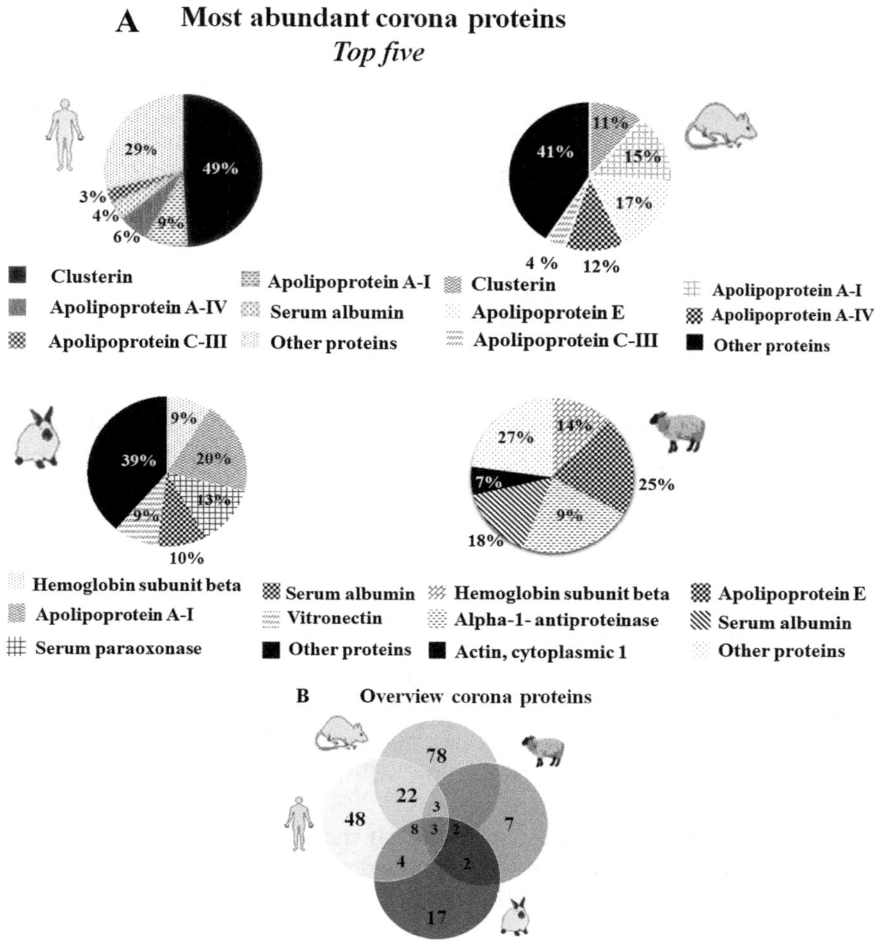

Figure 1.1 (A) Composition of the protein corona from different animal sources as determined by LC-MS exemplarily shown for un-functionalized PS-NPs. The five most abundant corona proteins based on all identified proteins are summarized. (B) Venn diagram showing common proteins from the list of all identified proteins per animal source for PS and PS-NH_2. Not shown are the number of common proteins from human and sheep (0) and mouse and rabbit (6). Reproduced from ref. 54 with permission from American Chemical Society, Copyright 2018.

cationic NPs are redirected to scavenger receptors. The adsorption of albumin was much better on the surface of anionic NPs compared to cationic NPs, leading to ~23% and 8% surface coverage, respectively. Albumin and fibrinogen are some of the most abundant proteins that constitute the PC composition on a variety of NPs. However, some proteins, such as apolipoproteins, show specific binding with other nanostructures, exhibiting a better adsorption affinity for liposomes and polymeric NPs, but a lesser affinity for metallic

NPs. A key factor required for the binding of proteins to NPs is hydrophobicity.[30] Therefore, polymeric and hydrophobic NPs facilitate the adsorption of proteins such as transferrin, haptoglobin, fetuin A, kininogen, histidine-rich glycoprotein, and intrinsic clotting pathway factors; however, most of these proteins can also show affinity for metal-based NPs[30] (Table 1.1).

Sempf *et al.* have studied hard corona formation on the surface of uncoated poly-lactic-*co*-glycolic acid (PLGA) NPs incubated in human plasma and observed that 15 proteins were localized in the hard corona, consisting of seven that are not found in abundance in plasma. Apolipoprotein E, vitronectin, histidine-rich glycoprotein and kininogen-1 were the proteins from plasma adsorbed on the PLGA NPs and thus provided the best signal.[61] Further, Darabi Sahneh *et al.* also developed an experimental model explaining the two phase dynamics of a PC complex. They suggested the use of two formulae that provide information about the metastable and stable states of a PC. Mathematically, they could predict that the PC is independent of simulation by adding suitable values of parameters as well as understand the NPs' interaction with a physiological environment, applicable to toxicology and nano-based drug delivery applications.[62]

1.4 Kinetics of Protein Corona Formation

During the beginning of PC formation, the most abundant proteins but those with less affinity for the NP surface are the first to be adsorbed and develop a layer that is recognized as the soft corona. Subsequently, high affinity proteins tend to replace the low affinity proteins from the medium and thus develop a hard corona layer[7,12,33,55] In the case of the hard corona, a direct interaction of proteins occurs with the NP surface; however, in the soft corona, proteins interact with the hard corona proteins through weak protein–protein interactions. The kinetics of the formation of the soft and hard PCs differs significantly. The development of the hard corona is very rapid and occurs within seconds or minutes, whereas a longer duration (hours or even days) is required for the development of the soft corona, as higher affinity proteins are being substituted by lower affinity proteins. This procedure relies on the concentration of protein as well as the biological medium constitution.[17] Several evidences indicate that even if plasma concentration is low, corona proteins can totally mask the NP surface and thus alter its behaviour and physico–chemical characteristics.[55] Interaction of soft corona proteins with hard corona proteins is possible as they desorb from the NP surface, which provides an opportunity for the interaction of other biomolecules from the medium. The aforementioned exchange is attributed to the competitive adsorption and desorption of proteins, which can be influenced by several parameters such as interaction period, protein concentration, and adsorption efficiency of proteins with the NP. This exchange could be explained by the "Vroman effect"[52,63] and is comprised of two phases.

Table 1.1 A list of common plasma proteins found in PCs developed on different types of NPs.

Type of NP	Proteins identified in corona	References
AuNPs	Albumin, fibrinogen chains, apolipoprotein A1, transport proteins, coagulation factors	52, 58 and 59
SiO_2 NPs	Ig, lipoproteins, complement proteins, coagulation proteins, acute phase proteins, cell proteins, serum proteins	52, 55 and 59
Magnetic NPs	Albumin, apolipoprotein A1, complement factors, vitronectin, hemoglobin	59
Polystyrene NPs	Coagulation factors, Ig, lipoproteins, acute phase proteins, complement proteins, plasminogen, anti-CD4, C4a, albumin	52, 58 and 59
Latex NPs	Albumin, apolipoproteins, Ig, hemoglobin, haptoglobins	52 and 59
Copolymer NPs	Albumin, apolipoproteins, fibrinogen, Ig, C4BP-α-chain	59
Supraparamagnetic iron oxide NPs	Albumin, α-1-antitrypsin, fibrinogen chains, immunoglobulin chains, transferrin, transthyretin	56, 58 and 59
Carbon nanotubes	Fibrinogen chains, immunoglobulin light chains, fibrin, albumin, ApoA1, component proteins, fibronectin	52 and 58
TiO_2 NPs, ZnO NPs, SiO_2 NPs	Complement proteins, immunoglobulin light chains, fibrin, albumin, fibronectin albumin, Ig, fibrinogen, transferrin, apolipoprotein A1	52 and 59
Citrate-coated AgNPs	Albumin, α-1-antiproteinase, α-2-HS-glycoprotein, apolipoprotein A1, serotransferrin, α-macroglobulin, α-fetoprotein, apolipoprotein B100, α-2-antiplasmin, complement C3, β-2-glycoprotein 1, fetuin-B, inter-α-trypsin inhibitor heavy chain H1, hemoglobin foetal subunit β, inter-α-trypsin inhibitor heavy chain H3, inter-α-trypsin inhibitor heavy chain H2, hemoglobin subunit α, complement factor B, hemopexin, serpin A3-6	60
AgNPs coated with polyvinylpyrrolidone	Albumin, α-2-HS-glycoprotein, α-1-antiproteinase, apolipoprotein A1, serotransferrin, α-2-macroglobulin, α-fetoprotein, apolipoprotein B100, complement C3, α-2-antiplasmin, inter-α-trypsin inhibitor heavy chain H1, fetuin-B, β-2-glycoprotein 1, hemoglobin foetal subunit beta, inter-α-trypsin inhibitor heavy chain H3, inter-α-trypsin inhibitor heavy chain H2, vitamin D-binding protein, transthyretin, hemoglobin subunit α, complement factor B	60

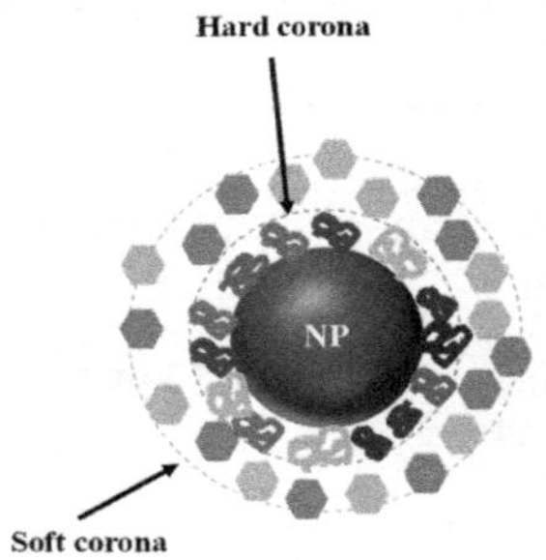

Property	Hard Corona	Soft Corona
Timescale	Hours	Seconds to minutes
Protein affinity	Low	High
Protein binding strength	Protein weekly bound	Protein strongly bound
Interaction point	Protein to protein	Proteins to NP
Type of interaction	Protein to protein	Van der waals
Dissociation constant	High	Low
Conformation changes	Low	High
Proteins	Apolipoproteins	Albumin, apolipoproteins, IgG

Figure 1.2 Schematic representation showing the formation and properties of the two types of protein coronas: hard (inner layer) corona and soft (outer layer) corona.

During the first phase, proteins with the greater association rates adsorb quickly while during the second phase, proteins with short residence times are replaced by proteins with slower association rates but longer residence times.[64] Figure 1.2 represents the typical characteristics of a PC formed over the surface of a NP. Jansch *et al.* investigated the Vroman effect on ultrasmall superparamagnetic iron oxide (USPIO) NPs and revealed that the Vroman effect on USPIO cannot be measured and the exchange of high affinity proteins with abundant proteins on USPIO was also not detected. Furthermore, as the incubation time increased, the amount of adsorbed fibrinogen and Ig over the NP surface was enhanced while the relative quantity of key proteins, such as apolipoproteins, fibrinogen, and albumin, remained constant.[65] Winzen *et al.* studied a complementary investigation of the hard and soft corona using human plasma and individual proteins considering HSA and apolipoprotein A-I (ApoA-I). They observed that there was no significant alteration in hard corona formation. Though, the soft corona acquired some extra proteins, which were due to the differential binding affinity and other physiological factors of proteins.[2]

1.5 Common Techniques to Characterize Protein Coronas on NPs

Characterization of PCs is essential for understanding the mechanism of formation, kinetics, and their functional aspects. The common techniques required for the detailed characterization of PCs include UV-visible spectroscopy, Fourier-transform infrared spectroscopy, dynamic light scattering, zeta potential, circular dichroism (CD) spectroscopy, and isothermal titration calorimetry (ITC). These techniques are required to confirm the layer of protein present on the NP surface and are considered as primary characterizations. Further, the complete characterization of the PC can be performed by using advanced techniques such as electrophoresis, chromatography, mass spectrometry, and nuclear magnetic resonance, *etc.* In this chapter, we will only discuss the primary characterization techniques for PCs; more detailed characterizations can be found in Mahmoudi *et al.*'s work.[23]

1.5.1 UV-visible Spectroscopy

Proteins absorb at a wavelength of 280 nm from UV-visible spectroscopy light.[66] Absorption of the UV light by proteins generally occurs due to the presence of aromatic amino acids, *i.e.*, phenylalanine, tyrosine and tryptophan. Binding of NPs to proteins changes the absorption spectra of NPs as well as the proteins, which is due to the electron transfer capabilities of the proteins with their optical properties.[13,67–70] Yadav *et al.*[68] utilized UV-visible spectroscopy to reveal the adsorption isotherm of lysozyme and BSA on anionic silica NPs and reported that adsorption of lysozyme can be correlated with the increase in the concentration. However, adsorption of BSA did not follow any adsorption pattern. In another study, it was observed that attachment of azurin protein on the surface of AuNPs makes a red shift in the surface plasmon resonance (SPR) of NPs, which was attributed to the energy transfer from azurin to the NPs' surface. Casals *et al.* reported that the SPR band of 10 nm AuNPs showed a red shift from 518 nm to 526 nm when incubated with a cell culture medium.[8] They reported that using UV-visible adsorption spectra, one can monitor the NPs' protein interaction by following the absorbance pattern of SPR. However, characterization by UV-visible spectroscopy alone is not enough because it is not very sensitive and, therefore, further characterization of protein-NP complexes is needed. Recently, we reported the formation of a PC on NPs and the observed changes were followed by different characterization methods including UV-visible spectrophotometry and transmission electron microscopy (TEM). In one of our studies, the typical absorbance of CeNPs (4+) was followed after dispersing them in MilliQ water and a DMEM cell culture medium (supplemented with 10% FBS). The results showed that CeNPs (4+) incubated in DMEM indicated two absorbance peaks at 280 and ~292 nm, which were ascribed to the presence of protein (from the cell culture medium) and CeNPs (4+), respectively.[71] Additionally, Vidic *et al.* have also reported PC formation over ZnMgO nanopowder when dispersed in a bacterial culture medium, luria bertani (LB), as well as the mammalian cell culture medium RPMI. The UV-visible absorption spectrum of ZnMgO dispersed in the LB medium displayed no optical features typical of ZnMgO nanopowder. The optical absorption recorded from the ZnMgO-LB suspension originates from components present in the LB medium such as amino acids, NaCl, and yeast extract. An intensity loss was observed in the spectrum after 24 hours, which indicates that the components were not present in free form; rather, they were attached to NPs. Similar to LB, NPs dispersed in the RPMI medium did not show any optical transitions characteristic of ZnMgO. After 24 hours of incubation of the ZnMgO NPs in complete RPMI medium, the absorbance intensity was reduced in the range of 200 and 300 nm except in the range of higher wavelengths. Thus, it can be concluded that this intensity enhancement is probably due to the existence of new electronic transitions originated after the formation of complexes between ZnMgO and components of the medium.[72]

1.5.2 Fourier-transform Infrared Spectroscopy

Fourier-transform infrared spectroscopy (FTIR) is a technique that basically works on the vibration energy of chemical bonds, which can be used to get information on proteins adsorbed on the surface of NPs. Kong *et al.* explained the use of FTIR for the identification of a protein's secondary structure.[73] They provided a detailed discussion on the prediction of protein structure on the basis of the vibrational energy of peptide bonds and suggested that a protein's secondary structure can be studied on the basis of infrared light absorption by amide bond I (1700–1600 cm^{-1}),[23] II (1540 cm^{-1})[74] and III. Out of all the amide bonds, vibrations in amide bond I have been reported to be the most crucial.[73] FTIR spectroscopy has been used for the evaluation of protein coatings on the surfaces of NPs and the difference between the vibrational energy of protein-coated NPs in comparison to uncoated the NPs has been reported by various researchers.[67,69,70,75] Ma *et al.* used FTIR to study the adsorption of HSA, globulin (HGG), and fibrinogen (HSF) over MSNPs. The pore size and morphology of MSNs influenced the adsorption of HAS and HSF while adsorption did not alter the conformation of HGG.[28] Additionally, Shang *et al.* have also demonstrated the changes observed from the conjugation of BSA with AuNPs at varying pH (3.8, 7, and 9) using the FTIR technique. They observed that bio-conjugation of BSA with AuNPs leads to a conformation change probably due to the considerable enhancement in sheet and turn structures in BSA and thus a decrease in the helical structure component.[76]

1.5.3 Dynamic Light Scattering and Zeta Potential

Dynamic light scattering (DLS) or photon correlation spectroscopy is used to study the dynamic behaviour of fluids on the basis of their Brownian motion.[77] Plenty of research has been done on the characterization of protein-NPs using DLS.[24,67,78,79] DLS can be used to study the hydrodynamic size distribution of NPs that can be calculated from the time depending on the scattering intensity. Similarly, Casals *et al.*[8] used DLS to measure the hard coronas of proteins on the surface of AuNPs. The zeta potential of any solution shows its stability on the basis of electrostatic repulsion between similarly charged particles and it varies due to variation in dispersions.[80] The zeta potential values offer information about the stability of the particle. An increase in zeta potential values is associated with greater electrostatic repulsion and, hence, enhanced colloidal stability. It is important to consider that the magnitude of a NP's surface charge relies on the solution pH. In this context, Patil *et al.* reported that a change in pH from acidic to alkaline induces a significant change in the zeta potential value of cerium oxide NPs.[80] It was found that with the incubation of cerium oxide with BSA, positively charged NPs exhibit a greater affinity for BSA adsorption compared to negatively charged NPs. It was also found that the cellular uptake of NPs varies with the surface charge of NMs. Zeta potential can be used to quantitatively measure the coating of proteins on the NPs' surface as with the growing protein coating, there will be a concomitant change in zeta potential.[8] Natte *et al.* investigated the adsorption of BSA on pristine silica NPs (SNPs) and silica polyethyleneglycol core–shell nanohybrids (SNPs@PEG) by measuring zeta potential.[81] Bare SNPs displayed a zeta potential of −21 mV, which remained the same when a lower BSA concentration was used to disperse the NPs. In contrast, a concentrated mixture of proteins lowered down the zeta potential up to −8 mV, similar to the zeta potential values of BSA dispersed in PBS. They found that the amount of BSA adsorption was dependent on the BSA concentration used as well as the incubation time with the NPs. Coating of PEG on the SNPs' surface was found to inhibit the short- and long-term formation of BSA PCs on SNPs, which was evident by the slight decrease in zeta potential due to the large polymer chain length of PEG.

1.5.4 Isothermal Titration Calorimetry

This is a physical technique, which provides information about the thermodynamic aspects of the interaction between small molecules (NPs) and biomolecules (protein, carbohydrates, DNA *etc.*) The thermodynamics of association between two molecules can be characterized by the association constant (K_a), enthalpy (ΔH_b), free energy (ΔG_b), and entropy (ΔS_b).[82] Many studies have utilized isothermal titration calorimetry (ITC) to determine the thermodynamic properties of protein–NP interactions.[83,84] In this technique, quantification of protein binding on the surface of NPs is achieved by titration of the protein in a solution of the NPs and the heat response is recorded simultaneously, which is then used to get a thermodynamic component measurement with the help of isothermal functions.[85] Huang *et al.*[83] used ITC to study the thermodynamic profile during the interaction of BSA with AuNPs, coated with different agents (negatively charged group, 11-mercapto-1-undecanesulfonate; a mixture of MUS and 1-octanethiol (OT); mixture of MUS and branched OT (3,7-dimethyloctane-1-thiol) (MUS/brOT) called MUS/OT). The ITC data obtained (Figure 1.3) suggested that the adsorption profile of BSA on MUS/OT was different from that of MPA (mercaptopropionic acid)/ OT in terms of its heat exchange profile. They observed that the complex of BSA with MUS/brOT was completely exothermic while MUS/OT was initially exothermic but later endothermic and consumed heat for the stabilization of the BSA-AuNP complex. This kind of variation in the thermodynamic profile could be due to a different kind of electrostatic interaction, which takes place because of the difference in the coating agent. Winzen *et al.* utilized ITC to study the affinity of the PCs around hydroxyethyl starch nanocapsules. They found that the presence of a large amount of HSA in the suspension showed a lower binding affinity to the NP surface, which can be attributed to soft

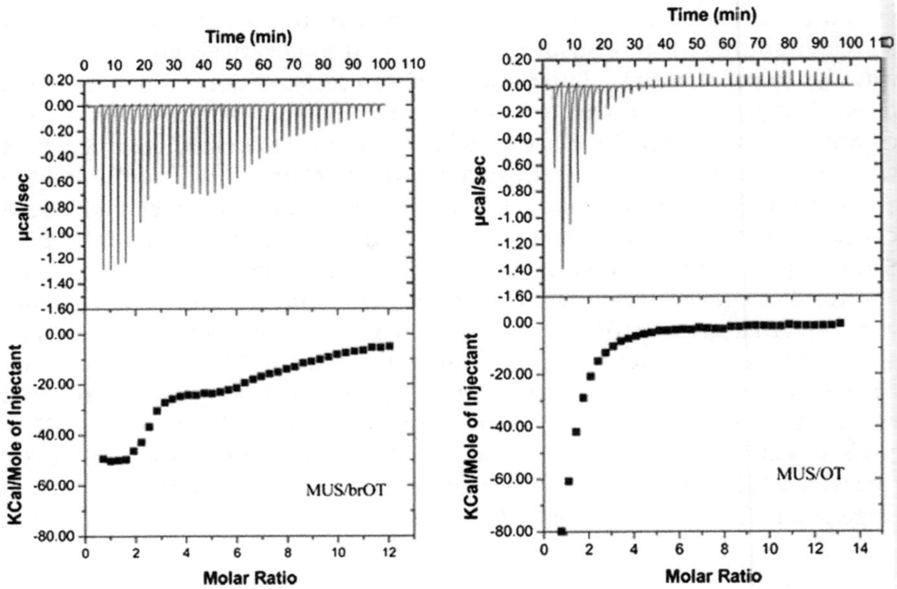

Figure 1.3 Isothermal calorimetric data from the titration of 500 M BSA into 8.4 M MUS/brOT and 7.7 M MUS/OT AuNP. Heat flow *versus* time during injection of proteins at 25 °C and heat evolved per mole of added proteins (corrected for the heat of protein dilution) against the molar ratio (protein–AuNP) for each injection are shown at the top and bottom, respectively. Reproduced from ref. 83 with permission from American Chemical Society, Copyright 2014.

corona formation, whereas the high amount of apolipoprotein A-I offers a greater binding ability due to the hard corona component.[2] Cedervall *et al.* used ITC techniques to examine the effect of different copolymer ratios and different rates of hydrophilicity/hydrophobicity of *N*-isopropylacrylamide (NIPAM)–*N*-*tert*-butylacrylamide (BAM) copolymer (NIPAM–BAM) NPs on the adsorption and desorption of HSA and fibrinogen.[12]

1.5.5 Circular Dichroism Spectroscopy

The structures of proteins vary after adsorption and binding to the surfaces of NPs. The proteins involved in the development of a hard corona generally express the secondary structure such as an α-helix, β-sheets or loops. The secondary structure of proteins in the PC can be obtained by circular dichroism (CD) spectroscopy.[86,87] CD spectroscopy relies on the principle of circularly polarized light that differentiates between the absorption of left- and right-handed polarized lights by proteins. Due to this, the light would have two dissimilar spin angular momentums, which can evaluate the secondary structures of the proteins.[88] Many reports are available that use CD spectroscopy for the determination of protein–NP interactions.[5,89–93] Ahmad *et al.*

have studied the absorption of BSA on the surface of $CoFe_2O_4$ NPs using CD spectroscopy. The results indicated two negative peaks at 208 and 222 nm, which correlates well with to the π–π* and n–π* transitions in the α-helix of BSA, respectively. These results further suggested that the interaction of BSA with $CoFe_2O_4$ NPs resulted in a decreased content of α-helix from 55.9% to 48.1% and an enhanced β-sheet content from 37.3% to 39.9%. The secondary structure of the protein has been modified, which suggests the aromatic amino acids are accountable for the binding of proteins to NPs.[93]

1.6 Applications of Protein Coronas

Considering the complexity with the retention of a protein's structure and function, conjugation of the desired protein with a NP becomes extremely challenging. Therefore, covalent or non-covalent binding of the required proteins to NPs leads to the development of novel functional hybrid systems, which offer myriad applications in biomedicines and catalysts. The following sub-topics will summarize the effective strategies developed utilizing protein–corona complex systems for biomedical applications.

1.6.1 Role of the Protein Corona Towards Targeted Drug Delivery

In order to develop an effective targeted drug delivery approach, it is expected that a longer circulation time of NPs in blood and accumulation of drug-loaded NPs at the target site are two of the core criteria. Conventional NPs' surfaces can be modified to increase their stability and prolong the circulation time of nanocarriers along with controlled bio-distribution.[94] A protein covering leads to an alteration of the NPs' hydrodynamic size, shape and, thus, the surface properties. Caracciolo *et al.* reported that after incubation of 3β-[*N*-(*N*′,*N*′-dimethylaminoethane)-carbamoyl]-cholesterol (DCChol)–dioleoylphosphatidylethanolamine DC-Chol-DOPE-DNA lipoplexes with a suspension containing FBS, the size of the NPs was altered from 244 ± 4 nm to 741 ± 1 nm. Therefore, the internalization mechanism of the NPs' entry into cells was also modified from clathrin-dependent to caveolae-mediated.[94] Such events can cause changes in the NPs' cellular internalization and co-localization based on the size. NPs can also be conjugated with different functional groups and moieties for selectively delivering drug(s) to the target site. Transferrin, insulin, folic acid, Apo A-1, and Apo E, *etc.*, are some of the commonly used targeting biomolecules to construct the drug delivery systems targeting cancer tissues. Additional small molecules such as anhydride, amine, hydroxyl, carboxyl, thiol, and epoxy could also be utilized to alter the NP delivery, as they facilitate association of NPs with endothelial cells, pancreatic cancer cells, and activated human macrophages.[95] Some apolipoproteins, such as Apo E, ApoA1, and ApoB-100, are also used to functionalize NPs to target the components of the nervous system.[96,97] The variety of functional groups present in the PC can be efficiently used for conjugation of the biomolecules mentioned with NPs. In a study, Zensi *et al.* have shown the covalent conjugation of HAS-coated NPs with ApoE, which was intravenously injected into SV 129 mice.[96] They sacrificed the mice after 15–30 mins of treatment and sections of different organs were analysed using TEM (transmission electron microscopy). Interestingly, NPs conjugated with ApoE were only found in the brain capillary endothelial cells and neurons while bare NPs were not taken up by the brain. Surface modifications of NPs can be achieved by the covalent conjugation[98] of NPs with the desired biomolecules. A famous illustration is NP albumin-bound (nab) technology, which consists of binding albumin to a NP's surface to impart effective tumor targeting with concomitantly reduced toxicity. Using this technology, an anticancer agent (abraxane) consisting of paclitaxel was designed. Figure 1.4 demonstrates a novel approach for NP decoration using different surface-modifying ligands specific to cells. Protein adsorption may affect the targeting ligands conjugated on the NP surface; however, several cases showing that functional proteins from the PC could facilitate the recruitment of NPs to the desired target site have been reported. Modification of the NP surface could also induce an alteration in the composition of the PC, which could lead to an interaction between the NP and cell membrane receptors, eventually favouring cellular uptake and other responses of NPs.

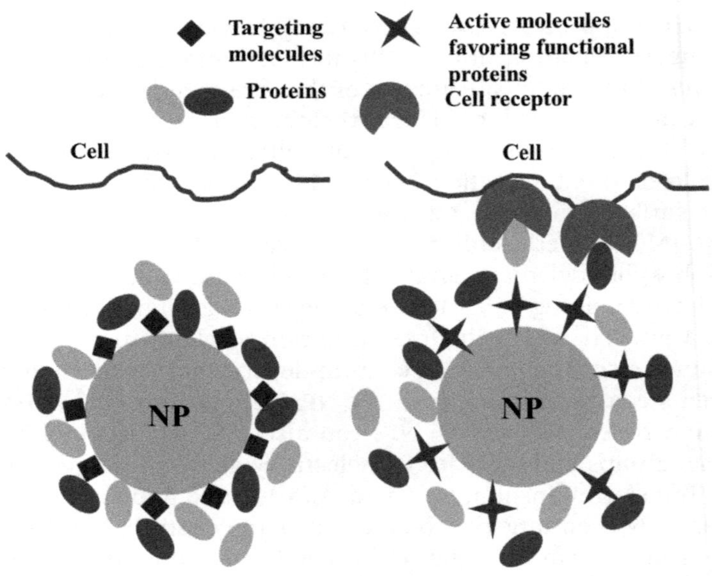

Figure 1.4 Protein-corona-based cell targeting strategy: protein corona formation over functionalized NPs leads to an alteration in the identification of the cell surface receptors (left image); however, protein corona formation along with the active molecules could facilitate the binding of the NPs to the desired targeted cells.

1.6.2 Engineering Nanoparticle–Protein Technologies for Early Cancer Detection

Since PC formation has been reported to alter the properties of NMs, efforts have been devoted to develop technologies that could be used for the sensitive detection of cancer. In this context, several strategies have been developed to probe personalized PCs and their utility in cancer research and management. Mahmoudi *et al.* have investigated the idea of personalized PCs (PPC) in great detail for cancer therapy.[33] PPC have shown a remarkable influence in cancer treatment where the expression pattern of proteins changes as per their physiological environment. Each cancer type is reported to have a variety of characteristic proteins and, therefore, considering the specific interaction between NPs and proteins to form the PC, patients with a specific cancer type may exhibit unique NP coronas. It is well-known that the physio–chemical characteristics of NPs play important roles in the development of PCs around them, therefore, the use of characteristic PCs in the diagnosis of specific diseases would require the synthesis of NPs with the required features.[99] In this context, a detailed understanding of the relationship between the physio–chemical properties of NPs and PC patterns would be the most essential step for developing strategies for diagnostic applications of nanotechnology. The development of PPC was recently validated in pancreatic cancer where clinically relevant proteins were found to be reduced, which affects the size, charge, and aggregation state of NPs.[24,100] PCs developed in pancreatic cancer displayed high negative zeta potentials; however, in other types of cancers, the charge did not vary to any significant magnitude. Thus, the NP and PC technology could work as a "nano-concentrator", which allows identification of the enhancement of a particular protein in the PC developed from cancer patients. In this perspective, the sequestration of proteins across the NP corona surface might be an outstanding opportunity to identify the low-abundant proteins and their connection with carcinogenesis. Another opportunity in which the NP-PC complex could be used as a "nano-constructor'' technology is to diagnose the elevated level of a particular protein in PCs from cancer patients' serum, although their concentration in blood serum is almost similar to that of healthy individuals. Utilizing the "nano-constructor" potential of PCs could be used to identify minor changes in protein concentrations even at the initial stage of disease progression. This strategy could also be used after chemotherapy or surgery, which allows the monitoring of the treatment's success in real time. In this context, blood levels of apolipoproteins could be used as an effective biomarker for the disease's identification, prognosis and to study the development of different types of cancers. The expression profiles of ApoA-1 and ApoC-III in small cell lung cancer cells vary noticeably and, therefore, can be used to distinguish between the non-small cell lung cancer cells from healthy lung tissues.[101] Blood levels of ApoE have been shown to follow the efficiency of chemotherapy as well as the prognosis of several diseases such as Alzheimer's.[102] Thus, the composition of PPC obtained from leukaemia patients, and its detailed characterization through mass spectroscopy, could be used to develop an appropriate strategy for the early detection of this disease.

1.6.3 Toxicity Reduction

It has been well documented that mammalian cells exposed to NPs undergo an alteration of their cellular physiology and related biochemical and signalling pathways. Despite their dependence on several parameters, NPs may induce toxicity to cells through compromising the membrane integrity of the cell and mitochondria, and frequent impairment of genetic materials. It has also been reported that bare NPs cause more damage to mammalian cells than those capped with biomolecules. Further, the physiological and immunological responses of cells could be altered by PC formation over NPs[103] due to the coating of the surface of NPs with proteins. The formation of a PC on a NP's surface could avoid the damaging effects caused by bare NPs has been investigated in several studies.[104,105] Such results may be due to the formation of a PC that leads to biocompatibility with the human body. BSA adsorption on naked poly-3-hydroxybutyrate-*co*-3-hydroxyhexanoate (PHBHHx NPs) has exhibited less cytotoxicity in comparison to NPs without corona, which was monitored in three different cell lines: A549, HepG2 and L929 cells.[106] Ge *et al.* have studied the spontaneous coating of human blood protein and subsequently followed the toxicity pattern of single-wall carbon nanotubes (SWCNTs) in human acute monocytic leukemia cells (THP-1) and human umbilical vein endothelial cells (HUVECs). The authors observed that bovine fibrinogen (BFG), BSA, transferrin (Tf), and γ globulin (Ig) capping on SWCNTs leads to decreased toxicity compared to bare SWCNT. The SWCNT–protein complex showed binding affinities of proteins in the order of BFG > Ig > Tf > BSA for THP-1 cells.[107] The likely reason behind the minimal toxicity observed by the BFG protein was ascribed to the compact rearrangement of this protein on the SWCNT surface and formation of ~4–5 layers of capping over SWCNT, imparting protection to cells from the direct exposure of SWCNT. Additionally, Žūkienė *et al.* have investigated that BSA coating on ZnO NPs displays less toxicity to CHO cells. It was observed that the binding of BSA to ZnO NPs triggers changes in the structure of BSA during PC formation at 4 °C. However, when corona formation was performed at 20 °C, additional changes in the conformation of BSA were observed (disulfide bridge breakage), which diminished the toxicity reduction effect of BSA-coated ZnO NPs.[108] Tenzer *et al.* have demonstrated that bare silica NPs show a haemolytic effect on red blood cells (RBCs); however, the formation of a PC on the NP surface prevents any damage to RBCs.[15] De paoli *et al.* reported that the interaction between carboxylated-multiwalled carbon nanotubes (CNT-COOH) and platelets is strongly dependent on the types of protein involved in the PC formation. Particularly, a PC composed of the various proteins indicates different influences: albumin PC reduces the platelet aggregation; however, histone H1 and gamma globulins induce platelet aggregation and fragmentation.[109] Wang *et al.* demonstrated that positively charged amino-modified polystyrene NPs (PS–NH_2–B) cause early cell membrane damage. The PC on the NPs' surface was explained by the "Trojan horse effect" and it was suggested that the NPs were first surrounded by the PC, which gives protection to cell membranes and also prevents intracellular damage until it reaches the lysosome, where the PC undergoes degradation.[104]

1.7 Limitations and Future Perspectives

With the current thrust towards the development of novel NMs for myriad biomedical applications, they are expected to be exposed to a biological milieu where they would interact with suspended proteins. Therefore, the NMs developed must not be considered as bare and the associated properties may not exactly translate in a biological system. Further, it remains to be seen whether the desired physico–chemical properties of NMs are intact even after PC formation, especially the pharmacokinetics and pharmacodynamics. Thus, the possible interactions of NMs with proteins and eventually the formation of the PC must be almost completely characterized. These efforts may include the interaction of proteins with the capping molecules present on the NP surface because the capping ligands offer promising control over their biological interactions. Although such efforts are less complicated than the characterization of a whole protein from the corona, there are limited efforts in this direction. Additionally, such a strategy may offer control over the NMs' interaction with proteins present in biological fluids and subsequently lead to the success of the clinical translation of nanomedicines. Active uptake of NMs is one of the major factors to improve the therapeutic efficacy, therefore, several studies have been performed to compare the uptake pattern. PC formation is also associated with the immune perturbation effect on macrophages. Such events suggest that corona formation could play an important role in potential immune and inflammatory diseases. PC formation may also significantly affect the physiological properties of NMs such as hydrodynamic size, agglomeration behaviour, surface charge, morphology, and sensitivity to pH and the presence of salt. Since all these factors are closely related with the toxicity of NMs, the safety data may also be influenced as a result of PC formation. It has been reported that PC formation over a NM surface leads to the decreased leaching of corresponding ions, thus significant inhibition in particle toxicity. Further, studies revealed that PC modulates the mechanism of uptake of NMs in mammalian cells such as endocytic rates and pathways, transport within the cells, trafficking rate of endosomes and lysosomes and, thus, the overall interaction and effect on the cells. For example, in the case of cell transfection, the presence of serum was found to induce suppression of transfection efficiency due to reduced endo/lysosomal escape of polyplexes with a PC, then PC-free polyplexes. Considering protein as one of the major components of biological fluids, there are other constituent molecules that need to be explored for their interaction with NMs of nanomedicinal potential. Among them, lipids and lipoproteins must be considered in future for their interactions with NMs and subsequent changes in the physiological properties of NPs and their effect on mammalian cells under *in vitro* and *in vivo* experimental conditions. On the basis of this discussion, it can be concluded that PC formation over a NM surface must be well characterized to realize the full potential of nanotechnology in biomedicines. More components of biological fluids may be considered for their interactions with NMs and subsequent biological effects. Although several efforts have been devoted to addressing these issues, more in-depth studies are needed to completely reveal the PC formation process and composition and translate their effects with biological interactions.

Acknowledgements

J. Shah would like to thank Department of Science and Technology, New Delhi, for providing an INSPIRE Junior Research Fellowship (INSPIRE-JRF). The financial assistance for the Centre for Nanotechnology Research and Applications (CENTRA) by The Gujarat Institute for Chemical Technology (GICT) is acknowledged. The funding from the Department of Science and Technology - Science and Engineering Research Board (SERB) (Grant No.: ILS/SERB/2015-16/01) to Dr. Sanjay Singh under the scheme of a Start-Up Research Grant (Young Scientists) in Life Sciences is also gratefully acknowledged. The Seed Grant (AU/SG/SAS/DBLS/17-18/03) support from Ahmedabad University provided to S. Singh is also acknowledged.

References

1. Y. Sun and Y. Xia, Shape-controlled synthesis of gold and silver nanoparticles, *Science*, 2002, **298**, 2176–2179.
2. S. Winzen, S. Schoettler, G. Baier, C. Rosenauer, V. Mailaender, K. Landfester and K. Mohr, Complementary analysis of the hard and soft protein corona: sample preparation critically effects corona composition, *Nanoscale*, 2015, **7**, 2992–3001.
3. S. Milani, F. Baldelli Bombelli, A. S. Pitek, K. A. Dawson and J. Rädler, Reversible *versus* irreversible binding of transferrin to polystyrene nanoparticles: soft and hard corona, *ACS Nano*, 2012, **6**, 2532–2541.
4. M. Lundqvist, J. Stigler, G. Elia, I. Lynch, T. Cedervall and K. A. Dawson, Nanoparticle size and surface properties determine the protein corona with possible implications for biological impacts, *Proc. Natl. Acad. Sci. U. S. A.*, 2008, **105**, 14265–14270.
5. W. Liu, J. Rose, S. Plantevin, M. Auffan, J. Y. Bottero and C. Vidaud, Protein corona formation for nanomaterials and proteins of a similar size: hard or soft corona? *Nanoscale*, 2013, **5**, 1658–1668.
6. V. Pareek, A. Bhargava, V. Bhanot, R. Gupta, N. Jain and J. Panwar, Formation and Characterization of Protein Corona Around Nanoparticles: A Review, *J. Nanosci. Nanotechnol.*, 2018, **18**, 6653–6670.

7. T. Cedervall, I. Lynch, M. Foy, T. Berggard, S. C. Donnelly, G. Cagney, S. Linse and K. A. Dawson, Detailed identification of plasma proteins adsorbed on copolymer nanoparticles, *Angew. Chem.*, 2007, **46**, 5754–5756.
8. E. Casals, T. Pfaller, A. Duschl, G. J. Oostingh and V. Puntes, Time evolution of the nanoparticle protein corona, *ACS Nano*, 2010, **4**, 3623–3632.
9. D. Dell'Orco, M. Lundqvist, C. Oslakovic, T. Cedervall and S. Linse, Modeling the time evolution of the nanoparticle-protein corona in a body fluid, *PLoS One*, 2010, **5**, e10949.
10. C. Röcker, M. Pötzl, F. Zhang, W. J. Parak and G. U. Nienhaus, A quantitative fluorescence study of protein monolayer formation on colloidal nanoparticles, *Nat. Nanotechnol.*, 2009, **4**, 577.
11. E. A. Vogler, Protein adsorption in three dimensions, *Biomaterials*, 2012, **33**, 1201–1237.
12. T. Cedervall, I. Lynch, S. Lindman, T. Berggård, E. Thulin, H. Nilsson, K. A. Dawson and S. Linse, Understanding the nanoparticle–protein corona using methods to quantify exchange rates and affinities of proteins for nanoparticles, *Proc. Natl. Acad. Sci. U. S. A.*, 2007, **104**, 2050–2055.
13. I. Delfino and S. Cannistraro, Optical investigation of the electron transfer protein azurin–gold nanoparticle system, *Biophys. Chem.*, 2009, **139**, 1–7.
14. Z. J. Deng, G. Mortimer, T. Schiller, A. Musumeci, D. Martin and R. F. Minchin, Differential plasma protein binding to metal oxide nanoparticles, *Nanotechnology*, 2009, **20**, 455101.
15. S. Tenzer, D. Docter, S. Rosfa, A. Wlodarski, J. Kuharev, A. Rekik, S. K. Knauer, C. Bantz, T. Nawroth, C. Bier, J. Sirirattanapan, W. Mann, L. Treuel, R. Zellner, M. Maskos, H. Schild and R. H. Stauber, Nanoparticle size is a critical physicochemical determinant of the human blood plasma corona: a comprehensive quantitative proteomic analysis, *ACS Nano*, 2011, **5**, 7155–7167.
16. M. P. Monopoli, C. Aberg, A. Salvati and K. A. Dawson, Biomolecular coronas provide the biological identity of nanosized materials, *Nat. Nanotechnol.*, 2012, **7**, 779–786.
17. C. D. Walkey and W. C. Chan, Understanding and controlling the interaction of nanomaterials with proteins in a physiological environment, *Chem. Soc. Rev.*, 2012, **41**, 2780–2799.
18. D. Docter, D. Westmeier, M. Markiewicz, S. Stolte, S. K. Knauer and R. H. Stauber, The nanoparticle biomolecule corona: lessons learned - challenge accepted? *Chem. Soc. Rev.*, 2015, **44**, 6094–6121.
19. C. Sacchetti, K. Motamedchaboki, A. Magrini, G. Palmieri, M. Mattei, S. Bernardini, N. Rosato, N. Bottini and M. Bottini, Surface polyethylene glycol conformation influences the protein corona of polyethylene glycol-modified single-walled carbon nanotubes: potential implications on biological performance, *ACS Nano*, 2013, **7**, 1974–1989.
20. A. L. Capriotti, G. Caracciolo, G. Caruso, P. Foglia, D. Pozzi, R. Samperi and A. Lagana, Differential analysis of "protein corona" profile adsorbed onto different nonviral gene delivery systems, *Anal. Biochem.*, 2011, **419**, 180–189.

21. D. Walczyk, F. B. Bombelli, M. P. Monopoli, I. Lynch and K. A. Dawson, What the cell "sees" in bionanoscience, *J. Am. Chem. Soc.*, 2010, **132**, 5761–5768.
22. H. D. Hill, J. E. Millstone, M. J. Banholzer and C. A. Mirkin, The role radius of curvature plays in thiolated oligonucleotide loading on gold nanoparticles, *ACS Nano*, 2009, **3**, 418–424.
23. M. Mahmoudi, I. Lynch, M. R. Ejtehadi, M. P. Monopoli, F. B. Bombelli and S. Laurent, Protein-nanoparticle interactions: opportunities and challenges, *Chem. Rev.*, 2011, **111**, 5610–5637.
24. M. A. Dobrovolskaia, A. K. Patri, J. Zheng, J. D. Clogston, N. Ayub, P. Aggarwal, B. W. Neun, J. B. Hall and S. E. McNeil, Interaction of colloidal gold nanoparticles with human blood: effects on particle size and analysis of plasma protein binding profiles, *Nanomedicine: Nanotechnol., Biol. Med.*, 2009, **5**, 106–117.
25. J. Piella, N. G. Bastus and V. Puntes, Size-Dependent Protein-Nanoparticle Interactions in Citrate-Stabilized Gold Nanoparticles: The Emergence of the Protein Corona, *Bioconjugate Chem.*, 2017, **28**, 88–97.
26. R. García-Álvarez, M. Hadjidemetriou, A. Sánchez-Iglesias, L. M. Liz-Marzán and K. Kostarelos, *In vivo* formation of protein corona on gold nanoparticles. The effect of their size and shape, *Nanoscale*, 2018, **10**, 1256–1264.
27. P. Foroozandeh and A. A. Aziz, Merging worlds of nanomaterials and biological environment: factors governing protein corona formation on nanoparticles and its biological consequences, *Nanoscale Res. Lett.*, 2015, **10**, 221.
28. Z. Ma, J. Bai, Y. Wang and X. Jiang, Impact of shape and pore size of mesoporous silica nanoparticles on serum protein adsorption and RBCs hemolysis, *ACS Appl. Mater. Interfaces*, 2014, **6**, 2431–2438.
29. P. Roach, D. Farrar and C. C. Perry, Interpretation of protein adsorption: surface-induced conformational changes, *J. Am. Chem. Soc.*, 2005, **127**, 8168–8173.
30. A. Gessner, R. Waicz, A. Lieske, B. Paulke, K. Mader and R. H. Muller, Nanoparticles with decreasing surface hydrophobicities: influence on plasma protein adsorption, *Int. J. Pharm.*, 2000, **196**, 245–249.
31. S. Dominguez-Medina, J. Blankenburg, J. Olson, C. F. Landes and S. Link, Adsorption of a Protein Monolayer *via* Hydrophobic Interactions Prevents Nanoparticle Aggregation under Harsh Environmental Conditions, *ACS Sustainable Chem. Eng.*, 2013, **1**, 833–842.
32. M. Hulander, A. Lundgren, M. Berglin, M. Ohrlander, J. Lausmaa and H. Elwing, Immune complement activation is attenuated by surface nanotopography, *Int. J. Nanomed.*, 2011, **6**, 2653–2666.
33. M. Mahmoudi, M. A. Sahraian, M. A. Shokrgozar and S. Laurent, Superparamagnetic iron oxide nanoparticles: promises for diagnosis and treatment of multiple sclerosis, *ACS Chem. Neurosci.*, 2011, **2**, 118–140.
34. M. S. Ehrenberg, A. E. Friedman, J. N. Finkelstein, G. Oberdorster and J. L. McGrath, The influence of protein adsorption on nanoparticle association with cultured endothelial cells, *Biomaterials*, 2009, **30**, 603–610.

35. I. Lynch, A. Salvati and K. A. Dawson, Protein-nanoparticle interactions: What does the cell see? *Nat. Nanotechnol.*, 2009, **4**, 546–547.
36. A. Gessner, A. Lieske, B. Paulke and R. Muller, Influence of surface charge density on protein adsorption on polymeric nanoparticles: analysis by two-dimensional electrophoresis, *Eur. J. Pharm. Biopharm.*, 2002, **54**, 165–170.
37. A. S. Fauci, *Harrison's Principles of Internal Medicine*, Mcgraw-hill, New York, 1998.
38. J. M. Yang, H. Yang and L. Lin, Quantum dot nano thermometers reveal heterogeneous local thermogenesis in living cells, *ACS Nano*, 2011, **5**, 5067–5071.
39. J. D. Hasday and I. S. Singh, Fever and the heat shock response: distinct, partially overlapping processes, *Cell Stress Chaperones*, 2000, **5**, 471–480.
40. M. Mahmoudi, A. M. Abdelmonem, S. Behzadi, J. H. Clement, S. Dutz, M. R. Ejtehadi, R. Hartmann, K. Kantner, U. Linne and P. Maffre, Temperature: the "ignored" factor at the nanobio interface, *ACS Nano*, 2013, **7**, 6555–6562.
41. M. Mahmoudi, S. E. Lohse, C. J. Murphy, A. Fathizadeh, A. Montazeri and K. S. Suslick, Variation of protein corona composition of gold nanoparticles following plasmonic heating, *Nano Lett.*, 2014, **14**, 6–12.
42. J. O'Brien and K. J. Shea, Tuning the Protein Corona of Hydrogel Nanoparticles: The Synthesis of Abiotic Protein and Peptide Affinity Reagents, *Acc. Chem. Res.*, 2016, **49**, 1200–1210.
43. V. H. Nguyen and B. J. Lee, Protein corona: a new approach for nanomedicine design, *Int. J. Nanomed.*, 2017, **12**, 3137–3151.
44. M. Kosmulski, pH-dependent surface charging and points of zero charge. IV. Update and new approach, *J. Colloid Interface Sci.*, 2009, **337**, 439–448.
45. G. Chen, X. Liu and C. Su, Distinct effects of humic acid on transport and retention of TiO2 rutile nanoparticles in saturated sand columns, *Environ. Sci. Technol.*, 2012, **46**, 7142–7150.
46. K. Pakarinen, E. J. Petersen, L. Alvila, G. C. Waissi-Leinonen, J. Akkanen, M. T. Leppanen and J. V. Kukkonen, A screening study on the fate of fullerenes (nC60) and their toxic implications in natural freshwaters, *Environ. Toxicol. Chem.*, 2013, **32**, 1224–1232.
47. B. Xie, Z. Xu, W. Guo and Q. Li, Impact of natural organic matter on the physicochemical properties of aqueous C60 nanoparticles, *Environ. Sci. Technol.*, 2008, **42**, 2853–2859.
48. A. B. Engin, M. Neagu, K. Golokhvast and A. Tsatsakis, Nanoparticles and endothelium: an update on the toxicological interactions, *Farmacia*, 2015, **63**, 792–804.
49. R. Gref, M. Luck, P. Quellec, M. Marchand, E. Dellacherie, S. Harnisch, T. Blunk and R. H. Muller, 'Stealth' corona-core nanoparticles surface modified by polyethylene glycol (PEG): influences of the corona (PEG chain length and surface density) and of the core composition on phagocytic uptake and plasma protein adsorption, *Colloids Surf., B*, 2000, **18**, 301–313.

50. J. L. Perry, K. G. Reuter, M. P. Kai, K. P. Herlihy, S. W. Jones, J. C. Luft, M. Napier, J. E. Bear and J. M. DeSimone, PEGylated PRINT Nanoparticles: The Impact of PEG Density on Protein Binding, Macrophage Association, Biodistribution, and Pharmacokinetics, *Nano Lett.*, 2012, **12**, 5304–5310.
51. S. M. Moghimi, I. Muir, L. Illum, S. S. Davis and V. Kolb-Bachofen, Coating particles with a block co-polymer (poloxamine-908) suppresses opsonization but permits the activity of dysopsonins in the serum, *Biochim. Biophys. Acta*, 1993, **1179**, 157–165.
52. P. Aggarwal, J. B. Hall, C. B. McLeland, M. A. Dobrovolskaia and S. E. McNeil, Nanoparticle interaction with plasma proteins as it relates to particle biodistribution, biocompatibility and therapeutic efficacy, *Adv. Drug Delivery Rev.*, 2009, **61**, 428–437.
53. D. Dutta, S. K. Sundaram, J. G. Teeguarden, B. J. Riley, L. S. Fifield, J. M. Jacobs, S. R. Addleman, G. A. Kaysen, B. M. Moudgil and T. J. Weber, Adsorbed proteins influence the biological activity and molecular targeting of nanomaterials, *Toxicol. Sci.*, 2007, **100**, 303–315.
54. L. K. Muller, J. Simon, C. Rosenauer, V. Mailander, S. Morsbach and K. Landfester, The Transferability from Animal Models to Humans: Challenges Regarding Aggregation and Protein Corona Formation of Nanoparticles, *Biomacromolecules*, 2018, **19**, 374–385.
55. M. P. Monopoli, D. Walczyk, A. Campbell, G. Elia, I. Lynch, F. B. Bombelli and K. A. Dawson, Physical-chemical aspects of protein corona: relevance to *in vitro* and *in vivo* biological impacts of nanoparticles, *J. Am. Chem. Soc.*, 2011, **133**, 2525–2534.
56. U. Sakulkhu, M. Mahmoudi, L. Maurizi, J. Salaklang and H. Hofmann, Protein corona composition of superparamagnetic iron oxide nanoparticles with various physico-chemical properties and coatings, *Sci. Rep.*, 2014, **4**, 5020.
57. C. C. Fleischer and C. K. Payne, Nanoparticle-cell interactions: molecular structure of the protein corona and cellular outcomes, *Acc. Chem. Res.*, 2014, **47**, 2651–2659.
58. P. P. Karmali and D. Simberg, Interactions of nanoparticles with plasma proteins: implication on clearance and toxicity of drug delivery systems, *Expert Opin. Drug Delivery*, 2011, **8**, 343–357.
59. S. R. Saptarshi, A. Duschl and A. L. Lopata, Interaction of nanoparticles with proteins: relation to bio-reactivity of the nanoparticle, *J. Nanobiotechnol.*, 2013, **11**, 26.
60. J. H. Shannahan, X. Lai, P. C. Ke, R. Podila, J. M. Brown and F. A. Witzmann, Silver nanoparticle protein corona composition in cell culture media, *PLoS One*, 2013, **8**, e74001.
61. K. Sempf, T. Arrey, S. Gelperina, T. Schorge, B. Meyer, M. Karas and J. Kreuter, Adsorption of plasma proteins on uncoated PLGA nanoparticles, *Eur. J. Pharm. Biopharm.*, 2013, **85**, 53–60.
62. F. Darabi Sahneh, C. Scoglio and J. Riviere, Dynamics of nanoparticle-protein corona complex formation: analytical results from population balance equations, *PLoS One*, 2013, **8**, e64690.

63. L. Vroman, A. L. Adams, G. C. Fischer and P. C. Munoz, Interaction of high molecular weight kininogen, factor XII, and fibrinogen in plasma at interfaces, *Blood*, 1980, **55**, 156–159.
64. T. M. Goppert and R. H. Muller, Polysorbate-stabilized solid lipid nanoparticles as colloidal carriers for intravenous targeting of drugs to the brain: comparison of plasma protein adsorption patterns, *J. Drug Targeting*, 2005, **13**, 179–187.
65. M. Jansch, P. Stumpf, C. Graf, E. Ruhl and R. H. Muller, Adsorption kinetics of plasma proteins on ultrasmall superparamagnetic iron oxide (USPIO) nanoparticles, *Int. J. Pharm.*, 2012, **428**, 125–133.
66. S. C. Gill and P. H. von Hippel, Calculation of protein extinction coefficients from amino acid sequence data, *Anal. Biochem.*, 1989, **182**, 319–326.
67. N. Jain, A. Bhargava, M. Rathi, R. V. Dilip and J. Panwar, Removal of Protein Capping Enhances the Antibacterial Efficiency of Biosynthesized Silver Nanoparticles, *PLoS One*, 2015, **10**, e0134337.
68. I. Yadav, V. K. Aswal and J. Kohlbrecher, Size-dependent interaction of silica nanoparticles with lysozyme and bovine serum albumin proteins, *Phys. Rev. E*, 2016, **93**, 052601.
69. X. Xu, Y. Wang, H. Wang, H. Su, X. Mao, L. Jiang, M. Liu, D. Sun and S. Hou, Synthesis of triangular silver nanoprisms and studies on the interactions with human serum albumin, *J. Mol. Liq.*, 2016, **220**, 14–20.
70. N. Dasgupta, S. Ranjan, D. Patra, P. Srivastava, A. Kumar and C. Ramalingam, Bovine serum albumin interacts with silver nanoparticles with a “side-on” or “end on” conformation, *Chem.-Biol. Interact.*, 2016, **253**, 100–111.
71. R. Singh and S. Singh, Role of phosphate on stability and catalase mimetic activity of cerium oxide nanoparticles, *Colloids Surf., B*, 2015, **132**, 78–84.
72. J. Vidic, F. Haque, J. M. Guigner, A. Vidy, C. Chevalier and S. Stankic, Effects of water and cell culture media on the physicochemical properties of ZnMgO nanoparticles and their toxicity toward mammalian cells, *Langmuir*, 2014, **30**, 11366–11374.
73. J. Kong and S. Yu, Fourier transform infrared spectroscopic analysis of protein secondary structures, *Acta Biochim. Biophys. Sin.*, 2007, **39**, 549–559.
74. D. A. Mbeh, T. Javanbakht, L. Tabet, Y. Merhi, K. Maghni, E. Sacher and L. H. Yahia, Protein Corona Formation on Magnetite Nanoparticles: Effects of Culture Medium Composition, and Its Consequences on Superparamagnetic Nanoparticle Cytotoxicity, *J. Biomed. Nanotechnol.*, 2015, **11**, 828–840.
75. L. Fageria, V. Pareek, R. V. Dilip, A. Bhargava, S. S. Pasha, I. R. Laskar, H. Saini, S. Dash, R. Chowdhury and J. Panwar, Biosynthesized protein-capped silver nanoparticles induce ROS-dependent proapoptotic signals and prosurvival autophagy in cancer cells, *ACS Omega*, 2017, **2**, 1489–1504.

76. L. Shang, Y. Wang, J. Jiang and S. Dong, pH-dependent protein conformational changes in albumin:gold nanoparticle bioconjugates: a spectroscopic study, *Langmuir*, 2007, **23**, 2714–2721.
77. W. Goldburg, Dynamic light scattering, *Am. J. Phys.*, 1999, **67**, 1152–1160.
78. P. M. Tessier, J. Jinkoji, Y.-C. Cheng, J. L. Prentice and A. M. Lenhoff, Self-interaction nanoparticle spectroscopy: a nanoparticle-based protein interaction assay, *J. Am. Chem. Soc.*, 2008, **130**, 3106–3112.
79. P. Del Pino, B. Pelaz, Q. Zhang, P. Maffre, G. U. Nienhaus and W. J. Parak, Protein corona formation around nanoparticles–from the past to the future, *Mater. Horiz.*, 2014, **1**, 301–313.
80. S. Patil, A. Sandberg, E. Heckert, W. Self and S. Seal, Protein adsorption and cellular uptake of cerium oxide nanoparticles as a function of zeta potential, *Biomaterials*, 2007, **28**, 4600–4607.
81. K. Natte, J. F. Friedrich, S. Wohlrab, J. Lutzki, R. von Klitzing, W. Osterle and G. Orts-Gil, Impact of polymer shell on the formation and time evolution of nanoparticle-protein corona, *Colloids Surf., B*, 2013, **104**, 213–220.
82. M. M. Pierce, C. Raman and B. T. Nall, Isothermal titration calorimetry of protein–protein interactions, *Methods*, 1999, **19**, 213–221.
83. R. Huang, R. P. Carney, K. Ikuma, F. Stellacci and B. L. Lau, Effects of surface compositional and structural heterogeneity on nanoparticle-protein interactions: different protein configurations, *ACS Nano*, 2014, **8**, 5402–5412.
84. M. De, C. C. You, S. Srivastava and V. M. Rotello, Biomimetic interactions of proteins with functionalized nanoparticles: a thermodynamic study, *J. Am. Chem. Soc.*, 2007, **129**, 10747–10753.
85. Q. Xiao, S. Huang, Z. D. Qi, B. Zhou, Z. K. He and Y. Liu, Conformation, thermodynamics and stoichiometry of HSA adsorbed to colloidal CdSe/ZnS quantum dots, *Biochim. Biophys. Acta*, 2008, **1784**, 1020–1027.
86. N. J. Greenfield, Using circular dichroism spectra to estimate protein secondary structure, *Nat. Protoc.*, 2006, **1**, 2876–2890.
87. N. Sreerama, S. Y. Venyaminov and R. W. Woody, Estimation of protein secondary structure from circular dichroism spectra: inclusion of denatured proteins with native proteins in the analysis, *Anal. Biochem.*, 2000, **287**, 243–251.
88. W. C. Johnson Jr, Secondary structure of proteins through circular dichroism spectroscopy, *Annu. Rev. Biophys. Biophys. Chem.*, 1988, **17**, 145–166.
89. M. Lundqvist, I. Sethson and B. H. Jonsson, Protein adsorption onto silica nanoparticles: conformational changes depend on the particles' curvature and the protein stability, *Langmuir*, 2004, **20**, 10639–10647.
90. S. H. Lacerda, J. J. Park, C. Meuse, D. Pristinski, M. L. Becker, A. Karim and J. F. Douglas, Interaction of gold nanoparticles with common human blood proteins, *ACS Nano*, 2010, **4**, 365–379.
91. L. Li, Q. Mu, B. Zhang and B. Yan, Analytical strategies for detecting nanoparticle-protein interactions, *Analyst*, 2010, **135**, 1519–1530.

92. G. Brancolini, D. B. Kokh, L. Calzolai, R. C. Wade and S. Corni, Docking of ubiquitin to gold nanoparticles, *ACS Nano*, 2012, **6**, 9863–9878.
93. F. Ahmad, Y. Zhou, Z. Ling, Q. Xiang and X. Zhou, Systematic elucidation of interactive unfolding and corona formation of bovine serum albumin with cobalt ferrite nanoparticles, *RSC Adv.*, 2016, **6**, 35719–35730.
94. G. Caracciolo, L. Callipo, S. C. De Sanctis, C. Cavaliere, D. Pozzi and A. Lagana, Surface adsorption of protein corona controls the cell internalization mechanism of DC-Chol-DOPE/DNA lipoplexes in serum, *Biochim. Biophys. Acta*, 2010, **1798**, 536–543.
95. R. Weissleder, K. Kelly, E. Y. Sun, T. Shtatland and L. Josephson, Cell-specific targeting of nanoparticles by multivalent attachment of small molecules, *Nat. Biotechnol.*, 2005, **23**, 1418.
96. A. Zensi, D. Begley, C. Pontikis, C. Legros, L. Mihoreanu, S. Wagner, C. Büchel and J. Kreuter, Albumin nanoparticles targeted with Apo E enter the CNS by transcytosis and are delivered to neurones, *J. Controlled Release*, 2009, **137**, 78–86.
97. J. Kreuter, T. Hekmatara, S. Dreis, T. Vogel, S. Gelperina and K. Langer, Covalent attachment of apolipoprotein AI and apolipoprotein B-100 to albumin nanoparticles enables drug transport into the brain, *J. Controlled Release*, 2007, **118**, 54–58.
98. L. Thiele, J. E. Diederichs, R. Reszka, H. P. Merkle and E. Walter, Competitive adsorption of serum proteins at microparticles affects phagocytosis by dendritic cells, *Biomaterials*, 2003, **24**, 1409–1418.
99. S. Palchetti, D. Pozzi, M. Mahmoudi and G. Caracciolo, Exploitation of nanoparticle–protein corona for emerging therapeutic and diagnostic applications, *J. Mater. Chem. B*, 2016, **4**, 4376–4381.
100. V. Colapicchioni, M. Tilio, L. Digiacomo, V. Gambini, S. Palchetti, C. Marchini, D. Pozzi, S. Occhipinti, A. Amici and G. Caracciolo, Personalized liposome-protein corona in the blood of breast, gastric and pancreatic cancer patients, *Int. J. Biochem. Cell Biol.*, 2016, **75**, 180–187.
101. J. Shi, H. Yang, X. Duan, L. Li, L. Sun, Q. Li and J. Zhang, Apolipoproteins as Differentiating and Predictive Markers for Assessing Clinical Outcomes in Patients with Small Cell Lung Cancer, *Yonsei Med. J.*, 2016, **57**, 549–556.
102. F. J. Wolters, P. J. Koudstaal, A. Hofman, C. M. van Duijn and M. A. Ikram, Serum apolipoprotein E is associated with long-term risk of Alzheimer's disease: The Rotterdam Study, *Neurosci. Lett.*, 2016, **617**, 139–142.
103. Y. K. Lee, E. J. Choi, T. J. Webster, S. H. Kim and D. Khang, Effect of the protein corona on nanoparticles for modulating cytotoxicity and immunotoxicity, *Int. J. Nanomed.*, 2015, **10**, 97–113.
104. F. Wang, L. Yu, M. P. Monopoli, P. Sandin, E. Mahon, A. Salvati and K. A. Dawson, The biomolecular corona is retained during nanoparticle uptake and protects the cells from the damage induced by cationic nanoparticles until degraded in the lysosomes, *Nanomedicine: Nanotechnol., Biol. Med.*, 2013, **9**, 1159–1168.

105. R. Tedja, M. Lim, R. Amal and C. Marquis, Effects of serum adsorption on cellular uptake profile and consequent impact of titanium dioxide nanoparticles on human lung cell lines, *ACS Nano*, 2012, **6**, 4083–4093.
106. Q. Peng, S. Zhang, Q. Yang, T. Zhang, X.-Q. Wei, L. Jiang, C.-L. Zhang, Q.-M. Chen, Z.-R. Zhang and Y.-F. Lin, Preformed albumin corona, a protective coating for nanoparticles based drug delivery system, *Biomaterials*, 2013, **34**, 8521–8530.
107. C. Ge, J. Du, L. Zhao, L. Wang, Y. Liu, D. Li, Y. Yang, R. Zhou, Y. Zhao and Z. Chai, Binding of blood proteins to carbon nanotubes reduces cytotoxicity, *Proc. Natl. Acad. Sci. U. S. A.*, 2011, 16968–16973.
108. R. Žūkienė and V. Snitka, Zinc oxide nanoparticle and bovine serum albumin interaction and nanoparticles influence on cytotoxicity *in vitro*, *Colloids Surf., B*, 2015, **135**, 316–323.
109. S. H. De Paoli, L. L. Diduch, T. Z. Tegegn, M. Orecna, M. B. Strader, E Karnaukhova, J. E. Bonevich, K. Holada and J. Simak, The effect of protein corona composition on the interaction of carbon nanotubes with human blood platelets, *Biomaterials*, 2014, **35**, 6182–6194.

CHAPTER 2

Biological Significance of the Nanoparticles Protein Corona

V. FOREST*

Mines Saint-Etienne, Université de Lyon, Université Jean Monnet, INSERM, U 1059 Sainbiose, Centre CIS, 158 cours Fauriel, CS 62362, F-42023 Saint-Etienne, France
*E-mail: vforest@emse.fr

2.1 Introduction

During the last decades, nanotechnology has evolved dramatically. Consequently, we are potentially exposed to increasing levels of nanoparticles. Indeed, they are included in many products that are becoming more and more a part of our everyday life (automobile industry, electronics, paints, cosmetics, food, hygiene products, and many more).[1] Many application fields benefit from their outstanding properties, including nanomedicine where nanoparticles are used for diagnosis and/or therapeutic purposes.

Despite these promising aspects, some concerns have arisen, especially regarding the impact of engineered nanoparticles on human health and on the environment. Indeed, while we are exposed more and more to these nanoparticles, their biological impact remains controversial.

Nanoparticles exhibit unique properties that are very different from their bulk counterparts because of specific physicochemical features. For instance, nanoparticles are characterized by a huge surface–volume ratio, making them extremely reactive. Consequently, the nanoscale that makes

Issues in Toxicology No. 40
Nanoparticle–Protein Corona: Biophysics to Biology
Edited by Ashutosh Kumar and Alok Dhawan

Published by the Royal Society of Chemistry, www.rsc.org

these materials very attractive in terms of applications can, on the other hand, be responsible for a different toxicological profile. It is, therefore, mandatory to thoroughly characterize the nanoparticles' physicochemical features and consider, in particular, as recommended by the OECD ISO/TR 13014:2012 standard, the following main parameters: size, shape, chemical composition, solubility, agglomeration/aggregation state, specific surface, surface chemistry, surface charge, *etc.*[2] These features define the "synthetic identity" of the nanoparticles (Figure 2.1A) and will have a strong impact on the nanoparticles' fate and biological outcomes.

But the picture is much more complex as pristine nanoparticles (*i.e.* as synthesized in biomolecule-free environments) will not keep a bare surface upon interactions with biological systems. Indeed, biomolecules will adsorb on the nanoparticle surface forming the so-called "protein corona" (as better defined in the next paragraph) of which its importance and relevance to the biological response are now unanimously acknowledged (Figure 2.1C).

The parameters involved in the formation of the nanoparticle protein corona as well as its biological consequences have been the subject of intensive research over the last two decades. However, because of its complexity and uniqueness, no firm conclusions can be drawn; only general principles have emerged.

The aim of this chapter is to discuss the up-to-date knowledge on the protein corona with a specific focus on the crucial issue of how it could affect nanoparticle biological impacts. This chapter is divided into four sections: the first one deals with the parameters influencing the protein corona formation, the second part reports on the consequences of the protein corona's presence on the nanoparticle's physicochemical properties, the third part focuses on the impact of the protein corona on the nanoparticle's biological fate and activity and the fourth section discusses issues associated with the protein corona. Finally, future directions are proposed.

2.2 Formation of a Protein Corona

The nanoparticle protein corona concept was introduced in the early 2000s by Cedervall *et al.*[3] It refers to the fact that upon contact with biological media, proteins, lipids, sugars and other macromolecules contained in these fluids adsorb onto the nanoparticle's surface, forming a crown, a "corona" in Latin. Because proteins are largely predominant in biological fluids, the corona consists mainly of a protein layer, which is why it is generally referred to as a "protein corona".

It is widely acknowledged that the adsorption of proteins on the nanoparticle's surface is mainly driven by electrostatic interactions although some other forces can contribute such as H-bonding, van der Waals interactions and salt bridge formations.[4–7] The high surface–volume ratio exhibited by nano-sized particles compared to their bulk counterparts (and therefore their high surface free energy) provides a driving force for enhanced interactions between nanoparticles and the surrounding molecules.[8,9] Consequently,

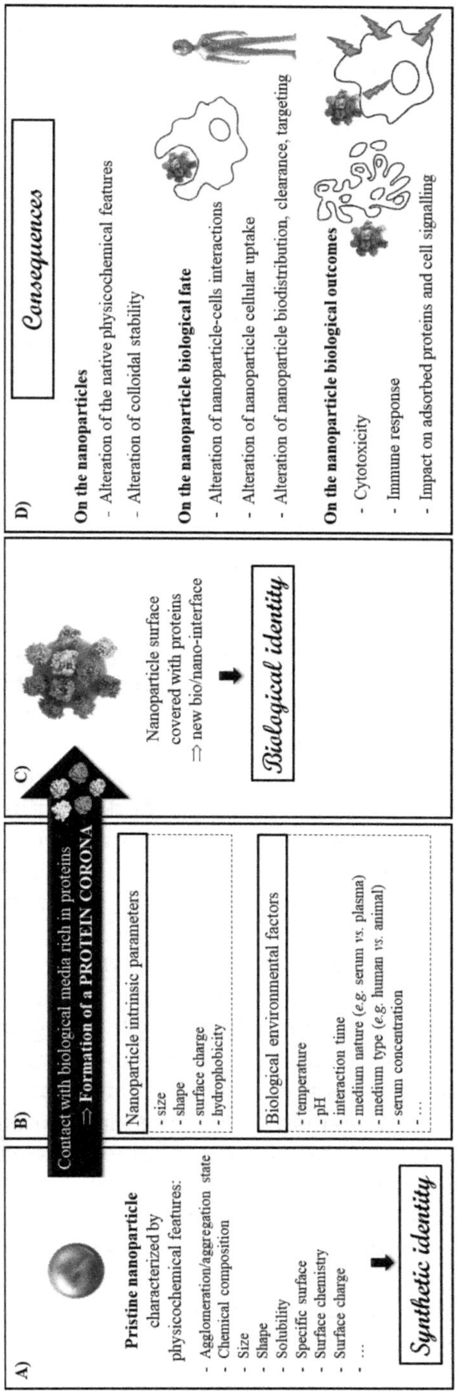

Figure 2.1 From the synthetic to biological identity of nanoparticles and the subsequent consequences.

nanoparticle–protein interactions either increase entropy or decrease enthalpy, both leading to a diminution of Gibbs-free energy, thus representing a favourable process.[10] Proteins can bind either directly to the nanoparticle's surface or indirectly through binding to proteins already adsorbed.[4]

The corona is a dynamic structure and its composition evolves over time. First, the proteins most abundant in the medium and with high mobility adsorb onto the nanoparticle's surface. They are then replaced by proteins exhibiting a higher affinity. This affinity-based competition is known as the "Vroman effect" since it was described by Vroman and Adams in the 1960s while they were characterizing the interactions of blood proteins with solid surfaces.[10,11] However, it seems to be a really fast process as Tenzer *et al.* demonstrated that an interaction between the nanoparticle's surface and plasma proteins is established as early as after 0.5 minutes of contact. This conclusion was reached for different types of nanoparticles: silica and polystyrene nanoparticles of various sizes and surface functionalizations. They also found that the composition of the generated corona did not change qualitatively but only quantitatively over time.[12]

It is also interesting to mention that although blood plasma is composed of thousands of proteins, some of which are more abundant than others, their abundance in the plasma does not correlate to their abundance in the protein corona.[13,14]

Based on the exchange rate of the proteins, a "soft" corona has been distinguished from a "hard" corona. The hard corona is composed of tightly bound proteins at the nanoparticle's surface characterized by a long exchange time (many hours), while the soft corona is represented by the second layer of proteins more loosely associated (not directly bound to the nanoparticle) that undergoes fast exchanges over time (seconds or minutes).[8,15] Because of these features, the soft corona is extremely difficult to characterize based on current available techniques, while the hard corona, which is more stable, is not removed during the analysis process and thus may be more easily studied.[10,16]

The dynamic nature of the corona is not only due to changes in its composition over time but also depends on changes in nanoparticle location. Indeed, nanoparticles can go through different physiological compartments, each of them having its own distinct set of proteins that interact in a unique way with the nanoparticle.[14] It was observed that after nanoparticle cell internalization, the protein corona, instead of being stripped off, is preserved, at least partially.[8] It was thus proposed that the nanoparticles carried a "memory" of the previous environments they went through.[10] The term corona "fingerprint" was coined to indicate the protein patterns of nanoparticles that characterize their corona, *i.e.* the relative abundance of each of these proteins.[17,18]

For a long time, the existence of the protein corona was overlooked and its biological consequences underestimated. Fortunately, many studies have been carried out to characterize this layer and better understand how it affects the biological fate of nanoparticles. It is now commonly admitted that the protein corona is of paramount importance as it induces vast biological outcomes (Figure 2.1D). The (patho)physiological effects triggered by nanoparticles are not attributed only to their physicochemical features anymore even if they will have an indirect influence as they impact the composition of the protein corona.

2.2.1 Parameters Involved in the Formation of the Protein Corona

Despite numerous works, no general rule can be established. Indeed, there is no universal corona; it is unique to each system. This can be easily explained by the fact that many factors are involved in its formation/composition, parameters related to the physicochemical properties of the nanoparticles but also to their biological environment (Figure 2.1B).[14,19] It is an interplay of different forces in a competing environment.[10]

2.2.1.1 Nanoparticles' Intrinsic Parameters

As illustrated in Figure 2.2, the main nanoparticle physicochemical parameters involved in the formation of the protein corona are the size, shape, surface charge and hydrophobicity.[6,9]

Size is one of the most important determinants of a protein corona's composition and therefore its biological effect.[20] Dobrovolskaia *et al.* found that more proteins were adsorbed on 30 nm than on 50 nm gold nanoparticles when incubated in human plasma.[21] Very small variations can have dramatic consequences as exemplified by Tenzer *et al.* who demonstrated that even a

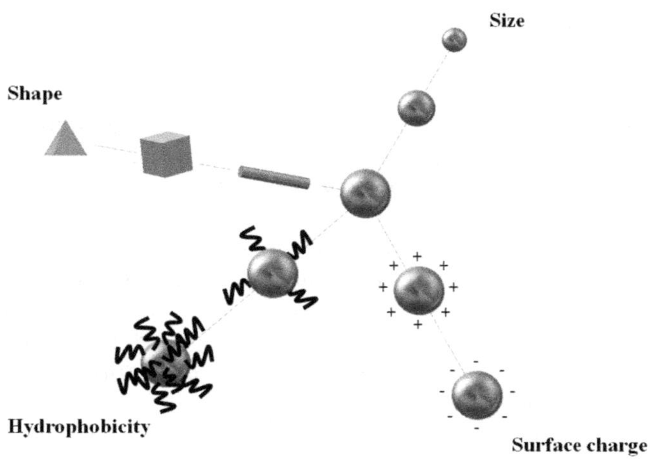

Figure 2.2 Main nanoparticle physicochemical properties influencing protein corona formation.

10 nm particle size difference significantly determined the protein corona of silica nanoparticles. Its composition was considerably affected quantitatively but not qualitatively.[13] The fact that particle size affects the amount of adsorbed proteins rather than the types of proteins was further confirmed by Dobrovolskaia *et al.*[22] However, it was not in agreement with results from Lundqvist *et al.* who exposed 50 and 100 nm polystyrene nanoparticles to human plasma and observed that the nanoparticle size affected the type of adsorbed proteins to their surface. For instance, apolipoprotein B-100 did not adsorb to 50 nm nanoparticles but bound to 100 nm nanoparticles.[5]

Globally, larger nanoparticles preferentially adsorb proteins with a high molecular weight, whereas proteins with a lower molecular weight accumulate on the surface of smaller nanoparticles.[23] This positive correlation between the nanoparticle size and the molecular weight of adsorbed proteins was also observed with ultrasmall superparamagnetic iron oxide nanoparticles (SPIONs, ranging from 3.5 to 15 nm), but the concentration and composition of the proteins within the corona were random.[23]

Related to the size, the curvature of the particles is a crucial parameter in the composition and conformation of the proteins forming the corona.[24] Indeed, the curvature of the particles influences the binding association constant of proteins with particles.[5,13] Thus, adsorption of proteins differs in planar and curved surface of a similar nature. Especially, nanoparticles' highly curved surfaces are characterized by low protein interactions; this increased curvature decreases protein adsorption to the point where it no longer occurs (for very small nanoparticles).[24,25]

Regarding the influence of surface charge, as a rule of thumb, charged particles tend to adsorb more proteins than neutral nanoparticles.[10,12,18,26] Logically, negatively charged particles attract positively charged proteins and *vice versa*; however, this simplified statement does not fit all circumstances.[10]

The hydrophobicity of a nanoparticle is also strongly correlated to its protein-binding properties.[3,6,26] In general, nanoparticles with hydrophobic surfaces tend to adsorb more proteins than those with hydrophilic surfaces.[13,14,27]

Deng *et al.* observed a shape-dependent adsorption of some proteins. In particular, clusterin and apolipoprotein D were found in the protein coronas of spherical nanoparticles but were not detected on nanorods or nanotubes.[28]

Nanoparticle surface chemistry has been shown to deeply influence the evolution and composition of a protein corona. As an example, the protein coronas formed at the surface of native, amine-modified and carboxy-modified silica nanoparticles were different in terms of composition and consequently induced different biological effects.[29,30] Therefore, modification of the nanoparticle surface through chemical functionalization is a way to change the biological outcomes of nanoparticles.[20]

The protein corona represents the new interface between nanoparticles and biological systems and will thus have an impact on the biological response. However, because the physicochemical properties of nanoparticles dictate the composition of this protein corona, the nanoparticle fingerprint encodes information about the size, surface charge, and agglomeration state of the nanoparticles and they will consequently have an indirect impact on the nanoparticle's fate. Of note, even small physicochemical changes in the corona proteins may elicit different biological responses. However, they are not the only parameters to consider as the formation of a protein corona also greatly depends on the biological medium.

2.2.1.2 Biological Environmental Factors

When designed for medical applications, such as drug delivery or imaging, nanoparticle administration often requires intravenous injections. On the other hand, while nanoparticles we are unintentionally exposed to can enter an organism through inhalation, ingestion or skin contact, because of their small size they can cross biological barriers and be found in the bloodstream. This is why many studies in the literature focus on the formation of protein coronas of nanoparticles exposed to blood (plasma or serum). Blood is a very rich biological medium, composed of about 3700 different proteins at various protein concentrations as high as 60–80 g L^{-1}.[8,17] As mentioned before there is no universal corona, especially considering its highly dynamic nature, but general trends can be mentioned: for instance, when nanoparticles are exposed to human blood, human serum albumin (HSA), immunoglobulin (IgG) and fibrinogen are found to be the major hard corona proteins binding firmly to the nanoparticle surfaces and showing a distinctive stability.[8]

Many parameters related to the experimental system deeply influence the formation and composition of the protein corona and subsequently the biological effects of the nanoparticles. Some of these factors have been ignored for a long time and can be responsible for the lack of consistency in results reported in the literature.[7]

The biological environmental factors influencing the formation of the protein corona especially include the nature of the biological medium (blood, serum, plasma, cell culture media, lung lining fluid, gastro-intestinal fluid, *etc.*),[31,32] the temperature,[33] the pH, the serum type (human *vs.* animal,[34] from a healthy donor or from a patient suffering from a disease[35,36]), the protein concentration,[6,37] the incubation duration,[12,38] *etc.*

Let's detail some examples. The influence of the protein concentration was illustrated by Monopoli *et al.* who demonstrated that increasing concentrations of plasma resulted in an increased thickness of the hard protein corona.[6]

Regarding the type of biological medium, it was shown that incubation of nanoparticles with human serum or human plasma resulted in the formation of protein coronas that elicit different cell responses (especially with respect to nanoparticle cellular uptake) due to their different compositions.[31]

The characterization of the protein corona is usually carried out after incubation of samples in human plasma. However, mice are often the first models used to test the efficiency of nanoparticles *in vivo* and the assessment of the protein corona of nanoparticles using human plasma could introduce unanticipated differences. This is especially relevant as attention has been paid to the correlation between the composition of the protein corona obtained from biological samples from human and mouse origins.[34]

Also, the composition of the hard corona varies among healthy individuals, as well as among patients with various diseases/medical conditions. It means that the same nanoparticles may have different protein coronas and thus different biological identities for different individuals.[36] This influence was confirmed *in vitro* with 10 nm gold nanoparticles; their incubation either with normal or malignant ovarian cell lysates resulted in different protein corona compositions.[35]

Beside studies performed exposing nanoparticles to blood to better understand what happens *in vivo*, especially relevant for studies in a nanomedicine context, many *in vitro* studies are also carried out to better understand the biological impact of nanoparticles at a cellular level. To that purpose, nanoparticles are incubated with cells in the presence of their culture medium, the composition of which has been shown to play a key role in the formation of protein coronas. Maiorano *et al.* compared two commonly used cell culture media, which differ in glucose, salt, and amino acid composition: Dulbecco Modified Eagle's Medium (DMEM) and Roswell Park Memorial Institute medium (RPMI) that were both supplemented with fetal bovine serum (FBS). They exposed various sized citrate-capped gold nanoparticles to these culture media and reported that the protein corona formed in DMEM was more abundant and stable compared to that formed in RPMI. It was also found that protein corona composition did not relate to the amount of serum proteins. Furthermore, the differences in the protein–nanoparticle complexes mediated different cellular responses. This result illustrates how crucial it is to take into account environmental factors in addition to nanoparticles' intrinsic parameters to better characterize and understand the protein corona composition and its subsequent impact on biological systems.[32]

Another example in support of this is the study from Lesniak *et al.* They exposed 40 and 100 nm polystyrene nanoparticles to sera with and without heat inactivation (HI), a process that destroys heat-labile complement proteins. It led to the formation of protein coronas with different compositions. In particular, higher amounts of proteins were adsorbed onto nanoparticles after exposure to HI serum compared to non-HI serum.[39]

Hence, systems are highly sensitive to even minor changes of some parameters. Experimental conditions are thus critical and should be tightly controlled to allow inter-study comparisons and avoid a lack of reproducibility. Consequently, little variations can be responsible for conflicting results. Therefore, by affecting the nature of the proteins binding to the nanoparticles, the parameters mentioned will have an influence on the nanoparticle itself and on its subsequent biological impact.

2.3 Impact of the Protein Corona on the Nanoparticle's Physicochemical Properties

The protein corona formed upon contact with biological media will serve as a new interface between nanoparticles and biological systems.[12] This protein layer defines the "biological identity" of nanoparticles, which could be significantly different from their original synthetic identity (Figure 2.1). Indeed, the pristine nanoparticle is "screened" by the proteins adsorbed on its surface and interactions with biological systems will be made through these proteins and not with bare materials.[17] This biological identity is what cells "see"[40] and it will determine the physiological behaviour of nanoparticles influencing their colloidal stability, targeting capability, kinetics of circulation, transport, cellular uptake and organ accumulation, degradation, drug release, signalling and toxicity.[8,41]

The formation of a protein corona can significantly alter a nanoparticle's physicochemical features especially surface charge and size. Indeed, hydrodynamic diameter includes both the nanoparticle size and the protein layer thickness, thus nanoparticle size tends to increase in biological fluids.

Generally, it is observed that the presence of a protein corona prevents agglomeration, or, in other words, increases colloidal stability. Indeed, many authors report that agglomeration is much diminished in a medium containing serum compared to a serum-free medium.[20,42–44]

Stabilization by the protein corona can be explained by electrostatic repulsion, which is achieved by rendering the surface charge negative (since serum proteins are mostly negatively charged).[20,45] And indeed, when nanoparticles are incubated with serum, such a change can occur that initially positively charged nanoparticles can turn negative.[5,46] It has been well documented with different types of nanoparticles such as polystyrene,[47] gold, silica[29,30,48] or magnetic nanoparticles.[49] And the higher the zeta potential, the longer the range of repulsive forces. Therefore, suspension stability is enhanced by increased zeta potential.[43]

Another explanation of the stabilizing effect of a protein corona against nanoparticle agglomeration lies in the fact that this protein layer provides a steric barrier,[45] preventing any nanoparticle–nanoparticle interactions.[42] This was exemplified by citrate-stabilized gold nanoparticles, which agglomerated immediately in phosphate buffered saline (PBS), while they were stable in plasma.[21]

Finally, by lowering the surface energy of the nanoparticle, the protein corona promotes its dispersion.[8]

However, it was also reported that the presence of a protein corona favoured agglomeration,[28,50] suggesting that additional factors may be involved such as the concentration of proteins, the nature of serum, *etc.* Protein concentration is a key parameter: at low plasma concentrations, proteins are unable to reach the nanoparticle surface quickly enough to prevent agglomeration.[14] The ability of plasma to stabilize nanoparticles is also nanoparticle-dependent. Polystyrene nanoparticles form dimers, trimers, and large agglomerates after introduction to plasma, even at a high protein concentration.[6,40] This may result from a failure of the protein corona to form sufficiently quickly on these nanoparticles. Alternatively, agglomeration may result from "bridging" between proteins on different nanoparticles.[14,40]

Oliveira *et al.* evaluated the impact of the protein corona on the colloidal stability of poly lactic-*co*-glycolic acid (PLGA) nanoparticles (bare or incubated in the presence of three media: BSA solution and plasma of mouse and human origins). The nanoparticles were stable in the BSA solution and mouse plasma, but not in human plasma. This unexpected behaviour in plasmas of different origins could be correlated to a significant variation of the amount of proteins adsorbed on the nanoparticles and suggests that inter-species variation could impact on the colloidal stability of the nanoparticles.[51]

Therefore, the protein corona alters the physicochemical features of the nanoparticles and it will have deep implications with respect to their behaviour in physiological systems.

2.4 Consequences of the Protein Corona on Nanoparticles' Biological Fate and Outcomes

Beyond the impact of the presence of a protein corona on the intrinsic properties of the nanoparticles, there will also be a strong impact on the nanoparticle–biological systems interactions and the subsequent cell response. As schematically summarized in Figure 2.1C and as detailed in the following paragraphs, the influence of the protein corona composition (in terms of protein types, their relative abundance, and conformation at the particle surface) will occur at different levels: nanoparticle–cell interactions, cellular internalization, *in vivo* biodistribution, toxicity, immune response and cell signalling.[17,52]

2.4.1 Impact of the Protein Corona on the Nanoparticles' Biological Fate

2.4.1.1 Impact of the Protein Corona on Nanoparticle–Cell Interactions

Among the numerous parameters that influence nanoparticle–cell interactions, the presence of a protein corona plays a key role. It was shown that direct physical associations between cells and nanoparticles are much stronger for the bare surfaces.[53] The privileged interaction of positively charged nanoparticles with a cell membrane, which harbours negative surface charges, is especially recognized. But, obviously, with adsorbed proteins altering the nanoparticle surface charge, the picture is more complicated than opposite surface charges mediating electrostatic interactions. The charge of the protein coat rather than the native particle surface charge will determine electrostatic binding.[54] Thus, as illustrated in Figure 2.3, because

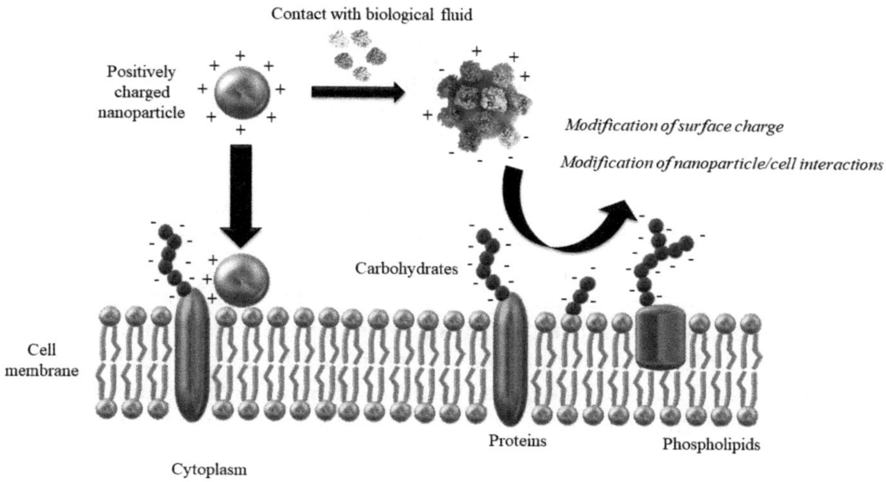

Figure 2.3 Influence of the protein corona on nanoparticle–cell interactions.

nanoparticle–cell membrane interactions mainly occur through electrostatic interactions, a protein corona, by inducing changes in the nanoparticle surface charge, will alter nanoparticle–cell interactions. This was experimentally proven by Lesniak *et al.* who observed that for identical particles and cells, under identical conditions, the interactions with cells and the biological outcomes could vary greatly in the presence or in the absence of a preformed corona in serum.[53]

Interaction with a cell membrane is the first step toward cell internalization, therefore, there is a close relationship between adhesion to a cell membrane and cellular uptake. Clearly, nanoparticles in serum-free conditions show a higher degree of adhesion on the cell membrane; the adhesion could, at least in part, contribute to the higher uptake efficiency.[53] On the contrary, the presence of a protein corona results in lower non-specific interactions between the pristine nanoparticle and the cell membrane leading to a decrease in cell uptake.[55]

2.4.1.2 Impact of the Protein Corona on the Nanoparticles' Uptake by Cells

An optimal understanding of how nanoparticles interact and are taken up by cells is of paramount importance both for nanomedicine and nanotoxicology. Indeed, when nanoparticles are used for biomedical purposes, therapeutic and/or diagnostic agents must generally enter the cells to reach their target. On the other hand, because of their size, high reactivity and large surface area, nanoparticles can interact with cell components, potentially inducing side effects and toxicity.[56] So depending on the context one may want to foster or avoid cell uptake. And this can be done through the modulation of the protein corona as it is now acknowledged that the cell internalization of nanoparticles strongly depends on their biological rather than synthetic identity and thus on the protein corona.[14,17] Therefore, a better knowledge of the specific role played by this factor is a crucial issue for the understanding of the biological fate of nanoparticles as well as potential adverse effects.

Usually, the protein corona has a protective role against cell uptake and nanoparticles exhibiting a protein corona are taken up to a lesser extent compared to their bare counterparts.[19,39,55,57–59] Indeed, it has been observed that nanoparticle uptake in serum-free conditions is, in most cases, higher than that measured for the same nanoparticle in the presence of serum.[53] The protein corona could decrease nanoparticle adhesion to the cell membrane due to the decreased surface-free energy after binding with blood proteins, thus decreasing particle–cell interactions and resulting in reduced uptake.[55,57]

But some studies have shown the contrary, *i.e.* the presence of the protein corona enhancing nanoparticle uptake by cells.[46,47,49]

These discrepancies can be explained by different cell internalization pathways. At this point, non-specific uptake must be distinguished from specific uptake.[54,56] Specific internalization is regulated by membrane receptors that are only activated by receptor-specific ligands to trigger internalization. Non-specific uptake is a random process without specific biomolecular control by the cell.[41] Therefore, as illustrated in Figure 2.4, in the absence of serum proteins, the uptake occurs through a direct interaction between the particles and the cell surface whereas in the presence of proteins, the uptake probably proceeds by a specific interaction between the adsorbed proteins and their

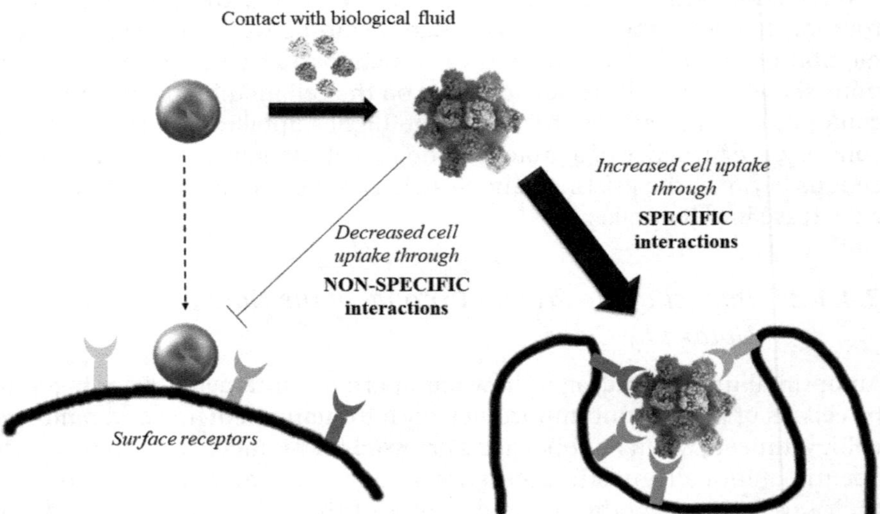

Figure 2.4 Influence of the protein corona on nanoparticle-specific (receptor-mediated) and non-specific uptake by cells.

receptors at the cell surface. In other words, in the presence of a corona, non-specific uptake seems to be decreased whereas specific uptake seems to be promoted.[41,43,46,55,60] Moreover, a protein-rich corona may interact with multiple receptors and multiple mechanisms may be involved simultaneously. In contrast, particles having one type of protein adsorbed on their surface may be restricted to a specific receptor.[49,54,56,60]

Cell internalization or endocytosis pathways will also be affected by the protein corona's composition. Endocytosis includes phagocytosis (mainly for large particles) and pinocytosis (rather for small particles, fluids and solutes). Pinocytosis comprises at least four distinct mechanisms: macropinocytosis, clathrin-mediated, caveolae-mediated, and clathrin- and caveolae-independent endocytosis, each of which depends on distinct components and mechanisms.[9,61] In most cases, the nanoparticle surface provides the driving forces (electrostatic, hydrophobic and hydrophilic (polar) forces) for the cellular internalization and decides the uptake pathway.[56] Thus the protein corona composition will directly influence the cellular uptake mechanism. In support of this conclusion, Lunov *et al.* demonstrated that when nanoparticles were incubated in a serum-free medium, the uptake mechanism was clathrin-mediated endocytosis for polystyrene nanoparticles functionalized with carboxylic groups and micropinocytosis for polystyrene nanoparticles functionalized with amine groups. However, their uptake mechanism shifted to phagocytosis when incubated in a medium containing serum.[62] Similarly, polystyrene nanoparticle uptake by macrophages in a protein-free medium was shown to change from clathrin-mediated endocytosis to phagocytosis when incubated in serum-enriched media.[58] Finally, Walkey *et al.* also observed that variations in serum protein adsorption correlated with differences in the mechanism and efficiency of nanoparticle uptake by a macrophage cell line.[18] In addition, an increased nanoparticle size due to the protein corona's presence can be potentially involved in endocytosis pathway shift.

Other factors need to be considered regarding the choice of endocytic pathway. Once again, even if the complexity of the protein corona makes it difficult to draw firm conclusions on its effect on nanoparticle cellular uptake, it is yet possible to say it is particle-property-dependent, cell-type-dependent and environment-dependent.[10] Indeed, it has been suggested that different cell types might have different uptake mechanisms, even for the same nanoparticles.[63,64] Cheng *et al.* also reported that the protein corona significantly decreased the internalization of gold nanoparticles in a particle-size- and cell-type-dependent manner.[57]

Nanoparticle agglomeration is another factor to take into account as it tends to lower cell internalization.[65,66] It is very likely because agglomeration rate and agglomerate size have a tremendous impact on particle transport and dramatically affect diffusion and gravitational settling.[29] As discussed before, many authors have reported that agglomeration is much diminished in a medium containing serum compared to a serum-free medium.[42,43] One possible explanation is that the adsorption of proteins affects the agglomerate size and surface charge of the nanoparticles, which alters the electrostatic binding affinity with cells.[56] It also alters the presentation of surface-adsorbed proteins to the cell's receptors. Agglomerates are large and require an increased recruitment of cell receptors to drive internalization. Agglomerates also possess an irregular morphology that produces multiple cell binding configurations with unequal uptake efficiencies. When agglomerates change the presentation of corona proteins to the cell membrane, this alters nanoparticle–cell interactions and changes the cell response.[65]

The type of medium should also be considered carefully. The previously mentioned study of Maiorano *et al.* on the comparison of the corona formed in DMEM or RPMI media also showed an impact on cell uptake. More specifically, internalization of protein–nanoparticle complexes in cells that were formed in RPMI media was notably higher than those formed in DMEM.[32]

Last but not least, the nature of the proteins composing the protein corona will dramatically affect cell uptake. While the protein corona decreases non-specific nanoparticle–cell membrane interactions (Figure 2.4), it can either promote or inhibit specific (receptor-mediated) interactions depending on its composition.[55] As an example, a positive correlation was observed between the amount of alpha-1 antitrypsin in the protein corona and cellular uptake.[67] But we should be cautious and keep in mind that the presence of a protein on the nanoparticle surface does not necessarily imply that it will interact with its corresponding cellular receptor. The protein orientation and conformation are critical factors.[14] Another example regarding variable cell uptake depending on the protein corona composition refers to opsonins and dysopsonins and will be discussed in the next section.

2.4.1.3 Impact of the Protein Corona on the Nanoparticles' Biodistribution, Clearance and Targeting

We have discussed the strong positive correlation between the plasma protein binding capacity of a nanoparticle and the rate at which it is taken up by cells. This is true both *in vitro* and *in vivo*, but *in vivo*, there is an additional layer of complexity. Nanoparticles can be taken up by cells from the reticuloendothelial system (RES), also known as the mononuclear phagocyte system (MPS). This system includes monocytes, macrophages and dendritic cells. This uptake results in the nanoparticles' removal from the blood circulation and rapid clearance.[14] Hence the notions of biodistribution, clearance and targeting are closely related, each of them being influenced by the protein corona's presence and composition.

Let's consider biodistribution first. It is clear that the *in vivo* location of a nanoparticle is related to its biological identity rather than its initial physicochemical properties, even though the latter will have an indirect influence as discussed before.[15] The impact of the protein corona on a nanoparticle's biodistribution will therefore be critical. A relevant example that binding of certain proteins can result in organ-specific distribution deals with apolipoproteins. Different apolipoproteins have been described to promote transport across the blood–brain barrier[3] by means of low-density lipoprotein (LDL) receptors.[6] The specific role of apolipoprotein E (apoE) was demonstrated by Yallapu *et al.* with enhanced transport across the blood–brain barrier of magnetic nanoparticles that had bound this protein.[49] Mahmoudi *et al.* rather emphasized the implication of ApoA-I and ApoA-II in their model of 3.5 nm sized and negatively charged SPIONs.[23] When incubated *in vitro* with FBS, nanoparticles adsorbed higher amounts of ApoA-I and ApoA-II compared to positively charged and plain SPIONs of a similar size. As these apolipoproteins can cross the blood–brain barrier through a saturable transport mechanism, once SPIONs were injected *in vivo*, ApoA-I and II were adsorbed on their surface and drove the nanoparticles to the brain vessels. This was demonstrated by magnetic resonance imaging only 5 min after injection.

Regarding nanoparticle clearance, it has been established that the protein corona may be responsible for recognition by the mononuclear phagocytic system leading to accelerated clearance.[26] In particular, this process is fostered when protein coronas are composed of opsonins. Opsonins are a subset of plasma proteins and include immunoglobulins (Ig), fibrinogen, and complement proteins. They promote the phagocytosis of nanoparticles by macrophages with rapid clearance of the nanoparticles from the bloodstream and accumulation in the liver and spleen for eventual elimination.[3,4,10,14,19,68] Adsorbed IgG facilitates nanoparticle uptake by interacting with Fc receptors (FcR) expressed on the surface of macrophages. FcR recognizes the Fc fragment of an adsorbed antibody. Accessibility of the Fc fragment is critical. Antibodies that adsorb to a nanoparticle specifically *via* their Fab fragment tend to present the Fc fragment to the environment. On the other hand, antibodies that adsorb non-specifically may have their Fc portion blocked by the nanoparticle surface. Both specific and non-specific adsorption of an antibody is possible, depending on the nanoparticle formulation.[14] On the contrary, adsorption of another type of proteins called dysopsonins like human serum albumin (HSA) or apolipoproteins can inhibit nanoparticle uptake and thus promote prolonged bloodstream circulation and the crossing of biological barriers.[3,4,10,14,19,38,68] Opsonization and dysopsonization are dynamic competing processes.[10]

In the context of nanomedicine, it is desirable to increase the nanosystem circulation time to allow it to reach its target and to benefit from the enhanced permeability and retention (EPR) effect for tumour delivery for instance.[69] Opsonization is a limiting factor for that. Hence, strategies to avoid the adsorption of proteins on the nanoparticle surface (opsonins in this particular example) have been developed, the most popular being the so-called PEGylation, consisting of coating the nanoparticle's surface with polyethylene glycol (PEG) polymers. These hydrophilic polymers provide steric stabilization and reduce the interactions of nanoparticles with blood proteins.[10,25,70] Consequently, PEGylation allows the modulation of the circulation time in the blood system as well as the biodistribution of nanoparticles. The reduction of unspecific cellular uptake of nanoparticles is referred to as the "stealth effect".[16]

An alternative to PEGylation was explored by Müller *et al.* through non-covalent coating of nanocarriers with surfactants based on structurally versatile poly-phosphoesters. They showed that this coating facilitated tailoring the protein adsorption on nanoparticles and their behaviour in a biological environment.[16]

If these strategies are efficient overall to prevent the formation of a protein corona, they cannot allow the achievement of 100% surface coverage and PEGylated nanoparticles cannot completely escape cellular uptake.[10] On the other hand, one should be careful with this process as making nanoparticles invisible to the immune system and more penetrating may also alter their toxicity profile as discussed in the following section.[27]

Also, excessive PEGylation can reduce the targeting potential of nanoparticles because of the strong inhibition of their uptake by the target cells. Thus, a balance between anti-opsonization strategies and target specificity should be reached.[20]

Indeed, targeting is another crucial issue in nanomedicine.[10,71] A typical example is nanosystems used for targeted drug delivery in the treatment of cancer; their efficiency will be directly related to their cellular uptake by tumour cells, sparing healthy cells to limit side-effects. A common strategy to achieve targeting at a cellular level is to functionalize the nanoparticle surface with biomolecules, whose receptors are over-expressed in the cells to be targeted. In this context, human transferrin has been widely used as a cancer targeting agent.[69]

But, as discussed before and as illustrated in Figure 2.5, targeting can be impacted by the presence of a protein corona.

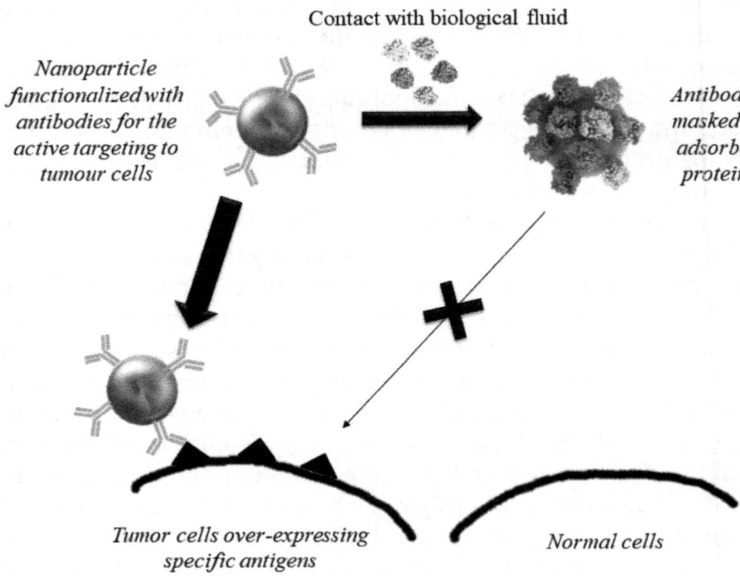

Figure 2.5 Impact of the protein corona on nanoparticle active targeting.

This effect is all the more important as nanovectors are administered through intravenous injections. Adsorbed plasma proteins can shield patterns grafted on the nanoparticle surface for targeting purposes. As an example, Salvati *et al.* showed that the targeting abilities of transferrin-functionalized silica nanoparticles disappeared after their incubation in biological fluids, due to the formation of a protein corona that prevented transferrin from binding its receptor. Consequently, the transferrin-related specificity was lost.[72]

Beside the screening of transferrin, adsorbed proteins can enter off-target cells through receptor-mediated endocytosis and trigger adverse effects.[72] Indeed, as discussed before, the protein corona can mediate a change in internalization pathways and different adsorption patterns could dictate the target.[49]

On the other hand, the formation of a protein corona is not always a bad thing as it may mitigate nanoparticle-induced cell toxicities.[4]

2.4.2 Impact of the Protein Corona on the Nanoparticles' Biological Outcomes

2.4.2.1 Impact of the Protein Corona on the Nanoparticles' Toxicity

Plasma protein binding capacity is correlated with cytotoxicity, very likely because of stronger nanoparticle–cell interactions.[14] As shown in Figure 2.3, pristine nanoparticles interact directly with cell membranes and can induce cell damages, potentially leading to cell death.[15] However, such toxicity is mitigated in the presence of serum in a reverse dose-dependent way: cell death decreases as serum concentration increases.[10]

In general, the presence of a protein corona reduces the cytotoxic effects of nanoparticles[20,29,68,73] as a consequence of a lower cell uptake.[32,41,53] Furthermore, many toxicity properties of nanoparticles derive from the reactivity of their surface in interfacing cellular membranes; the presence of a coating can mask these properties and can be considered as a factor that increases the safety of nanoparticles.[15,53] This statement can easily be generalized as it was observed with different types of nanoparticles such as carbon nanotubes,[73] silica[53] and graphene oxide nanoparticles.[74] The presence of a protein corona at the surface of these nanoparticles translated into a protective effect since cytotoxicity was mitigated in its presence compared to the cell damages observed in its absence (*i.e.* when cells were exposed to the same nanoparticles in a medium without proteins).[8]

Similarly, Docter *et al.* studied the influence of a preformed BSA corona on silica nanoparticles incubated with three different cell lines (A549, HepG2, and L929). They concluded that the toxicity of the nanoparticles was significantly lowered by the BSA corona.[59] It was hence proposed that, similar to PEGylation, this natural albuminization may improve the biocompatibility of nanoparticles, preventing unwanted biological effects.[27]

Another protective effect from the protein corona against cell toxicity lies in the fact that it represents a physical protective layer, especially for nanoparticles which degradation releases toxic by-products.[15] This was observed with zinc oxide (ZnO)[75] and silver (Ag)[76] nanoparticles in which the formation of a protein corona inhibited nanoparticle dissolution and consequent release of Zn and Ag ions that eventually became toxic for cells.

Lastly, the formation of a protein corona can increase the safety of nanoparticles by inhibiting the generation of radical oxygen species (ROS) by which several compounds exert their cytotoxic activity. For instance, zinc oxide nanoparticles are able to naturally trigger the formation of ROS; however, the presence of a protein corona at their surface inhibits this phenomenon.[75]

On the other hand, the composition of a protein corona may be responsible for adverse effects. For instance, as mentioned earlier, the protein corona may alter the *in vivo* biodistribution of nanoparticles like nanoparticles able to cross the blood–brain barrier when bound to apolipoproteins. This may lead to neurotoxic side effects.[23] Similarly, by interacting with cells, nanoparticle–corona complexes can initiate signalling processes.[77]

Proteins from the complement and coagulation systems are among the major components of the protein corona of various nanoparticles after exposure to blood. Activation of complement and coagulation is important both for opsonization and systemic toxicity as it can lead to severe anaphylaxis or thrombosis.[14,71] Indeed, the complement system is involved in the immune response. Upon binding to a nanoparticle surface, proteins from the complement can experience conformational changes resulting in their activation.[78]

Similarly, nanoparticles can interact with coagulation factors and hence manipulate the clotting events.[10] The protein corona has been linked to the pro-coagulant activity of nanoparticles.[20] Tenzer *et al.*[12] showed that when incubated with uncoated silica nanoparticles, human thrombocytes were activated. However, the same nanoparticles exhibiting a protein corona (obtained by previous incubation with human plasma) did not trigger this activation. Similarly, while uncoated silica nanoparticles had a significant haemolytic activity on red blood cells, the presence of a protein on the particle surface protected the cells from this damage.[12] Hemocompatibility is a critical issue for intravenously delivered nanoparticles. The haemolytic activity of nanoparticles can be dramatically decreased by the formation of a protein corona, reducing endothelial cell dysfunction, vascular thrombosis and various clinical manifestations related to haemoglobin release.[20]

2.4.2.2 *Impact of the Protein Corona on the Immune Response to Nanoparticles*

Interactions of the nanoparticles with the immune system are important to consider both for nanomedicine efficiency and safety reasons.[25] The presence of a protein corona at the nanoparticle surface mediates and has the ability to alter the immune response.[15]

When nanoparticles come into contact with human plasma, the majority of adsorbed molecules are immunoglobulins, complement factors and coagulation proteins. All these biomolecules have specific receptors on immune cells, which, upon their binding, results in cell activation.[4]

Nanoparticles bearing a protein corona display hydrophobic surfaces that are interpreted as danger signals by the immune system. Moyano *et al.* have shown in animal models that nanoparticle hydrophobicity dictated immune responses. They hypothesized that immune cells were "blind" or at least "short sighted" to bare nanoparticle surfaces,[79] while the protein corona has the ability to initiate alternative immune patterns. For instance, if the protein corona activates the components of the immune system like helper T lymphocytes type 1 (Th1), B lymphocytes and macrophages type 1 (M1), the entire panel of secreted molecules (immunoglobulins, cytokines and chemokines) generates an acute inflammation reaction but not a prolonged one and hence no neoplastic events. On the other hand, if the protein corona activates Th2, M2 and regulatory T lymphocytes (Tregs), then the array of secreted molecules will sustain a chronic inflammation and hence possible pro-tumoral activity.[80]

The complement system identifies and induces elimination of non-self-entities. The adsorption of complement proteins on the nanoparticle surface may lead to complement activation and inflammation.[15]

In addition, binding of proteins to the nanoparticle surface may result in abnormal unfolding of the protein with exposure to the environment of epitopes, which are normally hidden.[58] Such revealing of cryptic epitopes can cause an immune reaction potentially leading to immunotoxicity and inflammation.[10,20]

2.4.2.3 *Alteration of Protein Conformation Following Their Adsorption on a Nanoparticle Surface and Impact of the Protein Corona on Cell Signalling*

So far, we have considered the consequences of the presence of a protein corona on a nanoparticle's biological fate and outcomes. But let's flip the coin and focus on the consequences on the proteins themselves, in particular with regards to their structure and function. The binding of proteins onto a nanoparticle's surface may result in conformational changes to the secondary structure of adsorbed proteins consequently altering their biological activities and potentially leading to a complete loss of protein activity with the most flexible proteins being the most influenced.[8,45] It has also been shown that nanoparticles can impact protein fibrillation processes.[24]

The role of the protein corona in signalling processes has also begun to emerge.[64] A relationship was clearly established between the protein corona composition on silver nanoparticles and the proteomic or transcriptomic responses. The deregulated cellular pathways especially included oxidative stress response, mitochondrial energy metabolism, cell–cell contacts and kinase-dependent signalling.[81]

Because of all the biological potential consequences of the presence of a protein corona, one should be aware of its implication and should try to characterize it in the best way possible.

2.5 Issues Associated with the Protein Corona

2.5.1 Complexity of the Protein Corona

As discussed before, there is no universal protein corona. In addition, it evolves over time and depends on the nanoparticles' location, making its study a challenging task. For the same reasons, it is also hard to reliably predict the *in vivo* behaviour of nanoparticles, which is a crucial issue both in the context of nanomedicine and nanotoxicology.

Technically, the characterization of a protein corona is a multi-step process that has to face some hurdles. First, it is noteworthy to mention that only the hard corona can be studied, the soft corona being altered by the analytic process. Classically, the experimental approach consists of incubating nanoparticles with a biological milieu, then by means of (ultra)centrifugation, the unbound proteins are removed. Proteins adsorbed on the nanoparticle surface are eluted by the use of denaturing agents. The recovered proteins are finally analysed either through one- or two-dimensional gel electrophoresis or by proteomic analyses through mass spectrometry.[15,58,82]

Although they provide informative data, the set of techniques available to analyse the protein corona composition has some drawbacks. First, many techniques measure an ensemble average of all adsorbed proteins, and cannot distinguish individual proteins within the corona.[14] Moreover, because the corona is not at thermodynamic equilibrium, there could be statistical fluctuations in its composition and organization from particle to particle within the same sample. This suggests that nanoparticles within the same sample may have slightly different hard coronas and, consequently, different biological actions.[8,83] Furthermore, it cannot be excluded that some proteins present at a minor level could be responsible for major biological consequences and *vice versa*.[41]

2.5.2 Discrepancies Between *In Vitro* and *In Vivo* Data

Another issue with protein coronas is the gap between *in vitro* and *in vivo* data. Indeed, many studies performed *in vitro* have given informative data but they are not necessarily all relevant, not representative of what really happens *in vivo* due to differences in experimental conditions. Because *in vitro* models oversimplify and do not take into consideration important factors, they cannot reproduce accurately the complexity of living systems. On the other hand, *in vivo* studies are scarce, mainly because of the difficulty of recovering nanoparticles after administration.[17] In addition, the more complex sample preparation steps before analysis could also bring variations.[10]

Of interest is the work by Hadjidemetriou *et al.*[84,85] who characterized the protein corona formed *in vivo* on liposomes. The latter were injected intravenously into CD-1 mice and then recovered from the blood circulation. While the total amount of proteins bound to circulating liposomes was similar to that observed from *in vitro* incubations, the number of different proteins identified in the *in vivo* protein corona was considerably larger. This observation that the protein corona differs between *in vitro* (*i.e.* static-flow conditions) and *in vivo* (*i.e.* dynamic-flow environment) is in agreement with that of Palchetti *et al.* who reported that the protein corona formed on circulating liposomes was more negatively charged compared to the one formed on their static counterparts.[71] It was highlighted that the molecular complexity (*i.e.* the variety of protein species) of the protein corona formed around nanoparticles under dynamic flow is wider.[17]

Despite an increasing knowledge of nanoparticles, few nanoparticles are used in a clinical context.[17] This is partly due to the fact that most studies are performed *in vitro* while nanoparticles behave differently *in vivo*, in part because of a different protein corona. Further *in vivo* studies are necessary for a better understanding of nanoparticles' biological identity impact in living organisms.

2.6 Conclusion and Perspectives

Everyone agrees to recognize that the protein corona of nanoparticles plays a major role in their biological behaviour and outcomes. However, the exact nature of this role remains the subject of debate as conflicting results have been reported in the literature.[15] One reason is that many factors are involved in the formation and consequences of a protein corona. Thus, the biological effects triggered by nanoparticle–cell interactions cannot be fully evaluated in the absence of the context, *i.e.* without taking into account the combined contribution of the nanoparticle's physicochemical properties and the biological environment, both influencing the protein corona composition.[39,54] A better knowledge of the impact of protein coronas is critical for nanotoxicology and nanomedicine applications, especially for the safe and efficient design and applications of nanoparticles.[25]

Fortunately, the protein corona is now the subject of advanced research and is no longer an "ignored factor". Our knowledge on this topic is constantly increasing and becomes more accurate as new parameters previously overlooked are now taken into account.

So far, the research on protein coronas has been mainly focused on nanoparticle interactions with blood plasma or/and cell culture media containing serum.[45] Progress could be expected from the study of various biological fluids less studied but no less important such as lung surfactant or gastro-intestinal fluid. Indeed, depending on the way in which nanoparticles enter the body (*via* inhalation, oral ingestion, skin contact or injection), they will be exposed to very different environments, which will deeply impact the protein corona composition.[45] Because of the importance of the inhaled route, studies on nanoparticle contact with lung lining fluid are emerging. For instance, Konduru *et al.* reported that different nanoparticles (cerium oxide, barium sulfate, gold and zinc oxide nanoparticles) variably adsorbed proteins from the lung lining fluid. The amount of albumin in the protein corona varied as did nanoparticle uptake by macrophages. However, albumin coating did not affect the translocation of gold nanoparticles across the air–blood barrier. The extent to which the protein corona specifically affects the fate and biological effects of nanoparticles remains to be established.[86]

Furthermore, when nanoparticles move from one environment to another, such as from the lung to the blood, the protein corona can evolve considerably,[17] which can result in physiological responses different from those observed with the same nanoparticle directly injected into the blood.[14]

Another major factor has been largely neglected so far. It is the fact that physiological fluids are dynamic by nature.[12] And dynamic flows create shear stress on nanoparticles and provide a continual source of new biomolecules.[17] These conditions are usually not reproduced experimentally since current studies dealing with protein corona formation are generally conducted by static incubation or by gentle shaking at 37 °C from minutes to hours. Whereas body fluids are highly dynamic, the flow velocities in human ascending aorta can reach up to 60 cm s^{-1}, and particles are constantly conditioned by their microenvironment.[10] In this context, Palchetti *et al.*[71] demonstrated that the protein corona is influenced by exposure to dynamic flow (actually protein adsorption on a nanoparticle surface is largely amplified by dynamic flow), and that alterations are dependent both on incubation time and nanoparticle surface chemistry. Notably, changes in the protein corona composition induced by dynamic flow were found to affect cellular uptake of lipid nanoparticles by HeLa and MCF7 cancer cell lines.[71] A better understanding of the impact of the protein corona can come from experiments reproducing more faithfully "real-life" exposure conditions.

In the nanomedical field, many research teams focus their efforts on trying to predict the nanoparticles' biological fate and outcomes, in particular to design safe and efficient nanosystems.[17] In this context, computational methods can be very useful. Based on the assumption that nanoparticles of similar physicochemical properties will have similar bioactivity, a number of (quantitative) structure–activity relationship ((Q)SAR) models have been successfully developed for various nanoparticles, especially metal oxide nanoparticles.[67] Because the protein corona defines the biological identity of nanoparticles and affects their biological response, it is now accepted that the types of proteins and their abundance on the nanoparticle surface encode information that predicts nanoparticle bioactivity.[8] From here, correlations between the protein corona composition and cellular uptake have been explored with the ultimate goal to be able to predict a nanoparticle's behaviour from the knowledge of its protein corona composition. In this respect, Walkey *et al.* used nanoparticle properties (size, surface charge, surface area, agglomeration state, *etc.*) along with serum protein corona fingerprints to predict the associations between a library of 105 gold nanoparticles and A549 human lung epithelial carcinoma cells. That system was found to be 50% more accurate than a model based only on the physicochemical properties of the nanoparticles.[18] It means that the protein corona encodes more biologically relevant information about a nanoparticle than its physicochemical properties.[18] However, much more data are needed to develop reliable models.

The formation of a protein corona has long been considered (and sometimes is still considered today) as a problem, a detrimental phenomenon, leading to unpredictable biological identity and unwanted effects, such as particle recognition by immune system cells, macrophage activation and removal from blood circulation.[17,52] So strategies to prevent its formation were designed such as PEGylation as discussed before. However, this solution is not totally satisfactory and can also have drawbacks. Thus, as protein corona formation cannot be avoided, some researchers have proposed instead to exploit it. That means controlling it to eventually take advantage of it, for example for targeting purposes. Basically, it is founded on the idea of using the associated plasma proteins to target specific receptors on cells.[38] This approach involves the modification of the nanoparticle in a controlled manner to induce the binding of specific proteins.[15,16] There is now evidence that the corona proteins can be used for targeting specific cell types, and especially cancer cells.[87] For instance, the high affinity of vitronectin for the cationic-charged surface of lipoplexes was used to efficiently target the human breast cancer cells MDA-MB-435S, which overexpress two major vitronectin receptors. Internalization of naked lipoplexes (*i.e.*, without a protein corona, and thus without vitronectin on their surface) was lower.[87] Enrichment of apolipoproteins in a nanoparticle protein corona has also been used to target PC3 prostate carcinoma cells that express high levels of scavenger receptor class B type 1, thus facilitating their uptake by lipid transport mechanisms.[38] Prapainop *et al.* used the fact that nanoparticles with certain surface ligands bind to apo-B100, which induces unfolding and agglomeration resulting in cell-specific targeting to macrophages.[88] The ability to tune the formation of a protein corona and thus control its subsequent biological fate and consequences opens new perspectives in nanomedicine, such as decreasing toxicity and/or improving cellular uptake by specific target cells.[15]

Finally, it was recently reported that in patients, the blood composition can significantly vary depending on the disease. Significant changes in the expression of plasma proteins were observed due to the leakage of tissue biomarkers in blood. It consequently has an impact on the protein corona which composition may change depending on the disease type. This led to the introduction of the concept of personalized protein coronas.[20,36] The composition of the protein corona formed around lipid nanoparticles in human plasma drawn from pancreatic cancer patients was compared to that formed in plasma drawn from healthy volunteers. The protein corona of pancreatic cancer patients was found to be more enriched than that of healthy individuals.[10] Thus protein coronas formed in a biological environment may confer visibility to hidden biomarkers, which may correlate with the occurrence of pathology and can be used for diagnostic purposes. Hence, protein coronas on nanoparticles can be exploited to develop novel diagnostic tools for cancer.[20]

The stakes associated with a better knowledge of protein corona formation and biological consequences are huge. They concern both the nanotoxicology and nanomedicine fields. Progress in the understanding and control of protein coronas will allow the design of safer nanoparticles and will facilitate the use of safe and efficient nanomedicine approaches in clinics.

References

1. Nanotechnology Inventories, http://www.nanotechproject.org/inventories/, accessed 21 April 2018.
2. ISO/TR 13014:2012, ISO/TR 13014, http://www.iso.org/iso/catalogue_detail?csnumber=52334, accessed 28 November 2016.
3. T. Cedervall, I. Lynch, S. Lindman, T. Berggård, E. Thulin, H. Nilsson, K. A. Dawson and S. Linse, Understanding the nanoparticle-protein corona using methods to quantify exchange rates and affinities of proteins for nanoparticles, *Proc. Natl. Acad. Sci. U. S. A.*, 2007, **104**, 2050–2055.
4. G. Caracciolo, S. Palchetti, V. Colapicchioni, L. Digiacomo, D. Pozzi, A. L. Capriotti, G. La Barbera and A. Laganà, Stealth effect of biomolecular corona on nanoparticle uptake by immune cells, *Langmuir*, 2015, **31**, 10764–10773.
5. M. Lundqvist, J. Stigler, G. Elia, I. Lynch, T. Cedervall and K. A. Dawson, Nanoparticle size and surface properties determine the protein corona with possible implications for biological impacts, *Proc. Natl. Acad. Sci. U. S. A.*, 2008, **105**, 14265–14270.
6. M. P. Monopoli, D. Walczyk, A. Campbell, G. Elia, I. Lynch, F. B. Bombelli and K. A. Dawson, Physical-chemical aspects of protein corona: relevance to *in vitro* and *in vivo* biological impacts of nanoparticles, *J. Am. Chem. Soc.*, 2011, **133**, 2525–2534.
7. P. Foroozandeh and A. A. Aziz, Merging worlds of nanoparticles and biological environment: factors governing protein corona formation on nanoparticles and its biological consequences, *Nanoscale Res. Lett.*, 2015, **10**, 221.
8. M. P. Monopoli, C. Aberg, A. Salvati and K. A. Dawson, Biomolecular coronas provide the biological identity of nanosized materials, *Nat. Nanotechnol.*, 2012, **7**, 779–786.
9. A. E. Nel, L. Mädler, D. Velegol, T. Xia, E. M. V. Hoek, P. Somasundaran, F. Klaessig, V. Castranova and M. Thompson, Understanding biophysicochemical interactions at the nano–bio interface, *Nat. Mater.*, 2009, **8**, 543–557.
10. D. Chen, S. Ganesh, W. Wang and M. Amiji, Plasma protein adsorption and biological identity of systemically administered nanoparticles, *Nanomedicine*, 2017, **12**, 2113–2135.

11. L. Vroman and A. L. Adams, Findings with the recording ellipsometer suggesting rapid exchange of specific plasma proteins at liquid/solid interfaces, *Surf. Sci.*, 1969, **16**, 438–446.
12. S. Tenzer, D. Docter, J. Kuharev, A. Musyanovych, V. Fetz, R. Hecht, F. Schlenk, D. Fischer, K. Kiouptsi, C. Reinhardt, K. Landfester, H. Schild, M. Maskos, S. K. Knauer and R. H. Stauber, Rapid formation of plasma protein corona critically affects nanoparticle pathophysiology, *Nat. Nanotechnol.*, 2013, **8**, 772–781.
13. S. Tenzer, D. Docter, S. Rosfa, A. Wlodarski, J. Kuharev, A. Rekik, S. K. Knauer, C. Bantz, T. Nawroth, C. Bier, J. Sirirattanapan, W. Mann, L. Treuel, R. Zellner, M. Maskos, H. Schild and R. H. Stauber, Nanoparticle size is a critical physicochemical determinant of the human blood plasma corona: a comprehensive quantitative proteomic analysis, *ACS Nano*, 2011, **5**, 7155–7167.
14. C. D. Walkey and W. C. W. Chan, Understanding and controlling the interaction of nanoparticles with proteins in a physiological environment, *Chem. Soc. Rev.*, 2012, **41**, 2780–2799.
15. C. Corbo, R. Molinaro, A. Parodi, N. E. Toledano Furman, F. Salvatore and E. Tasciotti, The impact of nanoparticle protein corona on cytotoxicity, immunotoxicity and target drug delivery, *Nanomedicine*, 2016, **11**, 81–100.
16. J. Müller, K. N. Bauer, D. Prozeller, J. Simon, V. Mailänder, F. R. Wurm, S. Winzen and K. Landfester, Coating nanoparticles with tunable surfactants facilitates control over the protein corona, *Biomaterials*, 2017, **115**, 1–8.
17. G. Caracciolo, O. C. Farokhzad and M. Mahmoudi, Biological Identity of Nanoparticles *In Vivo*: Clinical Implications of the Protein Corona, *Trends Biotechnol.*, 2017, **35**, 257–264.
18. C. D. Walkey, J. B. Olsen, F. Song, R. Liu, H. Guo, D. W. H. Olsen, Y. Cohen, A. Emili and W. C. W. Chan, Protein corona fingerprinting predicts the cellular interaction of gold and silver nanoparticles, *ACS Nano*, 2014, **8**, 2439–2455.
19. P. Chandran, J. E. Riviere and N. A. Monteiro-Riviere, Surface chemistry of gold nanoparticles determines the biocorona composition impacting cellular uptake, toxicity and gene expression profiles in human endothelial cells, *Nanotoxicology*, 2017, **11**, 507–519.
20. P. Jain, R. S. Pawar, R. S. Pandey, J. Madan, S. Pawar, P. K. Lakshmi and M. S. Sudheesh, In-vitro in-vivo correlation (IVIVC) in nanomedicine: Is protein corona the missing link?, *Biotechnol. Adv.*, 2017, **35**, 889–904.
21. M. A. Dobrovolskaia, A. K. Patri, J. Zheng, J. D. Clogston, N. Ayub, P. Aggarwal, B. W. Neun, J. B. Hall and S. E. McNeil, Interaction of colloidal gold nanoparticles with human blood: effects on particle size and analysis of plasma protein binding profiles, *Nanomedicine*, 2009, **5**, 106–117.
22. M. A. Dobrovolskaia, B. W. Neun, S. Man, X. Ye, M. Hansen, A. K. Patri, R. M. Crist and S. E. McNeil, Protein corona composition does not accurately predict hematocompatibility of colloidal gold nanoparticles, *Nanomedicine: Nanotechnol., Biol. Med.*, 2014, **10**, 1453–1463.

23. M. Mahmoudi, S. Sheibani, A. S. Milani, F. Rezaee, M. Gauberti, R. Dinarvand and H. Vali, Crucial role of the protein corona for the specific targeting of nanoparticles, *Nanomedicine*, 2015, **10**, 215–226.
24. I. Lynch and K. A. Dawson, Protein-nanoparticle interactions, *Nano Today*, 2008, **3**, 40–47.
25. M. Neagu, Z. Piperigkou, K. Karamanou, A. B. Engin, A. O. Docea, C. Constantin, C. Negrei, D. Nikitovic and A. Tsatsakis, Protein bio-corona: critical issue in immune nanotoxicology, *Arch. Toxicol.*, 2017, **91**, 1031–1048.
26. D. E. Owens and N. A. Peppas, Opsonization, biodistribution, and pharmacokinetics of polymeric nanoparticles, *Int. J. Pharm.*, 2006, **307**, 93–102.
27. E. Casals and V. F. Puntes, Inorganic nanoparticle biomolecular corona formation, evolution and biological impact, *Nanomedicine*, 2012, **7**, 1917–1930.
28. Z. J. Deng, G. Mortimer, T. Schiller, A. Musumeci, D. Martin and R. F. Minchin, Differential plasma protein binding to metal oxide nanoparticles, *Nanotechnology*, 2009, **20**, 455101.
29. N. P. Mortensen, G. B. Hurst, W. Wang, C. M. Foster, P. D. Nallathamby and S. T. Retterer, Dynamic development of the protein corona on silica nanoparticles: composition and role in toxicity, *Nanoscale*, 2013, **5**, 6372–6380.
30. A. Kurtz-Chalot, C. Villiers, J. Pourchez, D. Boudard, M. Martini, P. N. Marche, M. Cottier and V. Forest, Impact of silica nanoparticle surface chemistry on protein corona formation and consequential interactions with biological cells, *Mater. Sci. Eng., C*, 2017, **75**, 16–24.
31. V. Mirshafiee, R. Kim, M. Mahmoudi and M. L. Kraft, The importance of selecting a proper biological milieu for protein corona analysis *in vitro*: Human plasma versus human serum, *Int. J. Biochem. Cell Biol.*, 2016, **75**, 188–195.
32. G. Maiorano, S. Sabella, B. Sorce, V. Brunetti, M. A. Malvindi, R. Cingolani and P. P. Pompa, Effects of Cell Culture Media on the Dynamic Formation of Protein–Nanoparticle Complexes and Influence on the Cellular Response, *ACS Nano*, 2010, **4**, 7481–7491.
33. M. Mahmoudi, A. M. Abdelmonem, S. Behzadi, J. H. Clement, S. Dutz, M. R. Ejtehadi, R. Hartmann, K. Kantner, U. Linne, P. Maffre, S. Metzler, M. K. Moghadam, C. Pfeiffer, M. Rezaei, P. Ruiz-Lozano, V. Serpooshan, M. A. Shokrgozar, G. U. Nienhaus and W. J. Parak, Temperature: the 'ignored' factor at the NanoBio interface, *ACS Nano*, 2013, **7**, 6555–6562.
34. D. Pozzi, G. Caracciolo, A. L. Capriotti, C. Cavaliere, G. La Barbera, T. J. Anchordoquy and A. Laganà, Surface chemistry and serum type both determine the nanoparticle-protein corona, *J. Proteomics*, 2015, **119**, 209–217.
35. R. R. Arvizo, K. Giri, D. Moyano, O. R. Miranda, B. Madden, D. J. McCormick, R. Bhattacharya, V. M. Rotello, J.-P. Kocher and P. Mukherjee, Identifying new therapeutic targets via modulation of protein corona formation by engineered nanoparticles, *PLoS One*, 2012, **7**, e33650.

36. M. J. Hajipour, S. Laurent, A. Aghaie, F. Rezaee and M. Mahmoudi, Personalized protein coronas: a "key" factor at the nanobiointerface, *Biomater. Sci.*, 2014, **2**, 1210–1221.
37. C. Gräfe, A. Weidner, M. V. D. Lühe, C. Bergemann, F. H. Schacher, J. H. Clement and S. Dutz, Intentional formation of a protein corona on nanoparticles: Serum concentration affects protein corona mass, surface charge, and nanoparticle-cell interaction, *Int. J. Biochem. Cell Biol.*, 2016, **75**, 196–202.
38. A. L. Barrán-Berdón, D. Pozzi, G. Caracciolo, A. L. Capriotti, G. Caruso, C. Cavaliere, A. Riccioli, S. Palchetti and A. Laganà, Time evolution of nanoparticle-protein corona in human plasma: relevance for targeted drug delivery, *Langmuir*, 2013, **29**, 6485–6494.
39. A. Lesniak, A. Campbell, M. P. Monopoli, I. Lynch, A. Salvati and K. A. Dawson, Serum heat inactivation affects protein corona composition and nanoparticle uptake, *Biomaterials*, 2010, **31**, 9511–9518.
40. D. Walczyk, F. B. Bombelli, M. P. Monopoli, I. Lynch and K. A. Dawson, What the cell 'sees' in bionanoscience, *J. Am. Chem. Soc.*, 2010, **132**, 5761–5768.
41. E. Brun and C. Sicard-Roselli, Could nanoparticle corona characterization help for biological consequence prediction?, *Cancer Nanotechnol.*, 2014, **5**, 7.
42. E. Izak-Nau, M. Voetz, S. Eiden, A. Duschl and V. F. Puntes, Altered characteristics of silica nanoparticles in bovine and human serum: the importance of nanoparticle characterization prior to its toxicological evaluation, *Part. Fibre Toxicol.*, 2013, **10**, 56.
43. A. C. Sabuncu, J. Grubbs, S. Qian, T. M. Abdel-Fattah, M. W. Stacey and A. Beskok, Probing nanoparticle interactions in cell culture media, *Colloids Surf., B*, 2012, **95**, 96–102.
44. J. S. Gebauer, M. Malissek, S. Simon, S. K. Knauer, M. Maskos, R. H. Stauber, W. Peukert and L. Treuel, Impact of the nanoparticle-protein corona on colloidal stability and protein structure, *Langmuir*, 2012, **28**, 9673–9679.
45. D. Docter, D. Westmeier, M. Markiewicz, S. Stolte, S. K. Knauer and R. H. Stauber, The nanoparticle biomolecule corona: lessons learned – challenge accepted?, *Chem. Soc. Rev.*, 2015, **44**, 6094–6121.
46. Y. Qiu, Y. Liu, L. Wang, L. Xu, R. Bai, Y. Ji, X. Wu, Y. Zhao, Y. Li and C. Chen, Surface chemistry and aspect ratio mediated cellular uptake of Au nanorods, *Biomaterials*, 2010, **31**, 7606–7619.
47. M. S. Ehrenberg, A. E. Friedman, J. N. Finkelstein, G. Oberdörster and J. L. McGrath, The influence of protein adsorption on nanoparticle association with cultured endothelial cells, *Biomaterials*, 2009, **30**, 603–610.
48. A. Kurtz-Chalot, J. P. Klein, J. Pourchez, D. Boudard, V. Bin, G. B. Alcantara, M. Martini, M. Cottier and V. Forest, Adsorption at cell surface and cellular uptake of silica nanoparticles with different surface chemical functionalizations: impact on cytotoxicity, *J. Nanopart. Res.*, 2014, **16**, 1–15.

49. M. M. Yallapu, N. Chauhan, S. F. Othman, V. Khalilzad-Sharghi, M. C. Ebeling, S. Khan, M. Jaggi and S. C. Chauhan, Implications of protein corona on physico-chemical and biological properties of magnetic nanoparticles, *Biomaterials*, 2015, **46**, 1–12.
50. S. Dominguez-Medina, L. Kisley, L. J. Tauzin, A. Hoggard, B. Shuang, A. S. D. S. Indrasekara, S. Chen, L.-Y. Wang, P. J. Derry, A. Liopo, E. R. Zubarev, C. F. Landes and S. Link, Adsorption and Unfolding of a Single Protein Triggers Nanoparticle Aggregation, *ACS Nano*, 2016, **10**, 2103–2112.
51. C. L. Oliveira, F. Veiga, C. Varela, F. Roleira, E. Tavares, I. Silveira and A. J. Ribeiro, Characterization of polymeric nanoparticles for intravenous delivery: Focus on stability, *Colloids Surf., B*, 2017, **150**, 326–333.
52. K. Hamad-Schifferli, Exploiting the novel properties of protein coronas: emerging applications in nanomedicine, *Nanomedicine*, 2015, **10**, 1663–1674.
53. A. Lesniak, F. Fenaroli, M. P. Monopoli, C. Åberg, K. A. Dawson and A. Salvati, Effects of the presence or absence of a protein corona on silica nanoparticle uptake and impact on cells, *ACS Nano*, 2012, **6**, 5845–5857.
54. V. Forest, M. Cottier and J. Pourchez, Electrostatic interactions favor the binding of positive nanoparticles on cells: A reductive theory, *Nano Today*, 2015, **10**, 677–680.
55. A. Lesniak, A. Salvati, M. J. Santos-Martinez, M. W. Radomski, K. A. Dawson and C. Åberg, Nanoparticle adhesion to the cell membrane and its effect on nanoparticle uptake efficiency, *J. Am. Chem. Soc.*, 2013, **135**, 1438–1444.
56. V. Forest and J. Pourchez, Preferential binding of positive nanoparticles on cell membranes is due to electrostatic interactions: A too simplistic explanation that does not take into account the nanoparticle protein corona, *Mater. Sci. Eng., C*, 2017, **70**, 889–896.
57. X. Cheng, X. Tian, A. Wu, J. Li, J. Tian, Y. Chong, Z. Chai, Y. Zhao, C. Chen and C. Ge, Protein Corona Influences Cellular Uptake of Gold Nanoparticles by Phagocytic and Nonphagocytic Cells in a Size-Dependent Manner, *ACS Appl. Mater. Interfaces*, 2015, **7**, 20568–20575.
58. S. R. Saptarshi, A. Duschl and A. L. Lopata, Interaction of nanoparticles with proteins: relation to bio-reactivity of the nanoparticle, *J. Nanobiotechnol.*, 2013, **11**, 26.
59. D. Docter, C. Bantz, D. Westmeier, H. J. Galla, Q. Wang, J. C. Kirkpatrick, P. Nielsen, M. Maskos and R. H. Stauber, The protein corona protects against size- and dose-dependent toxicity of amorphous silica nanoparticles, *Beilstein J. Nanotechnol.*, 2014, **5**, 1380–1392.
60. F. Zhao, Y. Zhao, Y. Liu, X. Chang, C. Chen and Y. Zhao, Cellular uptake, intracellular trafficking, and cytotoxicity of nanoparticles, *Small*, 2011, **7**, 1322–1337.
61. J. Wang, J. D. Byrne, M. E. Napier and J. M. DeSimone, More effective nanomedicines through particle design, *Small*, 2011, **7**, 1919–1931.
62. O. Lunov, T. Syrovets, C. Loos, J. Beil, M. Delacher, K. Tron, G. U. Nienhaus, A. Musyanovych, V. Mailänder, K. Landfester and T. Simmet, Differential

uptake of functionalized polystyrene nanoparticles by human macrophages and a monocytic cell line, *ACS Nano*, 2011, **5**, 1657–1669.
63. M. Mahmoudi, S. N. Saeedi-Eslami, M. A. Shokrgozar, K. Azadmanesh, M. Hassanlou, H. R. Kalhor, C. Burtea, B. Rothen-Rutishauser, S. Laurent, S. Sheibani and H. Vali, Cell 'vision': complementary factor of protein corona in nanotoxicology, *Nanoscale*, 2012, **4**, 5461–5468.
64. F. Marano, S. Hussain, F. Rodrigues-Lima, A. Baeza-Squiban and S. Boland, Nanoparticles: molecular targets and cell signalling, *Arch. Toxicol.*, 2011, **85**, 733–741.
65. A. Albanese and W. C. W. Chan, Effect of Gold Nanoparticle Aggregation on Cell Uptake and Toxicity, *ACS Nano*, 2011, **5**, 5478–5489.
66. F. Rancan, Q. Gao, C. Graf, S. Troppens, S. Hadam, S. Hackbarth, C. Kembuan, U. Blume-Peytavi, E. Rühl, J. Lademann and A. Vogt, Skin penetration and cellular uptake of amorphous silica nanoparticles with variable size, surface functionalization, and colloidal stability, *ACS Nano*, 2012, **6**, 6829–6842.
67. R. Liu, W. Jiang, C. D. Walkey, W. C. W. Chan and Y. Cohen, Prediction of nanoparticles-cell association based on corona proteins and physicochemical properties, *Nanoscale*, 2015, **7**, 9664–9675.
68. R. M. Pearson, V. V. Juettner and S. Hong, Biomolecular corona on nanoparticles: a survey of recent literature and its implications in targeted drug delivery, *Front. Chem.*, 2014, **2**, 108.
69. E. Mahon, A. Salvati, F. Baldelli Bombelli, I. Lynch and K. A. Dawson, Designing the nanoparticle-biomolecule interface for 'targeting and therapeutic delivery', *J. Controlled Release*, 2012, **161**, 164–174.
70. S. M. Moghimi, A. C. Hunter and J. C. Murray, Long-circulating and target-specific nanoparticles: theory to practice, *Pharmacol. Rev.*, 2001, **53**, 283–318.
71. S. Palchetti, V. Colapicchioni, L. Digiacomo, G. Caracciolo, D. Pozzi, A. L. Capriotti, G. La Barbera and A. Laganà, The protein corona of circulating PEGylated liposomes, *Biochim. Biophys. Acta*, 2016, **1858**, 189–196.
72. A. Salvati, A. S. Pitek, M. P. Monopoli, K. Prapainop, F. B. Bombelli, D. R. Hristov, P. M. Kelly, C. Åberg, E. Mahon and K. A. Dawson, Transferrin-functionalized nanoparticles lose their targeting capabilities when a biomolecule corona adsorbs on the surface, *Nat. Nanotechnol.*, 2013, **8**, 137–143.
73. C. Ge, J. Du, L. Zhao, L. Wang, Y. Liu, D. Li, Y. Yang, R. Zhou, Y. Zhao, Z. Chai and C. Chen, Binding of blood proteins to carbon nanotubes reduces cytotoxicity, *Proc. Natl. Acad. Sci. U. S. A.*, 2011, **108**, 16968–16973.
74. W. Hu, C. Peng, M. Lv, X. Li, Y. Zhang, N. Chen, C. Fan and Q. Huang, Protein Corona-Mediated Mitigation of Cytotoxicity of Graphene Oxide, *ACS Nano*, 2011, **5**, 3693–3700.
75. H. Yin, R. Chen, P. S. Casey, P. C. Ke, T. P. Davis and C. Chen, Reducing the cytotoxicity of ZnO nanoparticles by a pre-formed protein corona in a supplemented cell culture medium, *RSC Adv.*, 2015, **5**, 73963–73973.

76. S. Kittler, C. Greulich, J. Diendorf, M. Köller and M. Epple, Toxicity of Silver Nanoparticles Increases during Storage Because of Slow Dissolution under Release of Silver Ions, *Chem. Mater.*, 2010, **22**(16), 4548–4554.
77. Z. J. Deng, M. Liang, M. Monteiro, I. Toth and R. F. Minchin, Nanoparticle-induced unfolding of fibrinogen promotes Mac-1 receptor activation and inflammation, *Nat. Nanotechnol.*, 2011, **6**, 39–44.
78. Q. H. Quach and J. C. Y. Kah, Non-specific adsorption of complement proteins affects complement activation pathways of gold nanoparticles, *Nanotoxicology*, 2017, **11**, 382–394.
79. D. F. Moyano, M. Goldsmith, D. J. Solfiell, D. Landesman-Milo, O. R Miranda, D. Peer and V. M. Rotello, Nanoparticle Hydrophobicity Dictates Immune Response, *J. Am. Chem. Soc.*, 2012, **134**, 3965–3967.
80. C. Farrera and B. Fadeel, It takes two to tango: Understanding the interactions between engineered nanoparticles and the immune system, *Eur. J. Pharm. Biopharm.*, 2015, **95**, 3–12.
81. S. Juling, A. Niedzwiecka, L. Böhmert, D. Lichtenstein, S. Selve, A. Braeuning, A. F. Thünemann, E. Krause and A. Lampen, Protein Corona Analysis of Silver Nanoparticles Links to Their Cellular Effects, *J. Proteome Res.*, 2017, **16**, 4020–4034.
82. P. Aggarwal, J. B. Hall, C. B. McLeland, M. A. Dobrovolskaia and S. E. McNeil, Nanoparticle interaction with plasma proteins as it relates to particle biodistribution, biocompatibility and therapeutic efficacy, *Adv. Drug Delivery Rev.*, 2009, **61**, 428–437.
83. V. Forest and J. Pourchez, The nanoparticle protein corona: The myth of average, *Nano Today*, 2016, **11**, 700–703.
84. M. Hadjidemetriou, Z. Al-Ahmady and K. Kostarelos, Time-evolution of *in vivo* protein corona onto blood-circulating PEGylated liposomal doxorubicin (DOXIL) nanoparticles, *Nanoscale*, 2016, **8**, 6948–6957.
85. M. Hadjidemetriou, Z. Al-Ahmady, M. Mazza, R. F. Collins, K. Dawson and K. Kostarelos, In Vivo Biomolecule Corona around Blood-Circulating, Clinically Used and Antibody-Targeted Lipid Bilayer Nanoscale Vesicles, *ACS Nano*, 2015, **9**, 8142–8156.
86. N. V. Konduru, R. M. Molina, A. Swami, F. Damiani, G. Pyrgiotakis, P. Lin, P. Andreozzi, T. C. Donaghey, P. Demokritou, S. Krol, W. Kreyling and J. D. Brain, Protein corona: implications for nanoparticle interactions with pulmonary cells, *Part. Fibre Toxicol.*, 2017, **14**, 42.
87. G. Caracciolo, D. Pozzi, A. L. Capriotti, C. Cavaliere, F. Cardarelli, A. Bifone and A. Laganà, Cancer cell targeting of lipid gene vectors by protein corona, *Nanotechnology*, 2012, **3**, 354–357.
88. K. Prapainop, D. P. Witter and P. Wentworth, A Chemical Approach for Cell-Specific Targeting of Nanoparticles: Small-Molecule-Initiated Misfolding of Nanoparticle Corona Proteins, *J. Am. Chem. Soc.*, 2012, **134**, 4100–4103.

CHAPTER 3

Factors Affecting a Nanoparticle's Protein Corona Formation

PAL PATEL AND ASHUTOSH KUMAR*

Biological and Life Sciences, School of Arts and Sciences, Ahmedabad University, Central Campus, Navrangpura, Ahmedabad, 380009, Gujarat, India
*E-mail: ashutosh.kumar@ahduni.edu.in

3.1 Introduction

Nanoparticles (NPs) are particles having at least one dimension between 1 to 100 nm in size.[1] As nanoparticles have a large surface area compared to bulk particles, their reactivity is, equally, expected to be higher than their bulk counterpart.[2] Nanoparticles possess unique physiochemical properties, which make them attractive for various applications in personal care products, catalysis, imaging, optical, electronics, environmental and biomedical.[3] Nanoparticles are formed through natural phenomena (volcanic eruptions, forest fires), produced incidentally (by-products of industrial processes) and intensely by chemical synthesis.[4] In the fields of biomedical sciences, nanoparticles have shown numerous applications including drug and gene delivery,[5,6] tissue repair,[7] cancer therapy,[8] disease dignosis,[9] and imaging.[10]

As the nanoparticle size is very small, the electron presents on the surface of the nanoparticle, making it highly reactive. When the particle meets a biological system, the surface energy of the particle decreases by adsorbing

Issues in Toxicology No. 40
Nanoparticle–Protein Corona: Biophysics to Biology
Edited by Ashutosh Kumar and Alok Dhawan

Published by the Royal Society of Chemistry, www.rsc.org

biomolecules, mostly proteins.[11,12] Because of the interactions mentioned, a "bio–nano interface" is developed at the point at which two entities come into contact.[13–15] Proteins existing in a biological system are known to cover the nanoparticle's surface, thus resulting in the formation of a protein corona.[16] Blood plasma contains various proteins and lipids of different concentrations. After intravenous administration of a nanodrug, this leads to the interaction of nanoparticles with various physiological environments and protein mixtures.[17] Hence, a competition arises between the various biological molecules in order to be adsorbed on the nanoparticles' surface.[18]

Protein corona can exist in the form of soft and hard coronas. Initially, the most abundant proteins of the biological medium are adsorbed onto the nanoparticles' surface and form a soft corona. However, as observed over a period of time, some higher affinity proteins replace the soft corona and get adsorbed on the surface and form a hard corona. This is known as Vroman's effect.[19] Because of this effect, the corona is considered as a dynamic layer, which changes over time because of the adsorption of the ongoing protein and desorption from the surrounding biological environment.[18,19] Interestingly, the composition of the corona changes with time; however, the amount of adsorbed proteins remains relatively constant in a protein corona. For example, almost 300 different types of proteins have been reported to be adsorbed on amorphous silica NPs within less than 0.5 min of incubation with human plasma. Also, a time-dependent increase in the concentration of proteins was observed with no change in composition.[20]

Corona formation is affected by the physiochemical properties of nanoparticles as they are induced by conformational changes in the proteins.[21] The protein corona is highly dependent on the structure and composition of the nanoparticles such as, size, shape, surface charge, surface functional groups, hydrophilicity/hydrophobicity and nature of the physiological environment or biological environment, the exposure time and temperature. Therefore, the composition of the protein corona is different and exclusive to each nanoparticle.[21–23]

The size and interfacial composition are altered by the protein corona, which leads to a new biological identity for the nanoparticles, which influences the circulation lifetime of the nanoparticles, the signaling and kinetics, transport, accumulation, toxicity, cellular uptake and agglomeration of the nanoparticles. Therefore, the interaction between the nanoparticles and biological systems becomes vital to understand for their safe and effective working application.[11–14] However, it has also been reported that the interaction of nanoparticles with protein only leads to their increased size without an alteration in their functionalities.[24–26]

The toxicological effects of nanoparticles are greatly influenced by the adsorbed biomolecules present on the surface of the nanoparticles, which are responsible for the fate of the nanoparticles.[27] For example, protein adsorption can lead to the non-specific uptake of nanomaterials into cells by receptor-mediated endocytosis. Proteins can act as opsonins, so a major fate of nanoparticles in physiological media is opsonization. As a result, corona formation further impacts the cellular interactions of nanoparticles that help to determine their fate in humans.[28]

This chapter presents comprehensive information about the various factors influencing the nanoparticle–protein interaction including the physicochemical properties of the nanoparticles, the nature of the biological medium along with the exposure time and temperature.

3.2 Structure and Composition of the Protein Corona

Differences in nanoparticles' properties and environmental factors influence protein coronas. The phenomenon of adsorption occurring on the nanoparticles' surface is regulated by the binding affinity of proteins to nanoparticles and proteins.[19,21,22] In an interaction between nanoparticles and blood plasma, proteins are the biomolecules that are absorbed mainly on the surface of the nanoparticles; however, minor traces of lipids being absorbed have also been reported. Albumin, being the most abundant protein in blood, is often present in the fingerprints of protein coronas of nanoparticles. Other common proteins, such as complement proteins, fibrinogen, and immunoglobulin, are also found to be present in protein coronas and are known to significantly alter the properties of nanoparticles such as opsonization, allocation, and targeting efficiency. Protein patterns often contain recognizable sequence features. One of them is the binding site or active site of a protein or enzyme, as there are only a few amino acids that are essential for protein function and are found commonly in those sites.[28–30]

Protein coronas are classified into two parts based on protein binding affinity. One is the "hard corona" and the other is the "soft corona". The soft corona layer is formed when nanoparticles meet a biological system and proteins present in a higher abundance with a low affinity tend to get adsorbed on the nanoparticles' surface. High-affinity proteins replace this over time and form the hard corona.[31]

The hard corona protein often binds irreversibly to the nanoparticles; however, binding of the proteins with a low affinity, which forms the soft corona, is reversible. More often than not, this binding is replaced over the period or is detached during experiments (washing). Proteins forming the hard corona are directly absorbed on the nanoparticles' surface.

Binding of the proteins that form the soft corona can be done *via* one of two ways: either the protein gets adsorbed on the nanoparticle surface and binds to the nanoparticle or the different proteins interact together and simultaneously bind to the nanoparticle. Proteins constituting the hard corona layer are considered the prime component for the identity of nanoparticles in the biological system because of their long residence time[13–15,32–34] (Figure 3.1).

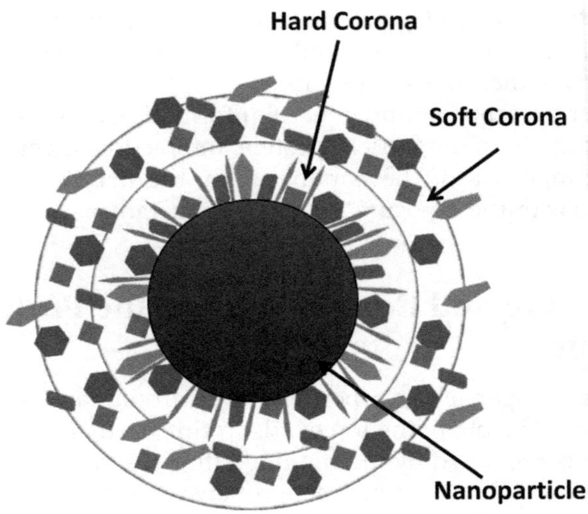

Figure 3.1 Schematic representation of the protein corona theory. A hard and soft layer of proteins covers the surface of the nanoparticle. Reproduced from ref. 35 with permission from Elsevier, Copyright 2014.

3.3 Formation of the Protein Corona

It has been reported that when nanoparticles come into contact with a biological medium, it results in a competition between various biomolecules such as protein and lipids, to interact with nanoparticles to form a protein corona.

In the case of intravenous administration of nanoparticles, proteins present in the blood plasma are adsorbed on the surface of nanoparticles, resulting in the formation of a dynamic layer called the protein corona.[21] The composition of the layers of a protein corona is observed to change with time due to proteins constantly being adsorbed and desorbed on the surface. The composition of the protein corona formed around each nanoparticle is different and is highly influenced by the various physiochemical parameters of each nanoparticle. The decreased surface energy of the nanoparticle leads to the formation of the protein corona. The interaction of a protein on the surface of a nanoparticle leads to an alteration of the properties of nanoparticles such as size, surface chemistry and surface charge. This alteration thereby affects the biological properties of the nanoparticles such as their cellular uptake, toxicity and bio-distribution. Several forces, such as van der Waals forces, electrostatic forces, electrosteric forces, solvation and solvophobic forces, play a vital role in the formation of protein coronas and in turn determine the structures of protein coronas.[4,15,21,29,35]

1 *Van der Waals forces*: These forces are asserted due to the fluctuations of electrons, which induce a small dipole in the nanoparticles, thereby inducing a dipole moment in the atoms of the adjacent nanoparticles as well and thus leads to creating an attractive force.[13,36]

2 *Electrostatic interactions*: Electrostatic interactions occur due to the electric charges present on nanoparticles. The electrostatic forces between nanoparticles of the same charge are repulsive while those of opposite charges are attractive in nature.[37] The disulfide bonds of bovine serum albumin (BSA) get adsorbed on the surface of gold nanoparticles (AuNPs) when bovine serum albumin is allowed to interact with gold nanoparticles. Hydrogen bonding, van der Waals forces, electrostatic and hydrophobic interactions were opined to play a vital role in driving the binding of proteins to gold nanoparticles.[38]
3 *Electrosteric forces*: Electrosteric forces are based on the steric repulsion between molecules adsorbed on nanoparticles, which is necessary to prevent their aggregation, agglomeration and coagulation in solutions. Electrosteric repulsion is observed when there is attachment of various polymers or surfactants on the surfaces of nanoparticles.[39] In the presence of natural organic matter, the stability of several nanoparticles, such as silver (Ag), titanium dioxide (TiO_2), aluminium oxide (Al_2O_3), zinc oxide (ZnO), iron (Fe) nanoparticles, fullerenes and carbon nanotubes (CNTs), increases as the formation of protein coronas on the surfaces of the nanoparticles stabilizes the electrostatic repulsion by providing a steric barrier.[40]
4 *Solvation and solvophobic forces*: Solvation is defined as the process of surrounding a solute with solvent.[41] The water molecules present in the medium bind to the hydrophilic nanoparticles with sufficient energy to form steric layers on the surface of the nanoparticles, making adhering of nanoparticles to one another difficult. Therefore, solvation forces lead to increased nanoparticle stability *via* hydrophilic repulsion.[13] 'Solvophobic' forces enable the nanoparticles to aggregate when the attractive interaction due to hydrogen bonding or other polar interactions between solvent particles becomes stronger than the interaction between solvent and solute.[42] Solvation forces were observed to play a significant role in interactions between SiO_2 NPs in water[13] (Figure 3.2).

3.4 Protein Conformation

Nanoparticles entering a biological fluid tend to undergo surface modification due to dynamic and physicochemical interactions with biological components, especially proteins. This occurs due to the protein being adsorbed on the surfaces of nanoparticles. At the time of adsorption, proteins may pass through the structural rearrangement that is called "conformational changes." Protein conformation is referred to as the dimensional shapes or structures of a reversible protein formed by the arrangement of an amino acid sequence in the polypeptide chain. Conformational changes of the protein are more likely to be thermodynamically positive if a charged sequence (hydrophobic) in the protein interacts with the other charged (hydrophobic) nanoparticle surface; however, the changes in

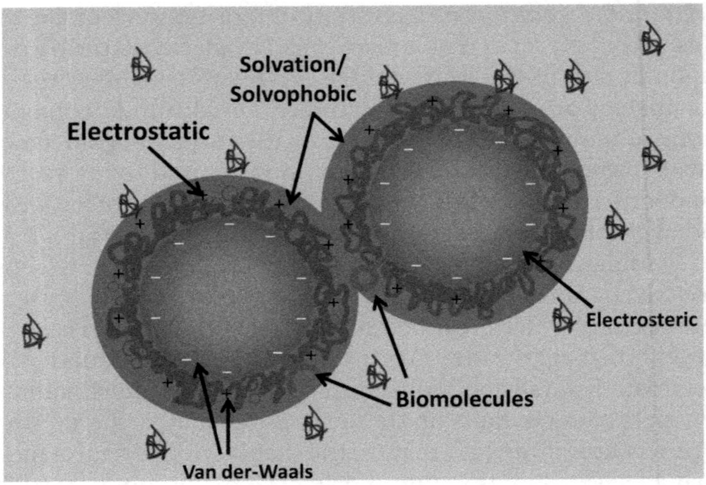

Figure 3.2 Schematic representation of an interaction at the nano–bio interface. Adapted from ref. 13 with permission from Springer Nature, Copyright 2009.

protein conformation are usually irreversible after desorption. The protein structure and nanoparticle surface property determine the conformational changes in the protein.[14,43,44]

The change in the conformation of adsorbed proteins is observed to be higher in the presence of hydrophobic nanomaterials compared to that in hydrophilic nanomaterials.[45] For example, quantum dots grafted with mercaptoundecanoic acid were taken into consideration to study the above. It was observed that they denature and inactivate the enzyme chymotrypsin effectively. Further, similar particles were grafted with a structurally similar but hydrophilic polyethylene glycol (PEG) derivative and they observed that the denaturation was not to the same extent.[21,46,47] Another example is the conformational change observed in iron-saturated human transferrin protein when allowed to interact with superparamagnetic iron oxide nanoparticles (SPIONs).[31] When various sizes of bare SPIONs and PVA-coated SPIONs were incubated with iron-saturated human transferrin protein, an irreversible change was observed in the conformation of the iron-saturated human transferrin protein due to the release of iron. More conformational changes were reported on the surfaces of the bare nanoparticles than the ones with the PVA coating.[31,48]

Compared to a planar surface, the high curvature of nanoparticles helps proteins to maintain their original structure. Various nanoparticles' surfaces and proteins have been reported to be perturbed because of corona formation. For example, lysozyme adsorbed onto silica nanoparticles or bovine serum albumin (BSA) adsorbed on AuNPs' surfaces showed a rapid conformational change at both secondary and tertiary levels of proteins. Most of the studies have reported a significant loss of α-helical content and increased sheet and turn structures in proteins when they are adsorbed onto nanoparticles.[21,22,49] Also, surface charges of nanoparticles not only lead to conformational changes in proteins but also in their biological effects. BSA adsorbed on positively charged polystyrene nanoparticles undergoes conformational changes compared to BSA adsorbed on negatively charged polystyrene nanoparticles, which retains its native structure. BSA-adsorbed anionic polystyrene nanoparticles are reported to inhibit cellular binding, but their cationic counterparts enhance cellular binding and toxicity. Similarly, negatively charged nanoparticles, such as citrate-coated gold nanoparticles and quantum dots with a carboxylate group on the surface inhibited cellular binding after adsorbing BSA or serum proteins.[30,50,51]

3.5 Time Evolution of the Protein Corona

Coronas are formulated with a long and short time of exchange because of the allocation of proteins and lipids present in the biological diversity. The lifetime of a hard corona is reported to be longer than a soft corona. Hard coronas are known to have a lifetime adequate for them to affect many biophysiological phenomena. Therefore, the hard corona is known to determine the biological identity of the particle.[21]

3700 proteins and more present in blood plasma compete to be absorbed on the surface of a nanoparticle. However, the competition doesn't have an impact on the composition of the corona over time as evidently, the protein corona is not a static layer. The composition of a protein corona is determined by the kinetic rate of adsorption and desorption of proteins and lipids being adsorbed on the surfaces of the nanoparticles.[21] In many cases, it is observed that proteins with a higher abundance in the blood plasma tend to form soft coronas. However, over a period, they are replaced by proteins with lower concentrations but a higher affinity.[52] For example, Casals and co-workers were able to demonstrate the time evolution of the protein corona formed on different-sized AuNPs ranging from 4 to 40 nm in cell culture media with 10% fetal bovine serum (FBS). They revealed that there is a reduction of surface charge when there is an increment in the thickness of the protein corona layer.[31] Albumin was reported to be the most abundant protein found on the surface of nanoparticles, which suggests it has an improved biocompatibility with nanomaterials. Furthermore, it has been reported that loosely bound proteins due to several factors like stability will over time evolve to form an irreversibly bound protein.[31,53]

The evolution of protein coronas around nanoparticles through the transfer of nanoparticles from one biological fluid (plasma) into another (cytosolic fluid) as an expressive model for cellular uptake was illustrated by Lundqvist and co-workers. This study took into consideration different nanoparticles such as 9 nm silicanano particles, 50 nm polystyrene nanoparticles and 50 nm carboxyl-modified polystyrene nanoparticles. It was observed that the protein corona evolves considerably in the second biological medium while the final corona conserves a "fingerprint" of its history or the first medium. This study can be advantageous in determining the transport pathways of nanoparticles, their fate and behavior.[31,54] Nagayama and co-workers investigated qualitative and quantitative changes in the adsorption pattern of serum proteins over a period of 5 min to 360 min. At different times, they used sodium dodecyl sulfate-polyacrylamide gel electrophoresis (SDS-PAGE) and western blotting to assess 50 nm lecithin-coated polystyrene nanoparticles. As per the quantitative study, it has been revealed that the total amount of adsorbed proteins increases over time. On the other hand, variation in protein adsorption is indicated by the qualitative study where adsorption of some proteins increases over time whereas the adsorption of others decreases over time. This study further indicates the elements – complement C3, IgG, apolipoprotein E (ApoE), and immunoglobulin A (IgA) – that showed an increase over time while the concentration of albumin remained constant as per this qualitative study.[55]

3.6 Properties Affecting the Formation of a Nanoparticle Protein Corona

Different protein corona profiles are produced due to different nanoparticle characteristics.[29] Through nanoparticle–protein complexes, the physicochemical and biological properties of nanoparticles, such as cellular behavior, bioavailability and biological responses, can be controlled. The different physiochemical properties of nanoparticles –size, shape, surface area, surface charge (zeta potential), surface chemistry (hydrophilicity/hydrophobicity), route of administration of nanoparticles (biological environment) and exposure time – affect the composition, thickness, and conformation of a protein corona. Primary knowledge of the role of each physicochemical parameter affecting the protein corona is key in the development of new targeting therapies for drug delivery and safety profiles[21] (Figure 3.3).

3.6.1 Size of nanoparticles

Nanoparticle size plays a critical role in corona formation. Similar nanoparticles of different sizes have been reported to have protein coronas with different compositions. Where nanoparticle size is concerned, the change of composition and the organization of proteins in the corona become crucial. Due to the highly curved surfaces of nanoparticles, protein–protein interactions often decrease.[19] Proteins that are adsorbed on the highly curved surfaces of nanoparticles tend to undergo conformational changes more than proteins adsorbed on less curved surfaces. Protein bindings are affected by the size and curvature of nanoparticles. For example, IgM-dependent complements are activated on dextran nanoparticles for which the optimal size range is 250 nm; however, larger nanoparticles fail to attract IgM and therefore activation doesn't occur to the same extent.[56] The effect of size and surface area of nanoparticles on the protein adsorption of NIPAM/BAM (50 : 50) copolymer nanoparticles ranging in size between 70 to 700 nm showed

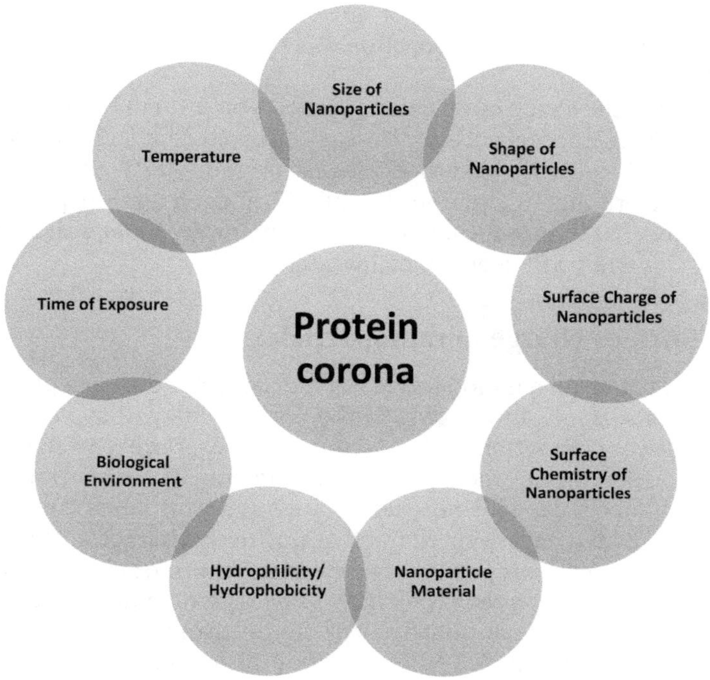

Figure 3.3 Factors influencing the composition of a protein corona.

that the amount of absorbed plasma protein increases with increasing surface area. The surface area available for protein binding increases with the decrease in the nanoparticles' size.[11,22]

Another study taking into consideration the different sizes of AuNPs demonstrated that the increased diameter of AuNPs enhances the molecular interaction between the plasma protein and the gold nanoparticles. Polyacrylic acid AuNPs having a size of 12 nm bind to fibrinogen with a higher affinity than polyacrylic acid AuNPs having a size of 7 nm.[57,58] The amount of proteins adsorbed on the surface is influenced by the size of the AuNPs. Nanoparticle size mainly determines the curvature of the nanoparticles that have different protein binding affinities. There is a major reason behind the influence of the size of AuNPs – an increase in the size of AuNPs leads to an increased binding affinity with fibrinogen.[59]

3.6.2 Shape of nanoparticles

The shape of the nanoparticles determines the amount of the protein, the type of protein adsorbed on the surface of the nanoparticles as well as the biological response of the nanoparticles. For example, the shape of AuNPs has been reported to have a vast effect on the interaction with cells. Spherical AuNPs have a higher association with cells compared to rod-shaped AuNPs.

Another study showed the effect of AuNPs' morphology on protein adsorption indicating that protein adsorption was higher on nanorods compared to nanospheres.[60]

The effect of the shape of TiO_2 nanoparticles on protein binding was investigated by Deng and co-worker.[57] It was observed that TiO_2 nanorods and nanotubes differentially adsorbed plasma proteins. IgM and IgG proteins bound mainly to the TiO_2 nanorods while fibrinogen was the prime protein bound to the TiO_2 nanotubes. However, the potential biological implication of this observation still needs to be investigated.[61]

3.6.3 Surface charge of nanoparticles

The surface charge of a nanoparticle is a crucial factor during a protein interaction on its surface. An increase in the surface charge has been observed to have a direct correlation with the protein adsorption. The charge on the nanoparticle determines the type of protein adsorbed on its surface. The surface of a nanoparticle with no charge binds fewer proteins compared to a nanoparticle with a negatively charged or positively charged surface.[62]

A protein corona's composition is different for unmodified, carboxyl- and amine-modified polystyrene particles. Polystyrene nanoparticles reveal that positively charged nanoparticles prefer to adsorb proteins with isoelectric points (pI) less than 5.5, such as albumin, while the negative surface charge enhances the adsorption of proteins with pI more than 5.5, such as IgG.[63] Surface modification of AuNPs with citrate or hexadecyltrimethyl ammonium bromide increases the net surface charge on the AuNPs, which produces an electrostatic attraction to the oppositely charged groups in serum proteins.[59]

Additionally, vesicles containing cationic lipids have been reported to undergo significant plasma protein binding because of the interactions between the cationic lipids and negatively charged plasma proteins.[64] It has been demonstrated that the surface charge of gold nanoparticles denatures the adsorbed proteins. AuNPs with different charged ligands showed that in the presence of positively and negatively charged ligands, proteins get denatured but with neutral ligands, the protein retains its native structure.[11]

3.6.4 Surface chemistry of nanoparticles

The different surface chemistry of nanoparticles leads to a variation in the composition of the protein coronas on the nanoparticles. Functionalization of nanoparticles through different groups can be employed to decrease protein adsorption. A polyamine 908 coating on polystyrene nanospheres reduces the adsorption of fibronectin.[62] In another study, the functionalization of carbon nanotubes (CNTs) and SiO_2 nanoparticles with pluronic F127 led to a reduction in serum protein adsorption.[22] In summary, different coating materials, such as PEG, poloxamer, poloxamine, pluronic F127, dextran, polysorbate and poly(oxyethylene), have been reported to alter corona formation leading to altered uptake, biodistribution and effects of the nanoparticle. Many studies have established that non-functionalized nanoparticles are stabilized using a bovine serum albumin/human serum albumin (BSA/HSA) coating, which prevents agglomeration of nanoparticles.[62]

3.6.5 Nanoparticle materials

Different protein adsorption patterns can be seen in different types of nanoparticles. The study of Deng and co-workers demonstrated the binding of a human plasma protein to commercially available metal oxide nanoparticles like TiO_2, SiO_2 and ZnO with a similar surface charge. It was reported that the proteins bound to the TiO_2 and SiO_2 NPs were similar. However, a notably different protein corona was observed on the surface of ZnONPs, indicating that the nanoparticle's surface can highly influence protein binding.[61] In a further study, the TiO_2 and SiO_2 NP protein coronas were found to have proteins like clusterin, apolipoprotein D and, also, alpha-2-acid glycoprotein. Interestingly, like transferrin, Ig heavy chain alpha and haptoglobin (alpha) were found in the protein corona of ZnO NPs.[31] Another study demonstrated the binding of a plasma protein to single-walled carbon nanotubes (SWCNT) and nano-sized silica and their different patterns of adsorption. Single-walled carbon nanotubes (SWCNT) were found to have the most abundant protein, serum albumin, in their protein corona; however, albumin was absent in the protein corona of silica NPs.

3.6.6 Hydrophilicity/hydrophobicity

The amount of adsorbed protein and the composition of the protein corona are affected by the hydrophobicity. As compared to the hydrophilic surface, the hydrophobic surface enhances the protein adsorption and increases the rate of nanoparticle opsonization. There is a higher chance of the protein being denaturized when it interacts with the charged or hydrophobic surface compared to the neutral and hydrophilic surface of nanoparticles.[33,43] In an isothermal titration calorimetry (ITC) study, Cedervall and co-workers revealed that increasing the hydrophobicity of polystyrene nanoparticles promoted more protein adsorption from plasma compared to their hydrophilic counterparts.[27,47] They found that hydrophobic nanoparticles adsorb more albumins with a shorter resident time than hydrophilic ones. Furthermore, this suggests that hydrophobic polystyrene nanoparticles have more protein binding sites.[32] Similarly, Moghimi and their colleague Patel earlier reported an important observation: more proteins bind with cholesterol-free liposomes as compared to cholesterol-rich liposomes.[65] Along with this, liposomes made of natural saturated lipids with a long carbon chain (C16) have increased protein adsorption as compared to their C14 counterparts.[66] Hence, it is proved that the protein binding affinity of a nanomaterial increases with a higher charge density and hydrophobicity.[67]

3.6.7 Biological environment/media composition

The composition of the biological medium is very important to determine the composition of a protein corona.[20,31] It has been demonstrated that nano–bio interactions between different cell culture media (Dulbecco Modified Eagle's Medium [DMEM] and Roswell Park Memorial Institute medium [RPMI] supplemented with 10% fetal bovine serum [FBS]) and differently sized citrate-coated AuNPs affects the formation of protein coronas and their cellular response. These media are commonly used in cell cultures and strongly differ in salt, amino acid and glucose composition.[68] A range of techniques like spectroscopy, electrophoresis and microscopy can be used to describe the biomolecular body formed by the dispersion of various sized nanoparticles in the different cell culture media. These techniques characterized corona composition, exchange kinetics of different classes of proteins, along with the physical status of AuNPs in the biological medium over time. DMEM significantly increases the formation of protein coronas in a time-dependent manner while RPMI showed different dynamics with a reduced protein coating.[22,68] For the characterization of protein–NP complexes, sodium dodecyl sulfate polyacrylamide gel electrophoresis (SDS-PAGE) and mass spectroscopy were performed and the results revealed that the composition of the protein corona does not relate to the abundance of serum proteins. To evaluate the biological impact of the nanoparticles, viability assays on two cell lines (HeLa [human epithelial cervical cancer cell line) and U937 (human leukemic monocyte lymphoma cell line]) were performed in the DMEM and RPMI media, for 15 nm AuNPs. AuNP uptake and cellular distribution were addressed by using confocal microscopy and a significantly different cytotoxicity, cellular uptake and cellular distribution in different compositions of cell culture media were reported. Specifically, cellular internalization was reported to be higher in the RPMI medium resulting in higher cytotoxicity compared to the DMEM medium. Furthermore, the protein corona formed in DMEM was observed to be more stable and abundant compared to the protein corona formed in the RPMI medium. These different formations of protein corona mediated by the biological environment can have a crucial impact on cellular response. Therefore, before performing cellular experiments, a detailed understanding of the effects induced by cell culture media on nanoparticles is crucial for understanding and depicting their interactions with nanoparticles.[68,69]

Monopoli *et al.* showed that the concentration of proteins present in plasma affects the thickness of a protein corona. By using a centrifugal sedimentation technique to decrease the protein concentration and dynamic light scattering to analyze the hydrodynamic size of nanoparticles, it was observed that a decrease in the concentration of plasma reduces the hard protein corona. Hence, protein corona composition may vary between the *in vitro* and *in vivo* test. This concludes that only studying the *in vitro* behavior of nanoparticles in a living biological environment (*in vivo* study) is not enough.[19]

Different concentrations of a plasma protein adsorption pattern were observed for two different nanoparticles: sulfonated polystyrene nanoparticles and silica nanoparticles. A semi-quantitative determination of the hard corona composition using 1D-PAGE and LC-MS/MS was used to study the total binding isotherms for the nanoparticle, identifying the nature and amount of the most relevant proteins in the plasma. This showed that the amount of protein in the corona changed according to the biological environment. When *in vitro* studies were compared with *in vivo* studies, they observed significant evolution of the hard protein corona in the latter studies, mainly due to an increase in the amount of protein present in the latter system.[19]

3.6.8 Time of exposure

The exposure time of nanoparticles to the surrounding biological medium controls the adsorption of protein on the nanoparticles. When nanoparticles meet biological fluids, a protein corona is formed immediately. Nearly 300 proteins binding to the surface of a nanoparticle are detected after 0.5 minutes of interaction with human plasma.[20] The total protein quantity adsorbed is said to be enhanced with longer durations of exposure.[68] The exchange of adsorbed proteins over time significantly modifies the abundance of protein binding, resulting in the formation of various protein corona patterns. This can be ascribed to the favorable replacement of higher affinity proteins over their lower affinity counterparts or the favorable adsorption of proteins with a lower molecular weight. However, the composition of the protein corona remains the same.[30,68]

3.6.9 Temperature

Temperature changes also alter protein–NP interactions and were studied briefly in some earlier work. Mahmoudi *et al.* demonstrated the extent to which temperature regulates the pattern of a protein corona. He showed the change in the pattern of a protein corona when magnetic nanoparticles (FePt NPs) at the human body's temperature, ranging between 5 °C and 45 °C, were exposed to various concentrations of HSA as well as apo-transferrin (apo-Tf). Both proteins are known to show a higher affinity toward FePt NPs at temperatures ranging between 9–23 °C. Hence, the results concluded that a slight change in the temperature could lead to a significant change in the formation and composition of a protein corona.[70] On the other hand, proteins adsorbed onto copper nanoparticles (CuNPs) increased when the incubation temperature was increased from 15 °C, 27 °C and 37 °C to 42 °C. Therefore, the body temperature for healthy human beings is said to vary between the range of 35–39 °C. However, when a fever or disease changes the body temperature up to 41 °C, it leads to changes in the biodistribution and bioavailability of nanoparticles.[30,71]

3.7 Conclusions

This chapter has highlighted the factors affecting the formation of a nanoparticle protein corona. It is well established that adsorption of a protein on the surface of a nanoparticle is a competitive phenomenon. The protein layer on the nanoparticle surface forms either a soft or a hard corona depending on the binding of the protein. Factors affecting the binding of biomolecules to the surfaces of nanoparticles are mostly the various physiochemical parameters of the nanomaterials; however, other parameters, such as biological environment, exposure time, and temperature, also play an important role.

Acknowledgements

Funding received from the Department of Biotechnology, Government of India, under the project "NanoToF: Toxicological evaluation and risk assessment on Nanomaterials in Food" (grant number BT/PR10414/PFN/20/961/2014) and DST SERB Project "Nanosensors for the Detection of Food Adulterants and Contaminants" (grant number EMR/2016/005286) is gratefully acknowledged. Financial assistance by The Gujarat Institute for Chemical Technology (GICT) for the Establishment of a Facility for environmental risk assessment of chemicals and nanomaterials is also acknowledged.

References

1. M. Auffan, J. Rose, J. Y. Bottero, G. V. Lowry, J. P. Jolivet and M. R. Wiesner, Towards a definition of inorganic nanoparticles from an environmental, health and safety perspective, *Nat. Nanotechnol.*, 2009, **4**, 634–641.
2. H. Shi, R. Magaye, V. Castranova and J. Zhao, Titanium dioxide nanoparticles: a review of current toxicological data, *Part. Fibre Toxicol.*, 2013, **10**, 15.
3. D. M. Berube, E. M. Searson, T. S. Morton and C. L. Cummings, Project on emerging nanotechnologies-consumer product inventory evaluated, *Nanotechnol. Law Bus.*, 2010, **7**, 152.
4. V. A. Senapati, K. Kansara, R. Shanker, A. Dhawan and A. Kumar, Monitoring characteristics and genotoxic effects of engineered nanoparticle-protein corona, *Mutagenesis*, 2017, **32**, 479–490.
5. K. Cho, X. Wang, S. Nie, Z. G. Chen and D. M. Shin, Therapeutic nanoparticles for drug delivery in cancer, *Clin. Cancer Res.*, 2008, **14**, 1310–1316.
6. T. Tanaka, L. S. Mangala, P. E. Vivas-Mejia, R. Nieves-Alicea, A. P. Mann, E. Mora, H. D. Han, M. M. Shahzad, X. Liu, R. Bhavane, J. Gu, J. R. Fakhoury, C. Chiappini, C. Lu, K. Matsuo, B. Godin, R. L. Stone, A. M. Nick, G. Lopez-Berestein, A. K. Sood and M. Ferrari, Sustained small interfering RNA delivery by mesoporous silicon particles, *Cancer Res.*, 2010, **70**, 3687–3696.

7. C. Xu, L. Mu, I. Roes, D. Miranda-Nieves, M. Nahrendorf, J. A. Ankrum, W. Zhao and J. M. Karp, Nanoparticle-based monitoring of cell therapy, *Nanotechnology*, 2011, **22**, 494001.
8. X. J. Liang, H. Meng, Y. Wang, H. He, J. Meng, J. Lu, P. C. Wang, Y. Zhao, X. Gao, B. Sun, C. Chen, G. Xing, D. Shen, M. M. Gottesman, Y. Wu, J. J. Yin and L. Jia, Metallofullerene nanoparticles circumvent tumor resistance to cisplatin by reactivating endocytosis, *Proc. Natl. Acad. Sci. U. S. A.*, 2010, **107**, 7449–7454.
9. L. Yan, F. Zhao, S. Li, Z. Hu and Y. Zhao, Low-toxic and safe nanomaterials by surface-chemical design, carbon nanotubes, fullerenes, metallofullerenes, and graphenes, *Nanoscale*, 2011, **3**, 362–382.
10. W. J. Rogers, C. H. Meyer and C. M. Kramer, Technology insight: *in vivo* cell tracking by use of MRI, *Nat. Clin. Pract. Cardiovasc. Med.*, 2006, **3**, 554–562.
11. I. Lynch and K. A. Dawson, Protein-nanoparticle interactions, *Nano today*, 2008, **3**, 40–47.
12. M. P. Monopoli, C. Aberg, A. Salvati and K. A. Dawson, Biomolecular coronas provide the biological identity of nanosized materials, *Nat. Nanotechnol.*, 2012, **7**, 779–786.
13. A. E. Nel, L. Madler, D. Velegol, T. Xia, E. M. Hoek, P. Somasundaran, F. Klaessig, V. Castranova and M. Thompson, Understanding biophysicochemical interactions at the nano-bio interface, *Nat. Mater.*, 2009, **8**, 543–557.
14. L. Treuel and G. U. Nienhaus, Toward a molecular understanding of nanoparticle-protein interactions, *Biophys. Rev.*, 2012, **4**, 137–147.
15. P. Foroozandeh and A. A. Aziz, Merging worlds of nanomaterials and biological environment: factors governing protein corona formation on nanoparticles and its biological consequences, *Nanoscale Res. Lett.*, 2015, **10**, 221.
16. M. P. Monopoli, A. S. Pitek, I. Lynch and K. A. Dawson, Formation and characterization of the nanoparticle-protein corona, *Methods Mol. Biol.*, 2013, **1025**, 137–155.
17. C. Corbo, R. Molinaro, A. Parodi, N. E. Toledano Furman, F. Salvatore and E. Tasciotti, The impact of nanoparticle protein corona on cytotoxicity, immunotoxicity and target drug delivery, *Nanomedicine*, 2016, **11**, 81–100.
18. L. Vroman, A. L. Adams, G. C. Fischer and P. C. Munoz, Interaction of high molecular weight kininogen, factor XII, and fibrinogen in plasma at interfaces, *Blood*, 1980, **55**, 156–159.
19. M. P. Monopoli, D. Walczyk, A. Campbell, G. Elia, I. Lynch, F. B. Bombelli and K. A. Dawson, Physical-chemical aspects of protein corona: relevance to *in vitro* and *in vivo* biological impacts of nanoparticles, *J. Am. Chem. Soc.*, 2011, **133**, 2525–2534.
20. S. Tenzer, D. Docter, J. Kuharev, A. Musyanovych, V. Fetz, R. Hecht, F. Schlenk, D. Fischer, K. Kiouptsi, C. Reinhardt, K. Landfester, H. Schild, M. Maskos, S. K. Knauer and R. H. Stauber, Rapid formation of plasma protein corona critically affects nanoparticle pathophysiology, *Nat. Nanotechnol.*, 2013, **8**, 772–781.

21. M. Rahman, S. Laurent, N. Tawil, L. H. Yahia and M. Mahmoudi, in *Protein-Nanoparticle Interactions*, Springer, 2013, pp. 21–44.
22. B. Kharazian, N. Hadipour and M. Ejtehadi, Understanding the nanoparticle–protein corona complexes using computational and experimental methods, *Int. J. Biochem. Cell Biol.*, 2016, **75**, 162–174.
23. P. Patel, K. Kansara, R. Singh, R. K. Shukla, S. Singh, A. Dhawan and A. Kumar, Cellular internalization and antioxidant activity of cerium oxide nanoparticles in human monocytic leukemia cells, *Int. J. Nanomed.*, 2018, **13**, 39–41.
24. P. Patel, K. Kansara, V. A. Senapati, R. Shanker, A. Dhawan and A. Kumar, Cell cycle dependent cellular uptake of zinc oxide nanoparticles in human epidermal cells, *Mutagenesis*, 2016, **31**, 481–490.
25. N. Meghani, P. Patel, K. Kansara, S. Ranjan, N. Dasgupta, C. Ramalingam and A. Kumar, Formulation of vitamin D encapsulated cinnamon oil nanoemulsion: Its potential anti-cancerous activity in human alveolar carcinoma cells, *Colloids Surf., B*, 2018, **166**, 349–357.
26. P. Patel, K. Kansara, D. Shah, N. V. S. Vallabani, R. K. Shukla, S. Singh, A. Dhawan and A. Kumar, Cytotoxicity assessment of ZnO nanoparticles on human epidermal cells, *Mol. Cytogenet.*, 2014, **7**, P81.
27. M. Lundqvist, J. Stigler, G. Elia, I. Lynch, T. Cedervall and K. A. Dawson, Nanoparticle size and surface properties determine the protein corona with possible implications for biological impacts, *Proc. Natl. Acad. Sci. U. S. A.*, 2008, **105**, 14265–14270.
28. C. Sacchetti, K. Motamedchaboki, A. Magrini, G. Palmieri, M. Mattei, S. Bernardini, N. Rosato, N. Bottini and M. Bottini, Surface polyethylene glycol conformation influences the protein corona of polyethylene glycol-modified single-walled carbon nanotubes: potential implications on biological performance, *ACS Nano*, 2013, **7**, 1974–1989.
29. V. H. Nguyen and B. J. Lee, Protein corona: a new approach for nanomedicine design, *Int. J. Nanomed.*, 2017, **12**, 3137–3151.
30. B.-J. Lee and V. H. Nguyen, Protein corona: A new approach for nanomedicine design, *Int. J. Nanomed.*, 2017, **12**, 3137.
31. P. Foroozandeh and A. A. Aziz, Merging worlds of nanomaterials and biological environment: factors governing protein corona formation on nanoparticles and its biological consequences, *Nanoscale Res. Lett.*, 2015, **10**, 221.
32. T. Cedervall, I. Lynch, M. Foy, T. Berggard, S. C. Donnelly, G. Cagney, S. Linse and K. A. Dawson, Detailed identification of plasma proteins adsorbed on copolymer nanoparticles, *Angew. Chem.*, 2007, **46**, 5754–5756.
33. T. Cedervall, I. Lynch, S. Lindman, T. Berggard, E. Thulin, H. Nilsson, K. A. Dawson and S. Linse, Understanding the nanoparticle-protein corona using methods to quantify exchange rates and affinities of proteins for nanoparticles, *Proc. Natl. Acad. Sci. U. S. A.*, 2007, **104**, 2050–2055.
34. M. Mahmoudi, M. A. Sahraian, M. A. Shokrgozar and S. Laurent, Superparamagnetic iron oxide nanoparticles: promises for diagnosis and treatment of multiple sclerosis, *ACS Chem. Neurosci.*, 2011, **2**, 118–140.

35. J. Wolfram, Y. Yang, J. Shen, A. Moten, C. Chen, H. Shen, M. Ferrari and Y. Zhao, The nano-plasma interface: Implications of the protein corona, *Colloids Surf., B*, 2014, **124**, 17–24.
36. H. Y. Kim, J. O. Sofo, D. Velegol, M. W. Cole and A. A. Lucas, Van der Waals dispersion forces between dielectric nanoclusters, *Langmuir*, 2007, **23**, 1735–1740.
37. Y. Min, M. Akbulut, K. Kristiansen, Y. Golan and J. Israelachvili, The role of interparticle and external forces in nanoparticle assembly, *Nat. Mater.*, 2008, **7**, 527–538.
38. P. Wang, X. Wang, L. Wang, X. Hou, W. Liu and C. Chen, Interaction of gold nanoparticles with proteins and cells, *Sci. Technol. Adv. Mater.*, 2015, **16**, 034610.
39. A. Vonarbourg, C. Passirani, P. Saulnier and J. P. Benoit, Parameters influencing the stealthiness of colloidal drug delivery systems, *Biomaterials*, 2006, **27**, 4356–4373.
40. O. Lunov, T. Syrovets, C. Loos, J. Beil, M. Delacher, K. Tron, G. U. Nienhaus, A. Musyanovych, V. Mailander, K. Landfester and T. Simmet, Differential uptake of functionalized polystyrene nanoparticles by human macrophages and a monocytic cell line, *ACS Nano*, 2011, **5**, 1657–1669.
41. A. D. McNaught and A. D. McNaught, *Compendium of Chemical Terminology*, Blackwell Science Oxford, 1997.
42. D. Ronis, E. Martina and J. Deutch, The solvophobic effect in simple fluid mixtures, *Chem. Phys. Lett.*, 1977, **46**, 53–55.
43. A. Gessner, R. Waicz, A. Lieske, B. Paulke, K. Mader and R. H. Muller, Nanoparticles with decreasing surface hydrophobicities: influence on plasma protein adsorption, *Int. J. Pharm.*, 2000, **196**, 245–249.
44. L. Treuel, How protein adsorption shapes the biological identity of NPs—where do we stand, *J. Phys. Chem. Biophys.*, 2013, **3**, e113.
45. J. Buijs, C. Costa Vera, E. Ayala, E. Steensma, P. Hakansson and S. Oscarsson, Conformational stability of adsorbed insulin studied with mass spectrometry and hydrogen exchange, *Anal. Chem.*, 1999, **71**, 3219–3225.
46. C. D. Walkey and W. C. Chan, Understanding and controlling the interaction of nanomaterials with proteins in a physiological environment, *Chem. Soc. Rev.*, 2012, **41**, 2780–2799.
47. C. D. Walkey and W. C. Chan, Understanding and controlling the interaction of nanomaterials with proteins in a physiological environment, *Chem. Soc. Rev.*, 2012, **41**, 2780–2799.
48. M. Mahmoudi, M. A. Shokrgozar, S. Sardari, M. K. Moghadam, H. Vali, S. Laurent and P. Stroeve, Irreversible changes in protein conformation due to interaction with superparamagnetic iron oxide nanoparticles, *Nanoscale*, 2011, **3**, 1127–1138.
49. D. Huhn, K. Kantner, C. Geidel, S. Brandholt, I. De Cock, S. J. Soenen, P. Rivera Gil, J. M. Montenegro, K. Braeckmans, K. Mullen, G. U. Nienhaus, M. Klapper and W. J. Parak, Polymer-coated nanoparticles interacting with proteins and cells: focusing on the sign of the net charge, *ACS Nano*, 2013, **7**, 3253–3263.

50. C. C. Fleischer, U. Kumar and C. K. Payne, Cellular Binding of Anionic Nanoparticles is Inhibited by Serum Proteins Independent of Nanoparticle Composition, *Biomater. Sci.*, 2013, **1**, 975–982.
51. C. C. Fleischer and C. K. Payne, Nanoparticle-cell interactions: molecular structure of the protein corona and cellular outcomes, *Acc. Chem. Res.*, 2014, **47**, 2651–2659.
52. M. Mahmoudi, I. Lynch, M. R. Ejtehadi, M. P. Monopoli, F. B. Bombelli and S. Laurent, Protein-nanoparticle interactions: opportunities and challenges, *Chem. Rev.*, 2011, **111**, 5610–5637.
53. E. Casals, T. Pfaller, A. Duschl, G. J. Oostingh and V. Puntes, Time evolution of the nanoparticle protein corona, *ACS Nano*, 2010, **4**, 3623–3632.
54. M. Lundqvist, J. Stigler, T. Cedervall, T. Berggard, M. B. Flanagan, I. Lynch, G. Elia and K. Dawson, The evolution of the protein corona around nanoparticles: a test study, *ACS Nano*, 2011, **5**, 7503–7509.
55. S. Nagayama, K. Ogawara, Y. Fukuoka, K. Higaki and T. Kimura, Time-dependent changes in opsonin amount associated on nanoparticles alter their hepatic uptake characteristics, *Int. J. Pharm.*, 2007, **342**, 215–221.
56. M. A. Dobrovolskaia, A. K. Patri, J. Zheng, J. D. Clogston, N. Ayub, P. Aggarwal, B. W. Neun, J. B. Hall and S. E. McNeil, Interaction of colloidal gold nanoparticles with human blood: effects on particle size and analysis of plasma protein binding profiles, *Nanomedicine*, 2009, **5**, 106–117.
57. Z. J. Deng, M. Liang, I. Toth, M. J. Monteiro and R. F. Minchin, Molecular interaction of poly(acrylic acid) gold nanoparticles with human fibrinogen, *ACS Nano*, 2012, **6**, 8962–8969.
58. R. Purohit, N. S. Vallabani, R. K. Shukla, A. Kumar, A. Dhawan and S. Singh, BSA coated gold nanoparticles exhibit size dependent interaction with lung cancer (A549) cells, *Mol. Cytogenet.*, 2014, **7**, P83.
59. P. Wang, X. Wang, L. Wang, X. Hou, W. Liu and C. Chen, Interaction of gold nanoparticles with proteins and cells, *Sci. Technol. Adv. Mater.*, 2015, **16**, 034610.
60. M. C. Cox, K. J. Barnham, T. A. Frenkiel, J. D. Hoeschele, A. B. Mason, Q. Y. He, R. C. Woodworth and P. J. Sadler, Identification of platination sites on human serum transferrin using (13)C and (15)N NMR spectroscopy, *JBIC, J. Biol. Inorg. Chem.*, 1999, **4**, 621–631.
61. Z. J. Deng, G. Mortimer, T. Schiller, A. Musumeci, D. Martin and R. F. Minchin, Differential plasma protein binding to metal oxide nanoparticles, *Nanotechnology*, 2009, **20**, 455101.
62. P. Aggarwal, J. B. Hall, C. B. McLeland, M. A. Dobrovolskaia and S. E. McNeil, Nanoparticle interaction with plasma proteins as it relates to particle biodistribution, biocompatibility and therapeutic efficacy, *Adv. Drug Delivery Rev.*, 2009, **61**, 428–437.
63. A. Gessner, A. Lieske, B. Paulke and R. Muller, Influence of surface charge density on protein adsorption on polymeric nanoparticles: analysis by two-dimensional electrophoresis, *Eur. J. Pharm. Biopharm.*, 2002, **54**, 165–170.

64. N. Oku, Y. Tokudome, Y. Namba, N. Saito, M. Endo, Y. Hasegawa, M. Kawai, H. Tsukada and S. Okada, Effect of serum protein binding on real-time trafficking of liposomes with different charges analyzed by positron emission tomography, *Biochim. Biophys. Acta*, 1996, **1280**, 149–154.
65. S. M. Moghimi and H. M. Patel, Tissue specific opsonins for phagocytic cells and their different affinity for cholesterol-rich liposomes, *FEBS Lett.*, 1988, **233**, 143–147.
66. P. R. Cullis, A. Chonn and S. C. Semple, Interactions of liposomes and lipid-based carrier systems with blood proteins: Relation to clearance behaviour *in vivo*, *Adv. Drug Delivery Rev.*, 1998, **32**, 3–17.
67. M. De, C. C. You, S. Srivastava and V. M. Rotello, Biomimetic interactions of proteins with functionalized nanoparticles: a thermodynamic study, *J. Am. Chem. Soc.*, 2007, **129**, 10747–10753.
68. G. Maiorano, S. Sabella, B. Sorce, V. Brunetti, M. A. Malvindi, R. Cingolani and P. P. Pompa, Effects of cell culture media on the dynamic formation of protein-nanoparticle complexes and influence on the cellular response, *ACS Nano*, 2010, **4**, 7481–7491.
69. G. Maiorano, S. Sabella, B. Sorce, V. Brunetti, M. A. Malvindi, R. Cingolani and P. P. Pompa, Effects of cell culture media on the dynamic formation of protein–nanoparticle complexes and influence on the cellular response, *ACS Nano*, 2010, **4**, 7481–7491.
70. M. Mahmoudi, A. M. Abdelmonem, S. Behzadi, J. H. Clement, S. Dutz, M. R. Ejtehadi, R. Hartmann, K. Kantner, U. Linne, P. Maffre, S. Metzler, M. K. Moghadam, C. Pfeiffer, M. Rezaei, P. Ruiz-Lozano, V. Serpooshan, M. A. Shokrgozar, G. U. Nienhaus and W. J. Parak, Temperature: the "ignored" factor at the NanoBio interface, *ACS Nano*, 2013, **7**, 6555–6562.
71. A. Bhogale, N. Patel, J. Mariam, P. M. Dongre, A. Miotello and D. C. Kothari, Comprehensive studies on the interaction of copper nanoparticles with bovine serum albumin using various spectroscopies, *Colloids Surf., B*, 2014, **113**, 276–284.

CHAPTER 4

NP–Protein Corona Interaction: Characterization Methods and Analysis

DIVYA PANCHAL[a], RUCHIT PATEL[a], MANTHAN SIDDHESHWARI[a], EFFTESUM RAHAMAN[a], VAISHWIK PATEL[a] AND AJAY S. KARAKOTI*[a,b]

[a]Biological and Life Sciences, School of Arts and Sciences, Ahmedabad University, Ahmedabad, Gujarat, 380009 India; [b]School of Engineering and Applied Science, Ahmedabad University, Ahmedabad, Gujarat, 380009 India
*E-mail: ajay.karakoti@ahduni.edu.in

4.1 Introduction

4.1.1 Brief History of the Protein Corona

The protein corona, as suggested throughout this book, is the term used for layers of proteins on the surfaces of materials directly in contact with a milieu of proteins. These layers are formed from the specific as well as non-specific adsorption of proteins on the surfaces of materials and thus can be regarded as an upgraded name for protein adsorption. Adsorption of proteins on surfaces gained importance in the early 60 s and the field drew the attention of various researchers who characterized the adsorption of proteins on multiple surfaces. The field started to gain much more interest in

Issues in Toxicology No. 40
Nanoparticle–Protein Corona: Biophysics to Biology
Edited by Ashutosh Kumar and Alok Dhawan

Published by the Royal Society of Chemistry, www.rsc.org

the early 80 s and 90 s as the transformation from ultrafine to nanoparticle-based chemistry surfaced. Liposomes and polymeric nanoparticle interactions with protein surfaces got specific attention as strategies around drug delivery started to emerge to create adsorption-resistant nanoparticle surfaces. While the concepts emerge gradually over time, the main theme of protein corona formation remains unchanged. In this chapter, we will briefly introduce nanoparticles and the nanoparticle protein corona. Aspects of the protein corona will be discussed so that the user can correlate the basic properties of corona formation that are studied using various characterization techniques. This will be followed by a brief introduction to the sampling methods and principles of various analytical techniques excluding the optical and electron-microscopy-based imaging techniques. The reader is encouraged to refer to books and journals for details of each analytical technique as a full description is beyond the scope of this chapter. Finally, relevant characteristics and properties of the nanoparticle–protein corona interaction are discussed based on the characterization technique that will help the reader to understand the utility of such characterization techniques in practice.

4.1.2 Nanoparticles

Nanoparticles (NPs) are nanoscale materials with sizes ranging between 1 to 100 nm and are often perceived as the link between bulk materials and atomic or molecular entities. A more rational definition will be the length scales at which you see drastic changes in the physical and/or chemical properties of materials as a function of size. This drastic change in properties due to an increase in surface to volume ratio is usually ascribed to the quantum confinement of electrons or simply as the higher energy (or boundary conditions) required by the electron cloud to participate in processes that the bulk counterparts did with ease. The atoms on the surfaces are more reactive as there are multiple dangling bonds increasing the overall surface energy that need immediate passivation. A combination of high surface energy and much higher surface to volume ratio is also responsible, in many ways, for the challenges associated with their characterization. Surface energetics results in the formation of intentional or un-intentional surface layers of NPs. NPs conventionally comprise three distinct layers that include the surface layer, the shell layer or the interfacial layer, and the core. The surface layer may constitute a mixture of various metal ions, organic contaminants, polymeric compounds, stabilizers and other such molecules. The interfacial layer or the shell layer forms an essential portion of the NPs and is synthetically dissimilar as compared to its bulk. Composed of certain ions, organic and inorganic molecules, the shell layer substantially controls the physical and chemical properties of NPs. The bulk of NPs that form the core is usually referred to as a "nanoparticle" (NP).

The reactivity of NPs is often influenced not only by their unique size, shape and structure but also by the composition of the surface and intermediate layers. Thus, the dispersion of NPs in the medium of their end application can also influence the final properties. Biological applications of nanomaterials require the dispersion of NPs in biological media and pose a great challenge in developing a clear understanding of the properties. Biological media have their own characteristic properties due to the presence of numerous biomolecules that are known to drive and govern the reactivity of nanoparticles. Likewise, the nanoparticles introduced can, in turn, have several effects on the biological medium. Therefore, studying NPs with varying sizes and their corresponding interactions with different biological systems (or proteins present therein) is primarily important for their safe and efficient application.

4.2 Protein Corona

4.2.1 Formation and Composition

The reactivity of nanomaterials with biological fluids is predominantly affected by various interactions between the surface of NPs and the medium used for their dispersion. These interactions are thought to evolve over time and have led to the emergence of the idea of protein–nanoparticle complexes; hence forming the basis of several extensive studies conducted on this matter over the last two decades.

Blood, human serum and plasma are the biological fluids used for the protein corona study of the nanoparticles. Among them, blood is the most commonly used biological fluid. Instead of biological fluids, synthetic media like Dulbecco's Modified Eagle's Medium (DMEM), Modified Eagle's Medium (MEM) and Rosewell Park Memorial Institute Medium (RPMI) and many others are also used for dispersing and studying the behavior of nanoparticles. These media may be supplemented with fetal bovine serum (FBS), fetal calf serum (FCS), several antibiotics, Bovine Serum Albumin (BSA), Human Serum Albumin (HSA), fibrinogen, and specific proteins like apolipoproteins. The presence of these proteins and other ingredients gives researchers the freedom to study the effects of such species on nanoparticles for protein corona formation.

When introduced in a biological medium, the nanoparticle surface interacts with various biomolecules such as proteins, nucleic acids, lipids and even biological metabolites owing to its high surface area and its need for the passivation of surface dangling bonds. The interaction primarily results in the surface adsorption of proteins giving rise to the formation of a single or multiple layer(s) of adsorbed proteins termed as a "protein corona". As the surface of the NP is modified, its cellular or tissue responses, such as its cellular uptake, bio distribution and bioavailability, will depend on the composition of the corona. Several outcomes can be predicted depending on the amount and type of proteins that are adsorbed in a defined period of time as the nanoparticle obtains a new identity and its physico-chemical properties are modified by the adsorbed layer of proteins. Thus, it is very important to understand the composition of the corona and its characteristics, such as its static or dynamic nature, to be able to predict the biological outcome of the surface-modified NP.

For disease applications, through intravenous administration of nanoparticles, blood plasma will be one of the first components that the NPs will encounter immediately upon administration. Human plasma (HP) consists of as many as 3700 different proteins with varying concentrations. Along with proteins, plasma also contains lipids that compete with the proteins to adhere to the NP surface. As a result, there is a competition between different biological molecules to adsorb on the NPs' surface. Initially, the proteins that are most abundant in the medium and those with an increasing mobility will get adsorbed on the surface; however, after a significant time period, they will be replaced by proteins that have a lower abundance in the medium but much higher affinity to the NP surface. Protein–nanoparticle binding affinities as well as protein–protein interactions highly affect the adsorption of proteins on the surface of NPs and need to be characterized thoroughly. This arrangement or pattern of proteins adsorbed onto the NP surface forming successive protein layers gives rise to a "hard corona" and a "soft corona". Proteins that bind strongly due to a higher affinity and that do not desorb easily give rise to a hard corona while the proteins that stay loosely attached to the NP surface due to a lower affinity form the soft corona. Also, the exchange rates ("on" and "off" rates or association and dissociation rates) of the adsorbed proteins can define whether the corona is "hard" or "soft". A typical hard corona shows much slower exchange rates as compared to the soft corona, thus, determination of exchange rates is crucial for the prediction of hard and soft corona formation over NP surfaces.

NPs exhibit a "synthetic identity" while they are processed and modified for their administration into any biological medium. However, upon their interaction with biological components, this "synthetic identity" is transformed into a "biological identity" (Figure 4.1).

The structure and composition of the protein corona change over time due to the various physical and chemical properties of the NPs such as their size, shape, composition, surface functional groups, surface charges, and the nature of the biological medium (such as blood or any growth medium) as well as the duration of exposure of the NPs in the medium. This identity influences all the physiological responses including agglomeration, cellular uptake, circulation lifetime, target delivery, signaling, kinetics, transport, accumulation, and toxicity of the NPs. Thus, studying these properties as a function of the biological medium is crucial for its applications.

The protein corona is believed to be a dynamic entity as it continues to change gradually in order to attune its composition to the surrounding biological environment. Hence, the protein corona is considered to be

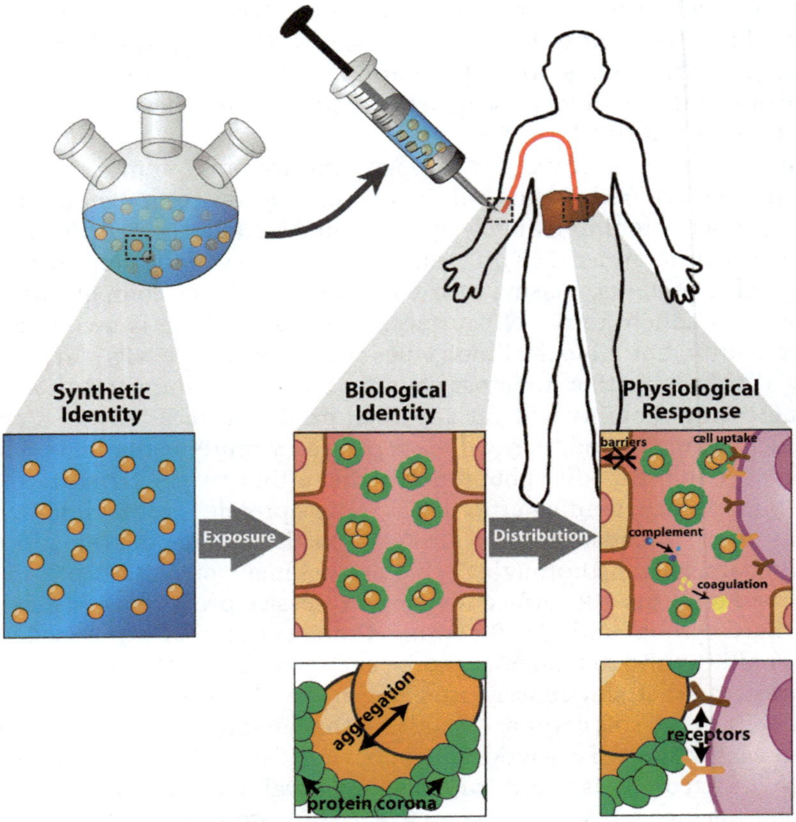

Figure 4.1 Correlation between synthetic identity, biological identity and physiological response. Reproduced from ref. 1 with permission from the Royal Society of Chemistry.

complex and there isn't a concept of a "universal" protein corona for all nanomaterials. Similarly, the thickness of the corona layer does not signify a relative abundance of these proteins in the biological medium. It can be inferred that the structure as well as the composition of the protein corona is exclusive to each nanoparticle and is governed by certain physicochemical factors (Figures 4.2–4.3). Different nanoparticles will have different ways of forming the corona, *i.e.*, irreversible corona formation, reversible corona formation, and temporary corona formation. Also, one study reports the formation of a corona in two stages: the early and late phase.

The surface chemistry of the particle and protein helical structures play a major role in the formation of the corona on the nanoparticle's surface. The changes in the formation of the corona depend on multiple factors, such as exposure period of any nanoparticle in the medium, the Vroman effect and association–dissociation values.

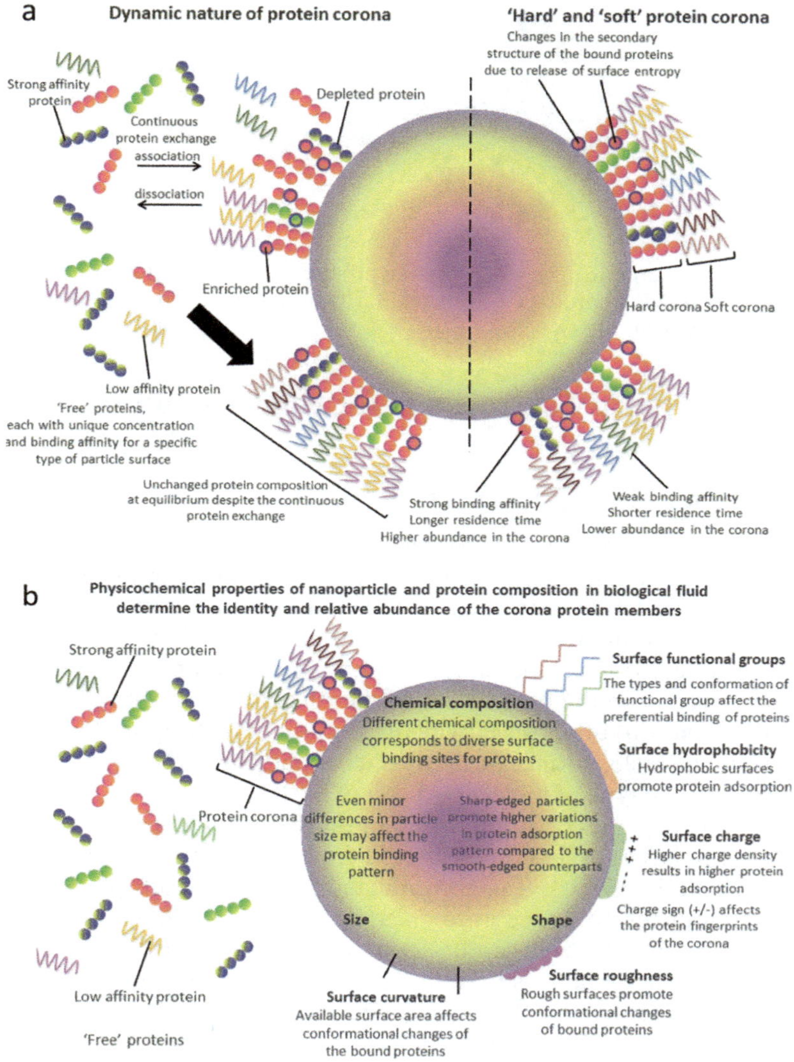

Figure 4.2 The nanoparticle–protein corona complex is the effective biological unit of a nanoparticle. (a) The competitive particle–protein interactions, variations in particle–protein association/dissociation rate along with the characteristic protein profile in the biological fluids result in the dynamic nature of a protein corona. The continuous bound protein–'free' protein exchange alters the protein corona composition prior to equilibrium, at which state the composition remains unchanged. The wide range binding affinities of proteins for the nanoparticle surface lead to the formation of hard and soft coronas. Protein members of the corona are enriched or depleted relative to their physiological abundance in the surrounding biological fluid. (b) The interconnected effects of a particle's physicochemical properties together with the protein profile in a biological fluid result in protein coronas with unique compositions. Reproduced from ref. 2 with permission from the Royal Society of Chemistry.

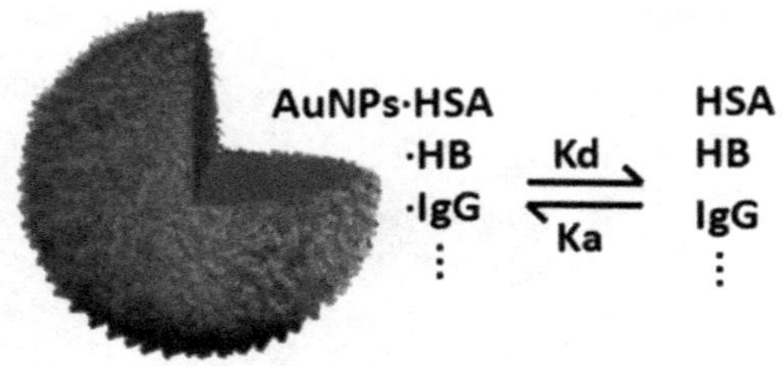

Figure 4.3 Schematic diagram of plasma protein adsorption and dissociation on a AuNP. Reproduced from ref. 3 with permission from the Royal Society of Chemistry.

4.3 Analytical Techniques for the Characterization of Protein Coronas

4.3.1 Dynamic Light Scattering (DLS)

DLS is often used to estimate the size distribution or size profile of NPs dispersed in a liquid medium. Also known as photon-correlation spectroscopy or quasi-elastic light scattering, it principally works under time-domain analysis, *i.e.*, examination of the temporal fluctuations in the scattered light intensity reflected from the nanoparticles. Fluctuations in intensity of the light scattered from the particles arises because of the continuous diffusion (Brownian motion) of particles. The measurements rely upon the size of the bulk of the particle, surface structures, concentration of the particle in the medium and the type of molecules present in the medium.

A laser or any monochromatic light source is incident on the sample through a polarizer. The light scattered from the particles passes through a second polarizer where it is accumulated by a photomultiplier, which, as a result, produces an image onto a screen – this phenomenon is termed a Speckle pattern. Polarizers can be set up in either a vertical–vertical geometry (second polarizer allows light in the same direction as the first) or vertical–horizontal geometry (second polarizer does not allow light in the same direction as the incident light). The particles scatter light in all directions (Rayleigh scattering – when the particles are small, *i.e.*, <1/10 of the wavelength of the incident light, the scattered light has the same energy as the incident light and is independent of the angle). When the size increases, Mie scattering occurs where the scattered light has a different energy and is angle-dependent.

The fluctuations in scattered light intensity are correlated against short decay intervals and a correlation between the scattered light against time is obtained through a correlation function. At short time intervals, correlation is high as the particles have not moved to a great extent. As the time interval increases, correlation decays exponentially and a time comes where there happens to be no correlation between the scattered intensity of the final and initial states. The correlation is high in large particles as they take longer to move out of the path of the incident laser light while the correlation will be low for smaller particles that move swiftly. Finally, the diffusion coefficient of particles is calculated by analyzing the decay rates from the correlation using deconvolution algorithms. Two of the common algorithms are cumulant analysis and Non-Negative Least Squares (NNLS – common examples of named NNLS algorithms include CONTIN, Regularization, GP and MNM, and REPES and DYNALS). The diffusion coefficient (D) can be used to calculate the hydrodynamic radius (R_H) of nanoparticles by the Stokes–Einstein equation:

$$D_t = \frac{k_B T}{6\pi \eta R_H}$$

where k_B = Boltzmann constant ($1.38064852 \times 10^{-23}$ J K^{-1}), T = temperature, η = absolute viscosity and R_H = hydrodynamic radius.

The results from the DLS are often a function of the sample consistency. It is often said that DLS works well for monodisperse samples as the scattering intensity varies exponentially with the size of the particles ($I^6 \alpha$ m). Sample preparation is crucial as the sample needs to be clear, homogeneous and without haze. In addition to this, DLS gives several types of particle sizes (weighted by intensity, volume, number and surface area) and often the user picks a size relevant to the size given by the TEM results. In a study carried out by the author of this book at the Environmental and Molecular Sciences Laboratory (EMSL) at the Pacific Northwest National Laboratory (PNNL) using standard gold nanoparticles (obtained from the National Institute of Standards and Technology (NIST)), it was observed that correlation of the particle size reported by DLS varied with the size of the particles (see Tables 4.1 and 4.2). With smaller particle sizes, the volume and surface area weighted results were found to be in agreement with other methods of size determination while with larger particle sizes, only intensity weighted results were closer to the actual size values. Significant underreporting of the sizes from DLS was found using number, intensity or surface area weighted particle size distributions. This was found for rather standard and nearly monodisperse gold nanoparticles with good scattering coefficients. For metal oxide nanoparticles or polymers with low scattering, the numbers may show significant differences based on the sizes of the nanoparticles and thus the user must exercise caution while working with such samples.

Table 4.1 Comparison of the sizes of standard gold nanoparticles obtained from NIST.

Size (nm)	AFM	SEM	TEM	DMA	SXA	DLS-NIST	DLS[a] (current study)
10	8.5 ± 0.3	9.9 ± 0.1	8.9 ± 0.1	11.3 ± 0.1	9.1 ± 1.8	13.5 ± 0.1	15.5 ± 0.1
30	24.9 ± 1.1	26.9 ± 0.1	27.6 ± 2.1	28.4 ± 1.1	24.9 ± 1.2	26.5 ± 3.6	30.8 ± 0.6
60	55.4 ± 0.3	54.9 ± 0.4	56.0 ± 0.5	56.3 ± 1.5	53.2 ± 5.3	55.3 ± 8.3	51.0 ± 1.2

[a]DLS size data was measured in the author's lab and compared with other data provided by NIST.

Table 4.2 Comparison of TEM size data from NIST samples with various formats of size data obtained from the DLS.

Size	TEM	Z-Ave			Lognormal distribution		
	NIST	NIST	EMSL	Intensity	Volume	Number	Surface area
10 nm	8.9	13.5	15.5	15.4	9.6	5.4	7.9
			15.5	15.4	8.4	3.9	6.5
			14.9	14.9	8.8	4.7	7.1
30 nm	26.9	26.5	31.2	31.2	24.5	17.6	21.5
			31.2	31.4	24.1	17.6	21.8
			30.0	30.2	24.9	19.6	22.9
60 nm	54.9	55.3	52.2	52.4	34.7	21.3	29.6
			51.3	51.3	35.2	21.1	29.2
			49.7	49.8	34.1	21.5	29.3

4.3.2 Z-potential

Z-potential is a function of the electric charge of NPs dispersed in any liquid medium, *i.e.*, the net charge that develops at the solid–liquid interface. The instrumentation used for size and surface charge estimation is the same. Z-potential refers to the electro kinetic potential generated on the NP dispersions under the influence of an electric field during electrophoresis. An electrical double layer (EDL) is formed at the NP–liquid interface as the net charge at the NP surface affects the distribution of the surrounding ions upon its administration into the liquid medium. As a result, two layers are formed: the inner layer called the stern layer is tightly bound to the NP surface and has an electric charge that is opposite to that of the NP, whereas, the external diffuse layer or the slipping plane contains a balanced ion distribution and is related to the mobility of nanoparticles in suspension under the influence of an electric field. The potential difference generated at the slipping plane between the surrounding medium and the stern layer is called the Z-potential.

The Z-potential is deduced from the electrophoretic mobility of charged particles under an applied electric field. The electrophoretic mobility (μ_e) of the particles is calculated as:

$$\mu_e = \frac{V}{E}$$

where V = particle velocity (μm s^{-1}), E = electric field strength (volt cm^{-1}) – both known quantities. The Z-potential is then calculated from the obtained μ_e by Henry's equation:

$$\mu_e = \frac{2\varepsilon_r \varepsilon_0 \zeta f(K_a)}{3\eta}$$

where ε_r = relative permittivity/dielectric constant, ε_0 = permittivity of vacuum, ζ = ZP, $f(K_a)$ = Henry's function, η = viscosity at experimental temperature.

When the thickness of the EDL is much smaller compared to the particle radius, the Henry's equation is then modified to the Helmholtz–Smoluchowski (HS) equation:

$$\mu_e = \frac{\varepsilon_r \varepsilon_0 \zeta}{\eta}$$

On the contrary, when the thickness of the EDL is much larger than the particle, the Henry's equation can be modified to the Hückel equation:

$$\mu_e = \frac{2\varepsilon_r \varepsilon_0 \zeta}{3\eta}$$

4.3.3 Centrifugation

Centrifugation works on the sedimentation principle where the action of centrifugal force is applied to separate the particles based on their size and density from the suspended liquid medium. During centrifugation, a powerful centrifugal force is applied to a suspension of mixed sizes of nanoparticles, which supports and accelerates the separation of particles from the fluid media based on density. Particles that have a higher density than that of the liquid medium migrate away from the centrifuge axis and sink while the less dense particles float at the top. The rate of separation of particles depends on the size and density of the particles. The larger the difference between the size and density of various components, the faster their separation.

The rate of centrifugation depends on the angular velocity of the rotor as well as its radius and can be expressed as revolutions per minute (RPM) or acceleration (g). The sedimentation velocity is a function of the size, and centrifugal acceleration, the density difference between the particle and the liquid and viscosity.

There are different types of centrifuges available based on their RPM such as micro centrifuges (12 000–13 000 RPM), high-speed centrifuges (up to 30 000 RPM) and ultracentrifuges (up to 70 000 RPM). The types of rotors that can be used are swinging bucket rotors, vertical tube rotors or fixed angle rotors. Density gradient centrifugation, differential centrifugation and ultracentrifugation are the types of centrifugation used based on the type of separation (size and density).

4.3.4 Chromatography

Chromatography is a common technique that involves separation of a mixture of compounds (also called solutes) into their individual constituents. All chromatography techniques usually contain two phases: the mobile phase and the stationary phase. The stationary phase is a stable structure onto which the sample is spotted or placed. This stationary phase is kept in the mobile phase, which is any solvent system that carries the sample up the stationary phase through capillary action to enable its separation. In high-performance liquid chromatography (HPLC), the solvent (mobile) phase is passed over the stationary phase (usually a column packed with packing material such as porous beads) under pressure and in a continuous manner to enable faster separation of solutes. The solute is injected into the solvent phase through a specially arranged injector. After each injection, some time is allowed for the individual components to pass over the separation column and separate into individual components before a fresh injection is carried out.

Chromatography techniques can be preparative, *i.e.*, separating the components for later use – a form of purification. Or, they can be analytical, *i.e.*, detecting the presence or measuring the relative proportion of any component in the sample mixture.

The separation is driven by differential affinities (strength of adhesion) of various components of a sample towards the stationary and mobile phase. Various chromatographic techniques use different types of interactions between the solute and the stationary (or mobile phase) phase such as adsorption, polarity, ionic charge, binding affinity or size. Specifically, for polarity-based separation, the solvent has a huge role to play and the strength of the solvent can be tailored by mixing solvents of different polarities. There are many types of chromatography, like paper chromatography, thin layer chromatography (TLC), ion exchange chromatography, gas chromatography, size-exclusion chromatography, liquid chromatography and affinity chromatography, where the principle of separation remains the same, it is just the basis of separation that is different. To study protein corona formation and interactions, the most commonly used chromatographic technique is size exclusion chromatography.

4.3.4.1 Size Exclusion Chromatography (SEC)

Size exclusion chromatography is the basic method to study the elution behavior of nanoparticle–protein, protein–protein and protein–ligand mixtures. Size exclusion chromatography uses porous particles to separate molecules of different sizes. SEC is also known as gel filtration or gel permeation chromatography. SEC provides information regarding the molecular weights of compounds or nanoparticles (with or without the protein corona). It is frequently used to characterize the molecular weights and differentiate macromolecules such as long chain amino acids, peptide complexes, DNA, RNA and polymers. When one uses any aqueous medium or gel-like material to filter, it can be called gel filtration. While, in gel permeation, an organic solvent is used in the mobile phase. Hence, the separation of molecules in gel permeation is also called fractionation. The sizes of the pores in the beads determine the size exclusion limit (or the resolution of the column).

A mixture of molecules when dissolved in the liquid phase (mobile phase) is applied to a chromatography column where beads form a solid support (stationary phase). The mass of beads in the column is often referred to as the column bed. The beads act as traps or sieves and function to trap the smaller molecules and hold them temporarily within the pore or bead, slowing down their egress from the pores. Larger molecules would pass around or only partially penetrate the pores or beads. So, the larger molecules elute first, and smaller molecules elute later. While, molecules that would pass through all the pores would elute last. Here, the particles would filter at different rates according to their sizes and molecular weights.

4.3.5 Nuclear Magnetic Resonance Spectroscopy (NMR)

Nuclear Magnetic Resonance Spectroscopy (NMR), a non-invasive and non-destructive technique, provides chemical information on the basis of the spin of the nuclei of the atoms of molecules and compounds. Before learning about NMR, we shall discuss the resonance condition. Resonance condition refers to two possible states that exist for either electro-paramagnetic (EPR) or nuclear magnetism in the presence of an external magnetic field.

In both EPR and NMR techniques, the lower-energy state possesses a nuclei spin parallel to the external magnetic field while in the higher-energy state, it is antiparallel (Figure 4.4). Thus any nucleus possessing a multitude of low and high energy states will show a transition from a lower-energy state to a higher-energy state when supplied with energy equivalent to the difference between the lower and higher energy states.

The data obtained in the NMR spectrum are generally helpful in the determination of molecular structures. It is a multidisciplinary technique that is applicable to the field of physics, chemistry and biology. In biochemistry and organic chemistry, the most commonly used NMR techniques based on the nuclei in focus are ^{13}C-NMR, ^{15}N-NMR and ^{1}H-NMR.

NMR can be generally used to study the physical, chemical and biological properties of molecules. It can determine folded–unfolded protein structures and their intermediates as well. NMR provides us with quantitative as

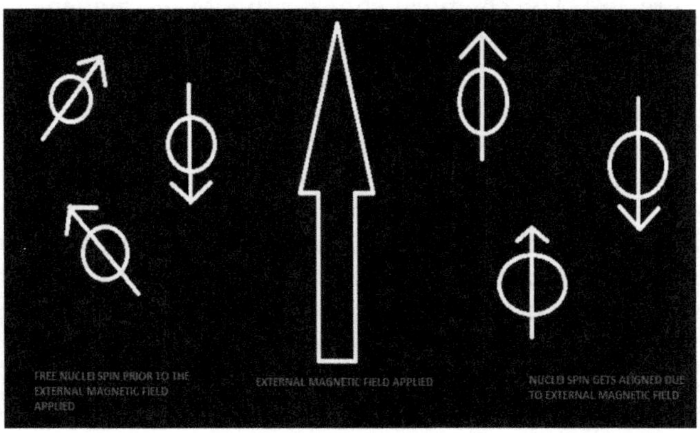

Figure 4.4 Nuclei spin and its effect after an external magnetic field is applied.

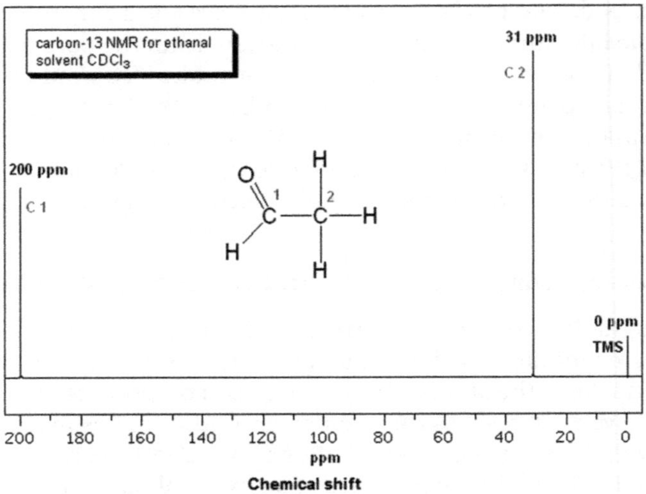

Figure 4.5 Diagrammatic representation of an NMR spectrum (carbon-13 NMR for ethanal; solvent $CdCl_3$).[100]

well as qualitative data. The spectral data are collectively studied to analyze complex proteins and their identities. For instance, ^{1}H-NMR provides data in one dimension generally and also gives structural clarification for any molecules.

4.3.5.1 *Chemical Shift*

The resonant frequency of the nucleus relative to the standard frequency in the magnetic field can be stated as the chemical shift in the NMR spectrum. The chemical shift (ppm) can be analyzed compared to a reference chemical's – tetramethyl silane (TMS) – frequency and the spectrophotometer frequency (Figure 4.5). The chemical shift for the lower frequency is recorded to be low and for the higher frequency, higher.

$$\text{Chemical shift} = \frac{\text{Observed frequency from TMS in Hz} \times 10^6}{\text{Spectrophotometer frequency in Hz}}$$

Solvents used in NMR: in proton NMR, a spectrum is obtained between the intensity and the relative shift in frequency of all the resonating nuclei in a molecule or compound that are generally dissolved in solvents. Hence, any NMR-active nuclei present in these solvents will also get recorded. Thus, to avoid mixing signals from the solvents and the molecules of interest, non-protonated or deuterated solvents are used.

The solvents used mostly for biochemistry and physical chemistry are carbon tetrachloride (CCl_4), carbon disulfide (CS_2), deuterochloroform ($CDCl_3$), hexa-deuterium-benzene (C_6D_6) and deuterium oxide (D_2O).

4.3.5.2 2D-NMR Technique

2D-NMR is a two-dimensional NMR technique providing spectra with data plotted for two different frequency axes. Here, the spectra are plotted like a grid map for both the x and y axis. It is considered to be the most suitable technique for analyzing complex proteins. 2D-NMR can be divided into the following: homo-nuclear and hetero-nuclear. The technique helps in obtaining the structural data that 1D-NMR can't provide. This includes many techniques as mentioned:

- Correlation Spectroscopy (COSY) – identifies all the coupled spins and their correlation.
- Heteronuclear Correlation Spectroscopy (HETCOR) – correlation between hydrogen and carbon generally.
- Heteronuclear Multiple/Single Quantum Coherence (HMQC/HSQC) – identifies the coupling amongst the hetero-nuclei.
- Nuclear Overhauser Effect Spectroscopy (NOESY) 3D-NMR technique – identifies neighboring spin systems and identifies chemical changes. One of the most powerful techniques to identify 3D structures from their conformations and protein molecules.

4.3.6 Isothermal Titration Calorimetry (ITC)

Calorimetry is the measurement of the change in temperature or enthalpy in a chemical reaction.

Isothermal titration calorimetry is the measurement of the change of enthalpy, entropy, binding affinity and stoichiometry between the biochemical compounds or biomolecules.

- Enthalpy (ΔH): the total amount of energy that is gained or released during a reaction.
- Entropy (ΔS): the measurement of the degree of disorder of a system; in this case, molecules, proteins or ligands move freely as individuals but are reduced to a very limited degree of freedom when bound with one another.
- Binding affinity: how strongly one molecule binds to the other. It is mainly denoted by K_d (dissociation rate) or K_b (association/binding rate).
- Stoichiometry (n): the number of reactants required to make a certain amount of product. In this case, the number of proteins or ligands required for binding as well as resulting in a change in enthalpy.

The device mainly consists of two compartments attached to heat sensors, a heat compensator attached to each of the chambers and software that measures the change and gives output data.

One of the chambers is used as a reference cell that contains the fluid medium (generally water) of interest and is maintained at a constant temperature. The second chamber serves as a sample cell where the desired reaction is to be carried out and is filled and calibrated for both the solutions to be at the same temperature (Figure 4.6a).

Very small amounts of proteins (or ligands) are injected through an injector inside the sample cell at predetermined time intervals; in addition to this, there is a stirrer attached with the injection that stirs the solutions and helps with the interaction and reaction process. The change in the heat dissipated is noted.

During the process, the heat sensors signal the heaters with the change in temperature, which in turn compensate for the heat and continuously try to keep the sample solution's temperature at its initial temperature (*i.e.*, same as the reference cell) and the energy or power required to do this is noted and converted as the "heat of interaction".

The heat sensors and compensators then bring the solution back to its initial temperature and the process is repeated until a point is reached where further addition of protein (or ligands) to the sample cell does not result in a variation in the temperature or it has reached a saturation point.

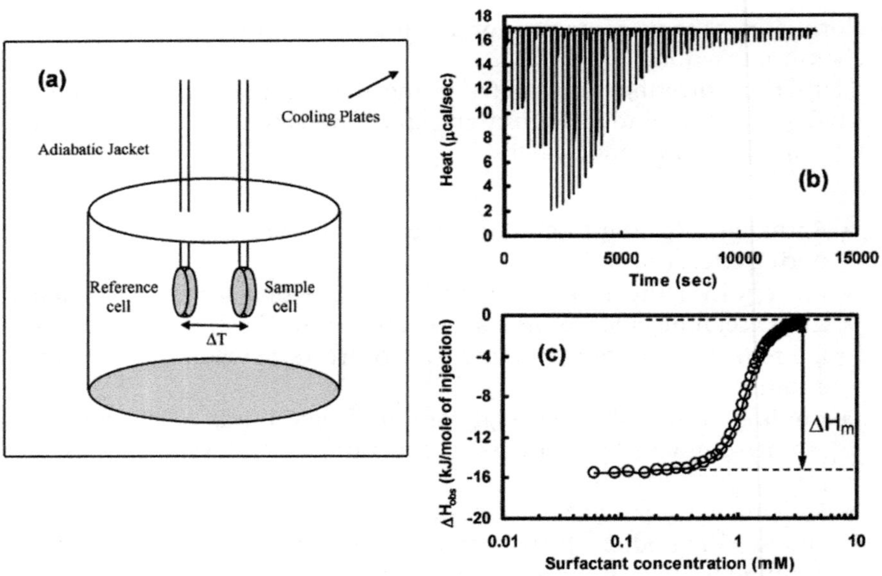

Figure 4.6 (a) Schematic of a commercial isothermal titration calorimeter (ITC); (b) raw heat signals for the titration of a non-ionic surfactant into water; (c) integrated heat corresponding to the enthalpy for each titration as shown in (b). Reproduced from ref. 4 with permission from the Royal Society of Chemistry.

A graph from the process is plotted according to the changes obtained in the variation of the gradual rise or drop of the temperature per micromolar of the solution against time taken.

4.3.6.1 Working

Whenever two molecules interact and bind with each other, there is a change in enthalpy, *i.e.*, either energy is absorbed (endothermic reaction) or released (exothermic reaction). In the case of proteins and nanoparticles, there are a lot of available binding sites where the protein can bind itself, which leads to the highest difference in the change in temperature. As the process progresses, the molar ratio between the proteins and the nanoparticle are gradually increased.

A gradual decrease in the change in temperature can be noted as the affinity for the available binding sites decreases and at last it reaches a point where the number of proteins and ligands introduced is equal or in excess of the number of sites available on the nanoparticles and the solution has reached a saturation point where the change in enthalpy is the least or is absent. These interactions or ΔH are then plotted on the graph in the form of spikes that go from the highest to the lowest. Each of these spikes denotes an integrated plot of molar ratio *versus* the interaction from which the K_d (dissociation rate) is derived (Figure 4.6). The molar ratio at the center of the graph gives us the stoichiometric value (n) and ΔH is the amount of heat released per mole of ligand or protein bound.

From this experiment, a number of conclusions can be drawn; firstly, the nature of the particles and the interaction between them, secondly the rate at which they readily interact with each other and lastly their thermodynamic characteristics.

ITC is highly reliable in the study and observation of the protein corona formations of nanoparticles and the nature of their reaction with respect to different kinds of proteins.[5]

4.3.6.2 Controls

Although ITC is less complicated to operate compared to other techniques where no specific sample preparation is needed, it should be ensured that the buffers that are used for the protein ligand binding studies are specific and compatible.

Different mixtures of inorganic salts for co-precipitation and buffer protonation, such as (1) tris base with nitric acid, (2) silver nitrate with sodium iodide or bromide, and (3) bovine carbonic anhydrase II with 4-carboxybenzenesulfonamide (CBS), are used to titrate the system as a universal standard. It is also suggested that measurements should be done in two ways: first as standard binding and then as protein–ligand binding so that they give precise results each time.[5] In addition to these, one of the most important issues that is usually neglected is the calibration of the instrument for validation, which can result in both systematic and statistical errors in data obtained during measurements.

4.3.7 Mass Spectrometry

Mass spectrometry (MS) is an analytical technique involving counting the ions and molecular weights of proteins according to the mass-to-charge ratio (m/z). It is a primary technique for the analysis of organic and inorganic components and has developed quite rapidly over the last 50 years.[6] On the basis of the operation, there are various mass spectrometers available in the research and analytics fields that can determine different characteristics of materials including the components of protein coronas.

To apply a force to charged particles (ions), mass spectroscopy analyzers use electrical and magnetic fields. The relationship between force, mass and the applied fields can be summarized in Newton's second law and the Lorentz force law:

$$F = ma \text{ (Newton's second law)}$$

$$F = e\,(E + v \times B) \text{ (Lorentz force law)}$$

where F is the force applied to the ion, m is the mass of the ion, a is the acceleration, e is the ionic charge, E is the electric field, and $v \times B$ is the vector cross product of the ion velocity and the applied magnetic field.

From Newton's second law, it is evident that the force will give an acceleration to the species depending upon its mass, and the Lorentz law suggests that the connected force is subject to the ionic charge of the species. Thus, it can be understood that mass spectrometers isolate particles as per their mass-to-charge ratio (m/z) instead of their mass alone.

4.3.7.1 *Instrumentation and Working*

Mass spectrometry has different variants but instrumentation in many forms is common with additional features in different mass spectrometers. The mass spectrometer is divided into three main parts: (1) ion source: sample is ionized, (2) analyzer: sample is characterized using the mass-to-charge ratio, and (3) detector system: for recognizing the particles and recording the relative abundance of every ionic species (Figure 4.7).

4.3.7.2 *Sample Preparation/Ionization*

The mass spectrometer can analyze samples in solid, liquid or gaseous states. Each type of sample can be ionized and this sample is further processed through the mass analyzer part of the mass spectrometer. There are a range of mass analyzers available that gather information-rich data in the form of mass spectra.

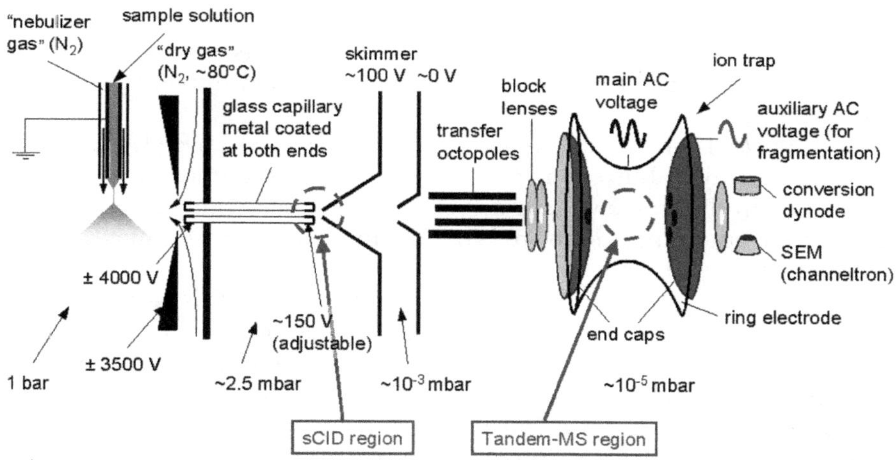

Figure 4.7 Schematic drawing of the mass spectrometer. Reproduced from ref. 7 with permission from the PCCP owner societies.

The samples in a mass spectrometer are introduced based upon the simple or complex nature of components. If the sample is simple or straightforward then it is directly infused with liquid or gas and then it is introduced into the mass analyzer section of the mass spectrometer, while if the sample is complex then primary separation like chromatography or electrophoresis[6] is done prior to the introduction of the sample to the analyzer.

In conventional MS, solid samples with low vapor pressure are directly inserted into an ionization chamber and are controlled with the help of a heating probe. Liquid samples are injected by hypodermic needles through a silicon-rubber-dam-like structure. Gas samples are leaked into the ionization chamber through metal or glass diaphragms.

After the ionization chamber, the samples are passed to the mass analyzer with the help of magnetic or electric fields. Molecular ions are formed when bombarded with a low energy electron beam of up to 10–15 eV. If the samples need to be fragmented further, they are bombarded with a higher energy electron beam (up to 70 eV), which results in fragmentation of the whole molecule/sample.

The ionization technique is divided further into two parts: (1) gas phase ionization (gas and vapor) and (2) desorption technique (liquid or solid). This technique is variable and depends on the physical state of the sample.

With all these segments, a mass spectrometer creates particles from the sample using an ionization source. These particles can be fragmented further into smaller fragments for higher resolution and separation efficiency. The fragments in the analyzer are identified based on their *m*/*z* ratio with the detector that changes the *m*/*z* response into electrical signals. The signals from the detector are transmitted to a PC and data are collected and analyzed with the help of instrument software. The heart of a MS system is its analyzer and various analyzers are currently in use for the analysis of samples. The most commonly used analyzers are Quadrupole, TOF, FT-ICR, and Orbitrap.

Quadrupole: There are four (quad) rods that use a selective electrical field to stabilize or destabilize ions passing through a radio frequency. Particles pass through the framework in a specific order based on the mass/charge proportion, but changes to the rods permit an extensive variety of m/z to pass through quickly, either constantly or in a progression.

TOF: Time-of-flight is a very simple type of analyzer. In this particular type, a sample after fragmentation is accelerated in the form of ions at the same potential with the help of an electric field and the time is calculated when it reaches the detector. This analyzer is generally used in TOF-MS.

FT-ICR: Fourier-transform ion cyclotron resonance mass spectrometry. As the name suggests, the sample fragments are moved in a cyclonic (circular) movement in the presence of a strong magnetic field. The cyclonic movement depends upon the mass-to-charge ratio. With the help of a Fourier transform, this recorded movement is changed or converted into a mass spectrum, which is later used to analyze data.

Orbitrap: Particles are electrostatically caught in an elliptical trajectory around an inner electrode axis. The terminal limits the particles so they circle around the inner electrode and move forward and backward along the axis. This wavering produces a graph, which is recorded by the instrument The frequencies of this graph depend upon the mass-to-charge ratio of the particles. Mass spectra are acquired by a Fourier change of the recorded graph. Orbitraps have a high mass precision, high affectability and a decent range.

4.3.7.3 Data Analysis

Data analysis is the most complicated part of mass spectrometry as each sample can produce multiple peaks upon ionization and fragmentation. A typical mass spectrum is shown in Figure 4.8. Several types of software are used to analyze the raw data and produce meaningful data that can be utilized for further studies. This spectrum has different peaks, as seen in Figure 4.8(R). The data compromises % abundance *vs.* the m/z ratio of the proteins detected.

4.3.7.4 Types of MS

For effective resolution and identification of complex molecules using mass spectrometers, it is often necessary to conduct a primary separation of molecules or ions of interest. This has resulted in giving rise to a number of hyphenated techniques or coupled mass spectroscopy systems. Common examples are introduced here for the benefit of the reader.

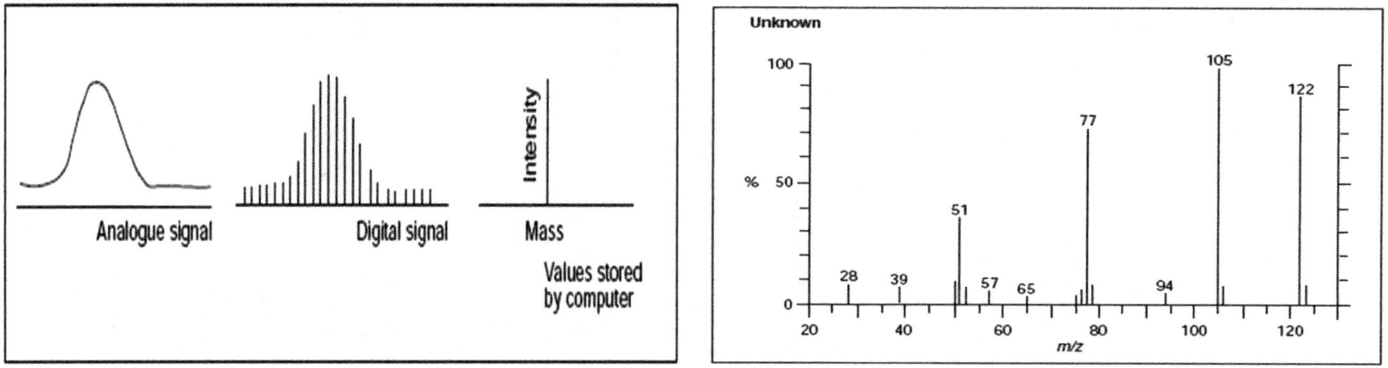

Figure 4.8 (L) Different types of signals received by detectors. (R) Representative mass spectrum of a sample after gathering information from the detectors. The data show different peaks against specific m/z ratios.[101]

Liquid chromatograph mass spectrometer: As the name suggests, LCMS is a technique that combines liquid chromatography with MS. The mobile phase or the starting phase is a liquid, which makes it different from GC-MS in which the mobile phase is a gas. A sample separated by LC is introduced into an ionization chamber where it is fragmented and ionized. After ionization, the sample is dried with the help of inert gases like nitrogen. Ions are then passed through the mass analyzer to obtain *m/z* separated data.

Inductive coupled plasma mass spectrometer: In ICP-MS, the sample is atomized and ionized by argon gas. In a high vacuum mass analyzer, these ions from the gas are introduced through cones (gaps). In view of the *m/z* proportion, the isotopes of the components can be distinguished.

Tandem mass spectrometer: In this technique, two or more quadrupoles are combined in tandem to generate a higher resolution. This technique is generally used for peptide sequencing. In tandem MS, a sample is volatilized and passed through the first quadrupole, which helps to separate specific ions. After initial separation, the ions are passed through a second quadrupole, which initiates further fragmentation and results in the high-resolution detection of fragments. Quadrupole–quadrupole, magnetic sector-quadrupole, and quadrupole-time-of-flight geometries, and nano liquid tandem mass spectrometry (Nano LC-MS-MS) are some variants available in tandem mass spectrometry.

Linear trap quadrupole (LTQ)-Orbitrap XL mass spectrometer: The LTQ Orbitrap XL hybrid ion trap-Orbitrap mass spectrometer is a Fourier-transform mass spectrometer (FTMS) variant of the Thermo Scientific LTQ XL direct particle trap and Orbitrap mass spectrometer. The instrument includes a higher-energy C-trap dissociation (HCD) impact cell. The high mass accuracy lowers false positive rates; the parallel acquisition mode optimizes productivity; the high MS and MS/MS sensitivity, wide dynamic range, and fast cycle times increase protein identification; these are some notable and exclusive features of the LTQ-Orbitrap XL mass spectrometer.

Matrix-assisted laser desorption ionization time-of-flight mass spectrometry (MALDI-ToF-MS): MALDI is widely used for the detection of glycoproteins, peptides, proteins, oligonucleotides and oligosaccharides. MALDI depends on the utilization of a matrix with a given sample/molecule of interest. The sample is embedded in a matrix that can be easily volatilized with a laser. The sample of interest is usually blended into a matrix with high absorption for laser energy. As the matrix absorbs most of the laser energy it can ionize large molecules from the sample of interest with minimum fragmentation. Most financially accessible MALDI mass spectrometers have a nitrogen laser with a wavelength of 337 nm. This technique is regularly combined with a time-of-flight detector to get appropriate charge–mass proportions.

4.4 Sampling and Preparation of Nanoparticles

Most studies carried out for studying protein corona formation include gold NPs, silver NPs, liposomes, polystyrene NPs, silica NPs, polymeric NPs (*e.g.*, *N*-isopropyl acrylamide, *N-tert*-butyl acrylamide), iron oxide NPs, gold nanorods, quantum dots (*e.g.*, dihydrolipoic acid-coated quantum dots (DHLA QDs) – CdSe/ZnS QDs), and others (*e.g.*, iron-platinum (FePt), cadmium-selenide (CdSe)/ZnS nanocrystals). Sampling and preparation of nanoparticles for analysis by some of the common techniques are described in this section.

The purpose of these experiments varies greatly depending on what physical and chemical aspects of the nanoparticles and protein coronas the researchers are interested in knowing. For example, the purpose could be to evaluate the evolution or development of protein coronas around nanoparticles with respect to time (temporal analysis), to check how the size and surface properties of nanoparticles affect the corona formation (kinetics), to discover what type and amount of proteins are adsorbed (proteomics), to study the interactions happening between the nanoparticles and the adsorbed proteins, and also the impact of these protein coronas on the NPs and their targets. These experiments provide a deeper understanding of nanoparticle-protein coronas (NP–PCs) as a biological entity.

In most experiments, the nanoparticles are used as-received or purchased, whereas, in other experiments, NPs of desired sizes as well as surface charges are synthesized to study how the size and charge variations affect protein corona formation. Following the synthesis or procurement, the NPs are incubated with certain buffers, a culture medium or human plasma for the formation of the protein corona. The dispersions are mostly made in a cell culture medium supplemented with 10% fetal bovine serum (FBS), glucose, sodium bicarbonate and penicillin/streptomycin, serum-free minimum essential medium, mammalian growth media (MGM), human serum albumin (HSA) and human plasma.[8–13] Nanoparticles are incubated with the media for the time period of the study deemed necessary for corona formation and later on analyzed using several comprehensive techniques.

Dynamic light scattering (DLS) and transmission electron microscopy (TEM) are widely used for analyzing the size profile of NP–PCs. Size measurements are carried out by DLS, which operates with a light source and has a fixed scattering angle of about 173°. The samples are diluted with either water, phosphate buffer saline (PBS) or the culture medium that is used, and analyzed mostly at a temperature between 25–37 °C. Z-potential measurements are carried out on the same instrument to determine the surface charge of the NPs. Malvern Zetasizer nano ZS, Zetasizer nano ZS90, Malvern Zetasizer 3000 Hsa and Brookehaven Zeta plus DLS detectors are generally used for analyzing size and surface charge.[8,9,14–18] The ionic strength and pH of the medium has a great impact on the charge of the nanoparticles and consequently their stability in a suspension medium. Thus, it is important to pay attention to the buffer strength and pH while preparing nanoparticle samples for studying these effects and the effect of serial dilution on the ionic strength must not be ignored while analyzing the results.

Differential centrifugation sedimentation (DCS) is thought to yield the "true size" of the NP–PC and the corona size (thickness of the adsorbed protein layer) using a simple core–shell density model. The total amount of protein necessary to form a shell is theoretically calculated, which yields a more direct estimation of the corona size than the hydrodynamic size obtained from DLS. DCS gives much more modulated evolutions of the corona layer with increasing exposure to the physiological medium.[10,19]

Purification of NP–PCs is carried out by centrifugation for the removal of weakly bound or unbound proteins. Centrifugation is considered to be the most conventional method to be used for separation of the NP–PCs from the excess proteins. Excessive washing in the centrifugation technique can also affect the formed complex and hence can cause unwanted damage to proteins and NPs as well. So, while using centrifugation, one must use it with care and also perform the same experiment with other methods such as size exclusion chromatography (SEC), magnetic separation and micro filtration.[20]

The pellet obtained after centrifugation is often resuspended in the dispersion medium or PBS. This procedure is continuously repeated at least 3–4 times for the complete removal of weakly bound or unbound proteins. Once the unbound and/or weakly bound proteins are removed, the sample is subjected to gel electrophoresis (SDS-PAGE) for the separation of the NPs and proteins. SDS-PAGE analysis of NP–PCs is carried out by suspending washed NP–PCs in SDS loading buffer and the solution is boiled for about 4–5 minutes at 100 °C. The sample is then loaded into a 4% stacking gel with 10% resolving gel and is subjected to an electrophoretic run on a 12% polyacrylamide gel at 120–130 V for around 60–80 minutes. The samples are often stained for visualization using suitable staining solutions. The molecular weights of proteins can be determined using protein molecular weight markers using software like ImageJ, GelAnalyzer *etc.*[8,9,15,16]

In cases where cells are experimented with to check the NP uptake or cellular internalization of NPs, flow cytometry is employed. To assess the NP uptake level in cells, NP dispersions are made by diluting NPs in the physiological medium used. For comparative studies, cells treated with NP–PCs, NPs and untreated cells are harvested by using proteolytic enzyme solution trypsin/EDTA and washed with the PBS for removal of culture media and proteolytic enzyme. Cells suspended in PBS are measured for uptake of NP–PCs in the flow cytometer. The cells are then fixed with formalin solution and again resuspended in PBS for flow cytometry measurements.[9,16,21,22]

4.5 Characterization of Protein Coronas

The synthetic identity of a nanoparticle, such as its size, shape, and surface chemistry, is affected by the dispersing biological environment and the duration of exposure. Exposure of NPs to any biological medium transforms its synthetic identity into a biological identity by forming a protein corona around the nanoparticles (NP–PC)s. The physiological response of protein-coated NPs is a function of various factors and greatly influences the composition as well as the structure of the NP–PCs. The factors can be broadly classified on the basis of thickness, conformation, surface charge, identity, quantity, arrangement, orientation, affinity, conformation, and aggregation of NP–PCs in the biological medium (Table 4.3).

By exploiting these parameters, numerous studies have been conducted utilizing many of the techniques mentioned earlier in the chapter. Employing these techniques, researchers have explored various aspects of protein coronas such as their time evolution in the biological environment, effect of the presence or absence of the protein corona and its impact on cells, how the NP size and surface properties determine the protein corona, relevance of physical and chemical characteristics of the protein corona, quantification of exchange rates and binding affinities of proteins, understanding the

Table 4.3 Different parameters of protein coronas and the purpose of the techniques employed to analyze them.[a]

Parameters of the NP–PC	Technique	Purpose
Thickness/size of the complex	Dynamic light scattering (DLS), differential centrifugal sedimentation (DCS), size exclusion chromatography and transmission electron microscopy (TEM)	Gives information about the size profile of the PC by measuring the overall size of the NP as well as the underlying surface
Surface charge	Zeta potential	Indicates whether the NP–PC complex is negatively or positively charged
Identity and quantity	Polyacrylamide gel electrophoresis (PAGE), liquid chromatography tandem mass spectrometry (LC-MS/MS)	The identity and quantity of adsorbed proteins gives information about different types of proteins and their interactions with the NPs
Affinity	Size exclusion chromatography, surface plasmon resonance (SPR), isothermal titration calorimetry (ITC)	To analyze the protein corona based on its association and dissociation rates
Molecular structure	Nuclear magnetic resonance spectroscopy (NMR)	To study protein–protein as well as protein–NP interactions (with proton NMR and using multiple spectra)

[a]Purified NPs are further analyzed based on the purpose of the study by using various analytical techniques such as MS, ITC, SEC and/or NMR.

kinetics of the protein corona, impact of NP–PCs on protein conformation as well as the effect of aggregation and the protein corona on the cellular internalization of NPs.

4.5.1 Surface Charge of the NP–PC

The surface charge of nanoparticles is considered an important factor while studying nanoparticle–protein interactions in any biological system. Mostly, the effect of protein binding on the surface charge of a nanoparticle–protein corona as a whole unit can be temporal (time-dependent) as well as dependent on the type and concentration of biological medium. The nanoparticles are either incubated with human plasma or serum or the particles are incubated with individual proteins and then examined for changes in their surface charge post-incubation for the desired time of the study.

There have been many studies where the nanoparticles have been conjugated with either amine ($-NH_2$) or carboxy ($-COO^-$) groups to observe the kind of interaction taking place at the bio–nano interface. In a study amine- and carboxy-modified SiO_2 NPs were examined following their incubation in mammalian growth medium for a period of 48 hours. For $SiO_2–NH_2$ conjugates, the Z-potential dropped from 8.7 ± 5.7 to around −10 ± 3.5 after 48 hours and for $SiO_2–COO^-$ conjugates, it increased from −10.4 ± 2.6 to −14.8 ± 3.2 after 48 hours, which suggested that PC formation proceeds with alteration of the surface charge by shifting the surface charge of the amine-functionalized particles from positive to negative.[11] In a study conducted on a defined set of four polystyrene NPs (PS–NPs) with average diameters of around 100 nm, incubated with serum, the corona formation resulted in a negative Z-potential for all PS–NPs, irrespective of their original Z-potentials. The Z-potential measurements obtained directly after serum incubation were as follows: PS-COOH (−22 ± 7), PS-NH_2 (−21 ± 12), PS-SO_3 (−24 ± 10), PS-PO_3 (−23 ± 9).[23] In another study, hydroxyethyl starch (HES) nano capsules modified with $-NH_2$ as well as −COOH incubated with plasma, human serum albumin (HSA) and apolipoprotein A-I (Apo A-I) showed similar Z-potential results. The Z-potential of HES-COOH dropped from −34 ± 3 to −13 ± 5 in plasma, −15 ± 4 in HSA and −8 ± 7 in Apo A-I, whereas the Z-potential of HES-NH_2 dropped from −27 (negative due to redispersion in water with SDS) to −11 ± 4 in plasma, −10 ± in HSA and −18 ± 8 in Apo A-I.[24] Citrate-capped AuNPs of different sizes (4–40 nm) were modified with mercaptoundecanoic acid (MUA) (negative surface charge) and aminoundecanethiol (AUT) (positive surface charge) and analyzed as a function of time. The Z-potential measurements showed no significant changes for AuNP-MUA post-incubation with CCM even after 48 hours. The Z-potential dropped from −56 mV to −54 mV, which shows that a soft PC might have formed on the AuNP-MUA surface, but it detached immediately upon washing of the NPs. The proteins were loosely bound and no evolution leading to the formation of a hard PC was observed. On the other hand, the AuNP-AUT conjugates, upon exposure to CCM, exhibited different results. The Z-potential of the synthesized conjugates was +64.4 mV due to the net positive charge on the conjugates and it dropped to around −22.3 mV after 48 hours, which showed that the formation of PC was driven by electrostatic interactions between the positively charged NPs and negatively charged serum proteins.[8]

As with time, the protein corona evolves with different plasma concentrations as well. Monopoli *et al.* (2011) showed the protein adsorption of two different NPs (sulfonated polystyrene and silica NPs) as a function of the plasma concentration. Following incubation of 200 nm SiO_2 NPs, the Z-potential was reduced to around −11 mV even at lower plasma concentrations. This suggests that an adsorbed protein layer forms in plasma, independent of its concentration. As observed from the results, even for the low plasma concentrations, formation of a protein layer occurs, which advances with increasing plasma concentrations (Table 4.4).[19]

These data sets suggest that, independent of their original surface charge, the NPs upon exposure to any biological medium show a transformation towards a negative surface charge, which is a measure of protein conjugation to the surface of NPs. Though, after washing the NP–PC conjugates, a slight increase in the negative Z-potential may occur, suggesting that some proteins get washed away, reducing the effect of PC on the Z-potential (Table 4.5).

Table 4.4 Z-potential measurements of SiO_2 nanoparticles incubated with increasing plasma concentrations.

% Plasma	Z-potential (mV)
0	−25.2 ± 0.2
3	−12.1 ± 2.1
6	−8.4 ± 0.7
20	−10.4 ± 0.6
55	−11.1 ± 1.0
80	−10.2 ± 0.3

Table 4.5 Z-potential measurements of untreated, treated and purified AuNPs.[25]

Size of AuNPs (nm)	Z-potential of untreated AuNPs (mV)	Z-potential after 24 hours' incubation (mV)	Z-potential after 24 hours' incubation and purification (mV)
5	−35 ± 1.6	−7.1 ± 0.8	−18.0 ± 2.0
15	−31.7 ± 7.2	−6.7 ± 1.1	−13.0 ± 3.0
80	−42.9 ± 6.2	−9.4 ± 1.0	−12.9 ± 2.8

All these observations are summarized and the resulting Z-potentials are plotted as a function of time (approximately 24 hours) in Figure 4.9.[8,9,11,19,23–26]

4.5.2 Kinetics of the NP–PC

Identification of adsorbed proteins on NPs surely provides adequate information about the amount and type of proteins associated with the NPs. However, it does not reveal the kinetics and equilibrium binding of the proteins associated with the NPs. In order to have a deeper understanding of the interactions between NPs and adsorbed proteins, information on the binding

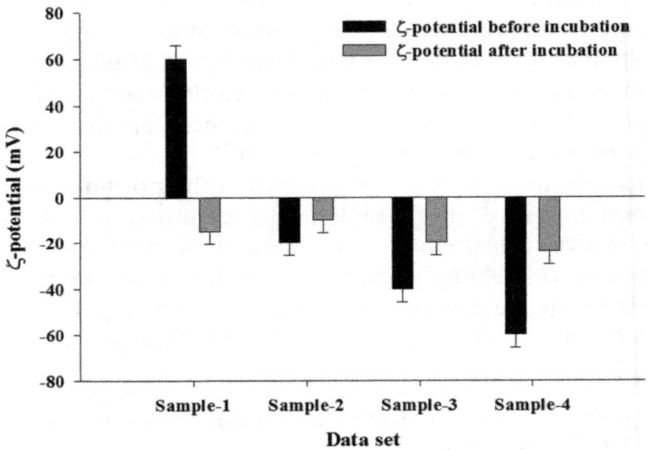

Figure 4.9 Changes in ζ-potential before and after media incubation (24 hours).

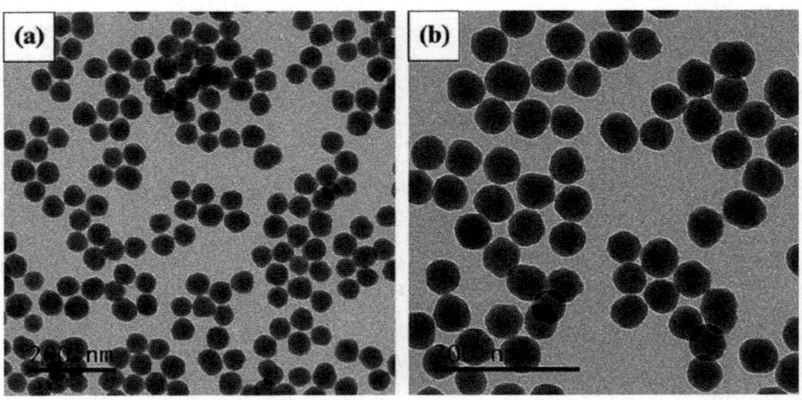

Figure 4.10 TEM images (a) and (b) of silica nanoparticles. Reproduced from ref. 39 with permission from the Royal Society of Chemistry.

affinities, and the association and dissociation rates of these adsorbed proteins is considered relevant. This yields the rates by which different proteins bind to and dissociate from the surface of NPs as well as the timescales on which the adsorbed proteins are replaced by free proteins. Cedervall *et al.* (2007) showed that surface plasmon resonance (SPR), isothermal titration calorimetry (ITC) and size exclusion chromatography (SEC) are preferable techniques to quantify the kinetics and equilibrium binding factors of the NP–PCs. Using a set of copolymer NPs of varying size and hydrophobicity, it was suggested that the exchange rates depend on the hydrophobicity of the particles. SPR studies on 70 nm *N*-isopropylacrylamide/*N*-*tert*-butylacrylamide (NIPAM/BAM) 85 : 15 (more hydrophilic) NPs with plasma found the dissociation rate constants to be $3.7 \times 10^{-4}\ s^{-1}$ and $6.1 \times 10^{-5}\ s^{-1}$ while for plasma proteins on the 70 nm NIPAM/BAM 50 : 50 (more hydrophobic) NPs, the dissociation rate constants were $2.0 \times 10^{-3}\ s^{-1}$ and $3.4 \times 10^{-5}\ s^{-1}$. Even for HSA and fibrinogen, the dissociation was faster from the more hydrophobic particles as compared to the more hydrophilic particles. Similar results were obtained with 200 nm NPs and all the SPR measurements were consistent with the ITC results showing that the exchange rates depend on both protein and NP identity.[27]

The distinct residence times of adsorbed proteins on nanoparticle surfaces may depend on the binding affinities of proteins. The kinetic on and off rates of any protein differ for specific NPs depending on the surrounding biological medium. Proteins that are present in higher concentrations bind to the NP surface owing to their high association rate constants, but can be replaced by proteins that have a higher affinity, known as the Vroman effect, despite their lower concentration in the medium and slower exchange rate. As it is known that the protein corona is dynamic, its composition can be determined accurately by taking account of the kinetic rates of adsorption and desorption of proteins.

Fluorescence correlation spectroscopy (FCS) has also been used to assess the association and dissociation behavior of the proteins. Protein corona formation was examined on FePt nanoparticles where time-resolved fluorescence quenching was used along with FCS in order to observe the association of proteins. The results stipulated that before successfully forming a stable corona, nanoparticles and proteins have to undergo numerous adaptations, such as the adsorbed proteins rearranging themselves to create space for new binding proteins.[28] These results show that the proteins with the highest mobility reach the nanoparticle surface at the early incubation stage but are later replaced by proteins with less mobility and a higher affinity for the nanoparticle surface. It can be interpreted that the soft corona, which develops initially upon the interaction of NPs with a biological medium, has faster exchange rates than the hard corona (where the already adsorbed proteins are replaced by low abundance proteins with a high affinity to the NP surface).

SEC can be useful to understand the association between lipids and copolymers. This method is used to understand the interaction behavior of the proteins and nanoparticles, as the elution behavior of the proteins changes on the basis of the interaction with the nanoparticles.[29] Nanoparticles with a 1:1 NIPAM–BAM ratio and diameters of 120 or 180–200 nm were incubated with plasma followed by centrifugal washing. The nanoparticles were separated based on their sizes (or molecular weights) using a sephacryl S-1000 SF column. An enzymatic assay was used to determine the presence of cholesterol in the eluted fractions suggesting that the plasma cholesterol binds to copolymers. SEC was also used to understand the protein exchange rates within the same experiment.[29] SEC can also be helpful to understand the kinetics of the nanoparticle and its associated proteins and/or ligands.[30]

A new approach using SEC has been designed to understand the strength of protein–nanoparticle interactions and also determine their lifetime after association with nanoparticles. Every single protein in the nanoparticle–protein complex would possess an elution profile that remains dependent on the dissociation rate, rate of flow, the time taken to run on the column and many other factors.[31,30]

ITC has been used extensively to study the binding interaction of surfaces of NPs with proteins. During many experiments with ITC, it was noted that instead of a quick change in the enthalpy as ideally suggested or assumed, a gradual change in the heat of the reaction was observed due to the interaction between proteins and nanoparticles. Observations made during several ITC studies are summarized in Table 4.6. A summary of important points for protein–nanoparticle interactions studied from ITC measurements are listed here:

(A) The adsorption rate of the protein on the particle depends largely on the serum composition and molar concentration of the particles (and proteins) in the system. A diluted serum or protein may not be able to produce a measurable change in the heat of the reaction. This is due to the vast number of binding sites compared to the adsorbing protein, resulting in a small number of interaction events adding up to the overall energy change of the system.
This process continues until a certain minimum amount of protein is present in the solution from where no further passive stabilization of the system can occur. Beyond this concentration of the protein, every incremental volume addition of protein to the nanoparticles results in a measurable heat change, which is recorded in the system. The first few interaction events between the proteins and the nanoparticles act as a self-calibrating method for the ITC before the actual measurable interaction takes place.
Thus, any change in the molar concentration of the protein or serum can change the behavior or rate of interaction between the particles.
(B) The data analysis from ITC usually considers that equilibrium is achieved between the adsorption and desorption of the proteins

following the Langmuir model. However, in reality the adsorption of proteins on hard surfaces may not be entirely reversible. This results in a slight discrepancy in the determined value of the dissociation constant (K_d – as K_d is dependent on the temperature) from the enthalpy change measured by ITC (ΔH_{itc}) when compared with the van't Hoff equations.[32]

Several studies were done to compare the binding affinities determined from van't Hoff's equations and ITC and were recently reviewed in a review article.[33] The investigations summarized therein led to the conclusion that the binding energy determined by ITC is a function of two different contributions given by:

$$\Delta H_{itc} = \Delta H_{bind} + \Delta H_{res}$$

where ΔH_{itc} = the total change in the enthalpy, ΔH_{bind} = the heat of the process or binding, ΔH_{res} = unrelated equilibrium, *i.e.*, it progresses independently of the process of binding.

Thus, ITC allows us to measure binding of proteins that may proceed without any change in enthalpy. One such example was given by studying the adsorption of RNAse A to cationic SPB where the ΔH_{bind} was measured to be zero and the entire binding process is driven by the positive changes in the entropy. Thus, ITC is useful for measuring reactions or biomolecular changes that take place through changes in the entropy without any enthalpy changes in the system.

These parameters are extremely useful in observing protein corona formation as a whole, *i.e.*, what the other driving forces that actually take part aside from the obvious NP–PC formations are.

(C) As is already known, all proteins have a tendency to bind with different nanoparticle surfaces, so when two similar types of proteins with a similar shape, size, and characteristics are put together in a NP-containing serum, a competitive binding is observed to take place.

The initial theories of competitive binding suggested that if there are two proteins – let's say A and B – then the protein with a high affinity (say protein A) will bind with the nanoparticles and will also hinder and affect the binding interaction between particle B and the nanoparticles.

Though recent studies suggest that when two proteins interact with the nanoparticles, both proteins interact independently of each other, *i.e.*, they bind with the nanoparticles irrespective of the presence of the other protein. After the first layer of the core corona formation occurs, both types of proteins form layers over the primary layer through protein–protein interactions. Very few proteins in these layers are capable of actually interacting with the nanoparticle at first, but then they undergo slight structural changes to bind with the nanoparticle.[34]

Throughout all these observations and conclusions, Langmuir's adsorption isotherm model plays a major role.

Table 4.6 ITC data measured for different nanoparticle and protein combinations.

Sr. No.	Nanoparticles	Size (nm)	Protein(s)/ components	Temperature	ΔH	ΔS	K_a	N	Reference
1	HES		Plasma	298.15	−285 ± 97	−845 ± 329	1.0 ± 0.4	182 000 ± 39 000	24
2	HES		HSA	298.15	−277 ± 43	−818 ± 147	0.8 ± 0.3	114 000 ± 16 000	24
3	HES		ApoA-I	298.15	−6010 ± 185	−20 000 ± 613	333 ± 124	10 ± 4	24
4	HES-COOH		Plasma	298.15	−306 ± 192	−1217 ± 318	1.0 ± 0.8	70 000 ± 13 500	24
5	HES-COOH		HSA	298.15	−308 ± 30	−928 ± 103	0.4 ± 0.2	156 000 ± 36 000	24
6	HES-COOH		ApoA-I	298.15	−5150 ± 787	−17 100 ± 2640	188 ± 79	6 ± 3	24
7	HES-NH_2		Plasma	298.15	−281 ± 139	−827 ± 468	1.2 ± 0.6	121 000 ± 29 000	24
8	HES-NH_2		HSA	298.15	−277 ± 45	−820 ± 152	0.5 ± 0.1	184 000 ± 33 000	24
9	HES-NH_2		ApoA-I	298.15	883 000 ± 24 000	2.4 ± 0.9 × 106	5.4 ± 0.1	8 ± 6	24
10	Polystyrene NH_2	58	BSA	300.15	−1.4 ± 0.7		0.40 ± 0.05		35
11	Polystyrene COOH	60	BSA	300.15	−1.4 ± 0.4		2.4 ± 0.9		35
12	Fullerenol		BSA	298	−130.4 ± 10.82	−355.8	1.276 ± 0.131	16.38 ± 0.994	36
13	Fullerenol		γ-Globulins	298	237.6 ± 54.8	862.6	0.512 ± 0.194	0.761 ± 0:156	36
14	Carbon		BSA	298	−1548		1.92	49	37
15	Carbon		HSA	298	−669.8		2.07	31	37
16	MPA -brOT		BSA	298.15	− 28.10 ± 0.80		7.86 ± 0.34	4.99 ± 0.15	38
17	Gold	40	HSA 7 μM	291.15	−19.6 ± 12.9	−6.2	1.4 ± 0.5	2.8	3
18	Gold	40	HSA 15 μM	291.15	−13.3 ± 8.4	−4.3	1.7 ± 0.3	2.8	3
19	Gold	40	HSA 50 μM	291.15	−6.7 ± 5.2	−2.1	1.5 ± 0.3	2.8	3
20	Gold	40	HSA 100 μM	291.15	−4.3 ± 10.3	−1.3	1.3 ± 0.1	2.8	3
21	Gold	15	HSA 50 μM	291.15	−24.9 ± 12.8		7.5 ± 1.3	1.5	3
22	Gold	25	HSA 50 μM	291.15	−17.4 ± 13.3		9.4 ± 0.8	1.9	3
23	Gold	40	HSA 50 μM	291.15	−6.7 ± 5.2		15.1 ± 3.1	2.8	3
24	Gold	70	HSA 50 μM	291.15	−8.3 ± 6.9		13.4 ± 1.2	4.7	3

4.5.2.1 Effect on Protein

Proteins are nowhere near stable due to the conformational change that is induced from the interactions between a NP and surrounding proteins. These changes in conformation result in a change in entropy to counterbalance the change as well as to accommodate a restricted conformation from a more free conformation in a biochemical system. A lot of research has been done to learn the specificity of different proteins. Many NPs are charged either positively, negatively or neutrally. It has been noted that the charges of ligands have a direct effect on proteins. It was found that proteins maintained their native structure in the presence of neutral ligands but proceeded to a denatured state in the presence of charged ligands. It was also observed that most of the conformational changes occurred due to loss of water molecules from the proteins. Another thing to note is that HSA binds with a high enthalpy change compared to other proteins like fibrinogen that bind without any enthalpy change. So, it can be concluded that competitive binding favors the HSA in comparison to other plasma proteins.

4.5.3 Size of the NP–PC Corona

Along with surface charge, exchange rates and composition, the protein corona shows high variability in the thickness of the corona layer thereby changing the hydrodynamic size of the NPs upon their interaction with any biological fluid. As the proteins conjugate with NPs, there is an increase in the thickness of the corona layer, which increases the size of the nanoparticle–protein corona complex and is greatly influenced by the nanoparticles' surface chemistry. This size profile is often analyzed by dynamic light scattering (DLS) measurements, transmission electron microscopy (TEM) (Figure 4.10) and sometimes with differential centrifugation sedimentation (DCS).

The binding of protein effectively increases the NP size over time. The time evolution of a protein corona was studied by Casals *et al.* with AuNPs ranging in size from 4 to 40 nm in CCM with 10% FBS. Amine-modified gold nanoparticles (Au-AUT conjugates) and carboxy-modified gold nanoparticles (Au-MUA conjugates) were studied, where the Au-MUA conjugates showed no relevant changes in the hydrodynamic size even after 48 hours and there was no evolution of a hard corona. On the other hand, the Au-AUT conjugates showed an increase in size after 48 hours of incubation indicating their successful evolution through a hard corona.[8] In a similar study with native, amine-modified as well as carboxy-modified SiO_2 NPs, SiO_2–NH_2 NPs had a larger hydrodynamic diameter (1036.3 ± 33.7 nm) than the native SiO_2 NPs (200 ± 3.6 nm) and the SiO_2–COO^- NPs (339 ± 6.3 nm) in media. Also, an unusual increase of 6812 nm was seen for SiO_2–NH_2 NPs after 1 hour of incubation, suggesting that SiO_2–NH_2 NPs had the most rapid agglomeration rate.[11]

The composition of a protein corona continuously changes due to the ongoing adsorption and desorption of proteins, which affects its size as well. Different sizes of NPs also appear to influence protein binding. In a study, it was shown that 30 nm Au colloids adsorbed a significantly more wide range of proteins than their 50 nm counterparts, suggesting that NP size plays an important role in corona formation.[40] A recent *in vivo* study on 40 nm and 70 nm Au nanorods (AuNRs) and 40 nm and 70 nm Au nanostars (AuNSs) showed that 70 nm AuNSs adsorbed a large number of proteins due to a higher surface area than the AuNRs, but the number of identified proteins was comparatively higher in 40 nm AuNSs. Here, the authors hypothesized that the presence of spikes on AuNSs may have affected the protein exchange and interaction. As the size of the spikes decreases, protein exchange becomes easier and hence different proteins can bind to their surface quickly leading to the formation of a complex corona.[41]

NP size distributions have also been shown to affect the binding affinities of proteins. A study of AuNPs of sizes ranging from 5 to 100 nm, incubated with human blood proteins (albumin, fibrinogen, gamma-globulin, histone, and insulin), showed that the size of the AuNPs affects the binding constant and Hill constant. It was observed that the binding constant increases with NP size whereas the Hill constant decreases with NP size.[42]

Figure 4.11 depicts the effect of the protein corona on the sizes of various nanoparticles before and after an incubation period of 24 hours.[8,9,11,24,25,43]

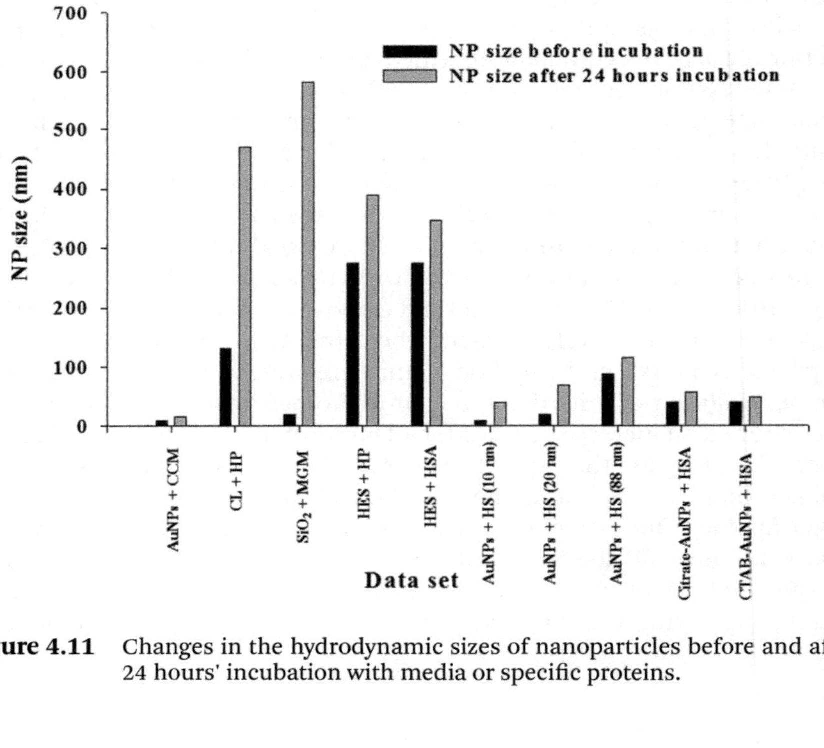

Figure 4.11 Changes in the hydrodynamic sizes of nanoparticles before and after 24 hours' incubation with media or specific proteins.

Table 4.7 Size exclusion chromatography for different nanoparticles and further techniques used for the separation of NP–PCs.

Nanoparticles	Proteins	Protein isolation technique	Technique for protein separation and determination	Reference
Liposomes (cationic liposome (CL))	Albumin, fibrinogen, apolipoproteins, IgG, α1-antitrypsin, α2-macroglobulin, IgM	Gel filtration	2D-PAGE	44
Solid lipid nanoparticles with Tween-80	Fibrinogen, IgG, IgM, apolipoproteins (including ApoE)	Gel filtration	2D-PAGE	45
Solid lipid nanoparticles with poloxamer 188	Fibrinogen, IgG, IgM, apolipoproteins (excluding ApoE), transthyretin, albumin	Gel filtration	2D-PAGE	45
Iron oxide nanoparticles	Albumin, IgG, IgM, fibrinogen	Gel filtration	2D-PAGE	46
SWCT AND DWCT (carbon nanotubes)	Albumin, fibrinogen, apo-lipoproteins, C1q	Affinity chromatography	SDS-PAGE with N-terminal sequencing	47

The thickness of the protein corona can be studied with SEC as well. Isolation of hard corona and soft corona can be done with SEC separately. Table 4.7 contains specific examples of nanoparticles that have been isolated or separated using SEC or a combination of SEC and other analytical methods.

4.5.4 Molecular Structure

NMR is used to characterize interactions between nanoparticles and proteins that are adsorbed onto the nanoparticles with hydrogen/deuterium exchange experiments.[48] Generally, silica, gold and other such nanoparticles are used for NMR experiments. Some lipoproteins and polymers are also used. NMR studies for understanding protein coronas are usually aimed at structural analysis. The majority of experiments using NMR are either for the structural or conformational analysis of proteins or the determination of bound/unbound proteins.

Chemical shift perturbation (CSP) is a technique that monitors protein–protein and protein–ligand interactions, and helps to identify the domain of the protein involved in protein–protein interactions in macromolecular complexes.[48] In CSP, two-dimensional NMR experiments are performed to determine differences or changes in the chemical environment of coupled hydrogen and nitrogen atoms of the protein upon formation of protein complexes. NMR and CSP can collectively provide information to study protein–protein or protein–particle interactions.[48]

Calculating the CSP can provide us with data on the bound and free proteins of each and every amino acid. With CSP and NMR data, we can label the amino acids and use the data qualitatively. The two-dimensional spectrum of human ubiquitin in the presence of gold nanoparticles showed properly folded proteins. If NMR fails to provide a spectrum, it could be due to an increase in the conformational flexibility of the loop within the complex.[48]

Residue-specific structural information regarding any particular protein bound on a nanoparticle can be obtained with NMR. Preferably, solid-state NMR helps to collect conformational information for peptides adsorbed on the nanoparticle. Human carbonic anhydrase I (HCAI) is more stable than human carbonic anhydrase II (HCAII) and provides a dynamic equilibrium between bound and unbound proteins in a mixture with silica nanoparticles.[49]

At equilibrium, the exchange between the two states is gradual and slow on the timescale of NMR, which allows the study of the solid surface on the protein structure under more specific conditions.[49]

For the first time, lipids were detected in the protein corona on the surface of nanoparticles in human plasma[29] using NMR spectroscopy. NIPAM–BAM 50:50 copolymer nanoparticles were mixed with plasma (HSA or BSA), and unbound proteins were repeatedly washed off with the help of centrifugation. NMR was able to determine the lipids on the surfaces of nanoparticles bound to the copolymer nanoparticles in plasma.

Biophysical NMR was used to study both the structural and stoichiometry data of a gold nanoparticle's surface. Amide chemical shifts are extremely sensitive to small changes in the chemical environment and a simple 2D HSQC spectrum can be used. NMR was used to examine the binding and unbinding of the proteins onto gold nano-clusters by studying the chemical shifts of the amine protons. Surface heterogenicity was also studied by analyzing the spectral changes upon adsorption. With NMR, we can probably study the adsorption capacity as well.[50]

In general, NMR spectroscopy is a high-resolution informative technique. It can be used to provide the structural analysis, conformational analysis and surface analysis of nanoparticles and proteins. Also, various methods and experiments can be tuned to study particular binding and unbinding of specific proteins. Protein corona formation studies can be studied with NMR; however, they seem to be limited in information and are more expensive than other complementary techniques. NMR needs certain backup data, references, and also the desired time and cost to perform such experiments. NMR isn't applicable when data analysis is required for larger particles and kinetics of protein adsorption cannot be understood by using NMR.

4.5.5 Protein Analysis and Identification

It has been discussed and explained in this chapter that nanomaterials interact promptly with proteins, peptides, and lipids present in media to form a PC on their surface. The proteins that are adsorbed on the nanomaterial to shape the PC are dependent on multiple factors including the natural media and the physicochemical qualities of the nanomaterial. In particular, research has shown that the PC may have an impact on nanomaterial uptake by cells and appropriation, change cytotoxicity, and increase inflammation and oxidative stress. The identity of these proteins is best analyzed by mass spectrometry.

Several studies have been carried out using various types of mass spectrometers to discover the exact constituents of protein coronas. Different metal, metal oxide and polymer nanoparticles tend to have variable amounts of constituent proteins in their coronas. Many studies suggest that as the size of the nanoparticle increases, the amount of protein found in the protein corona also increases. In this section, we will exemplify a few studies carried out on metals, metal oxides and polymer nanoparticles and list the results.

4.5.5.1 Metals

Metal-nanoparticle-based studies for the determination of protein coronas are the most common. Au- and Ag-NP-based studies have been done extensively in the literature. The most commonly used techniques for studying protein constituents in these experiments were liquid chromatography mass spectrometry (LCMS) and inductively coupled plasma MS (ICP-MS). Certain proteins, like serum albumin, fibronectin, thrombospondin, and apolipoprotein, are commonly found in various sizes of Au and AgNPs. In one study, almost 39 specific proteins were found to bind with AgNPs. AgNPs with a diameter of 110 nm were found to associate the largest number of constituent proteins compared to 20 nm AgNPs.[6] A systematic study was carried out to determine the size-dependent uptake of proteins on AuNPs of different sizes (4, 10, 13, 16, 24, and 40 nm). The study was based on the hypothesis that the 4 nm AuNPs are significantly smaller compared to the most abundant serum proteins, and, therefore, these particles will accommodate fewer proteins at their surface. The 10 nm AuNPs have a similar size compared to the most abundant proteins and they may assemble a proper PC, whereas the 40 nm NPs are above the size of larger serum proteins and, therefore, more prone to opsonization by immunoglobulin G (IgG), complement factors, and apolipoproteins.[51] At different sizes, a variation was observed in the formation of hard and soft coronas; on the basis of a study conducted on LCMS with AuNP, the results showed that 10–20 nm NPs demonstrated the fastest formation of hard PC.[51]

4.5.5.2 Metal Oxides

Metal oxide nanoparticle studies based on super paramagnetic iron oxide (SPIONs) were discussed in this chapter. Four different studies were analyzed that used inductively coupled plasma mass spectrometry (ICP-MS) and liquid chromatography mass spectrometry (LCMS) for protein data analysis. The data suggest that the constituent proteins of the PC increase with an increase in the sizes of the nanoparticles (Table 4.8). In a single study, a total of 52 and 58 plasma proteins were identified as being bound to $ZnO^{SM20(-)}$ and $ZnO^{SM20(+)}$, respectively. For $ZnO^{AE100(-)}$ and $ZnO^{AE100(+)}$, 58 and 44 proteins were bound, respectively, which indicated that positively charged nanoparticles have a greater ability to bind a higher number of proteins on their surfaces compared to negatively charged nanoparticles.[52]

4.5.5.3 Polymers

With polymers, the sizes of NPs varying from 50–100 nm were studied to ascertain the number of proteins adsorbed on the polymer surfaces as a function of size. Figure 4.12 shows data from various studies demonstrating a significant increase in the number of proteins bound to nanoparticles as their sizes increased. Studies also show that apolipoprotein was exclusively found in the maximum number of experiments conducted on polymer NPs.

4.5.5.4 Analytical Studies

Mass spectrometry, as discussed, has many variants and their use depends upon the outcome or desired results. Tables 4.8–4.10 consist of a summary of different studies using LCMS, ICPMS and other, newer, MS techniques that have been used for studying the composition of proteins on NP–PCs. A volatile buffer has to be used to avoid ion suppression and adduct formation.[55] Buffer, or other media, are generally used to keep nanoparticles stable in suspension. Various mediums, like PBS, FBS, HEPES and formic acid, are most commonly seen in studies done using LCMS. Data from Table 4.8 suggests that protein identification is one of the main reasons to use LCMS. In LCMS, samples are sorted by giving chromatographic treatment prior to introducing the sample to a mass analyzer, which can help to differentiate or sort the required components that are needed for a study. Table 4.9 summarizes various nanoparticles studied using ICP-MS. Mostly metal and metal oxide NPs are studied by ICP-MS such as AuNPs, AgNPs, SPIONs, and TiO_2 NPs. RPMI and FBS media are often used to introduce metals in ICP-MS. It must be noted that ICP-MS is predominantly used for determination of metal ion concentrations and studying the uptake of nanoparticles in cells post PC formation.

Table 4.8 Study of various nanoparticles using a liquid chromatograph mass spectrometer.

S. No	Nanoparticle	Medium	Study	Reference
1	Polystyrene	0.1% formic acid	Peptide probability and detection	56
2	TiO_2	PBS	Protein detection	57
3	CNTs (carbon nanotubes)	PBS	Characterization of the PC	58
4	Iron oxide	0.1% TFA	Identification of proteins present in solution	59
5	Au	0.25% protease max	Proteomics study of NPs	60
6	Ag and carbon nanotubes	FBS	Protein detection in the corona	6
7	Viral NPs (VNP) TMV	Formic acid	Protein detection in the PC	61
8	AuNPs	PBS	Amount of bound protein and composition of PC	62
9	CeO_2, $BaSO_4$ and ZnO	HPLC solvent (2.5% acetonitrile/0.1% formic acid)	Peptide detection	63
10	$PSOSO_3$ and SiO_2	PBS	Studying components of the hard corona	64
11	$PSOSO_3$ and hydrophilic SiO_2	Not specified	Protein identification	65
12	SPIONs	PBS	Interaction of different surface-coated SPIONs	66
13	Si	PBS	Protein identification	34
14	Lipid	0.1% Formic acid	Identification of proteins bound to liposome surface	67
15	Si	1% Formic acid	Qualitative and quantitative PC signatures	68
16	SiO_2	10% Serum (cMEM)	Molecular weights of different proteins	16
17	Si	Tyrode's buffer	Identification and quantification of biologically relevant corona	69
18	NIPAM/BAM copolymer	PBS	Protein identification	1
19	Vascular-targeted carrier (VTC)	PBS	Studying different molecular weight ranges	70
20	Silica-coated gold nanorods	PBS	Determination of the composition of the protein corona	71
21	Au	Iodoacetamide (IAA) and dithiothreitol (DTT)	Protein analysis	51
22	Super paramagnetic iron oxide nanoparticles (SPIONs)	PBS	Protein identification	72
23	Zinc oxide (ZnO)	PBS	Identification of bound proteins	52
24	SiO_2	Not specified	Protein identification	73
25	SAMNs (surface active maghemite nanoparticles)	50% acetonitrile/0.1% formic acid	Protein identification	74
26	Liposomes	HEPES	Protein identification	75
27	Polyacrylic acid coated cobalt ferrite (PAA, NPs) and SiO_2	Formic acid	Protein identification	76

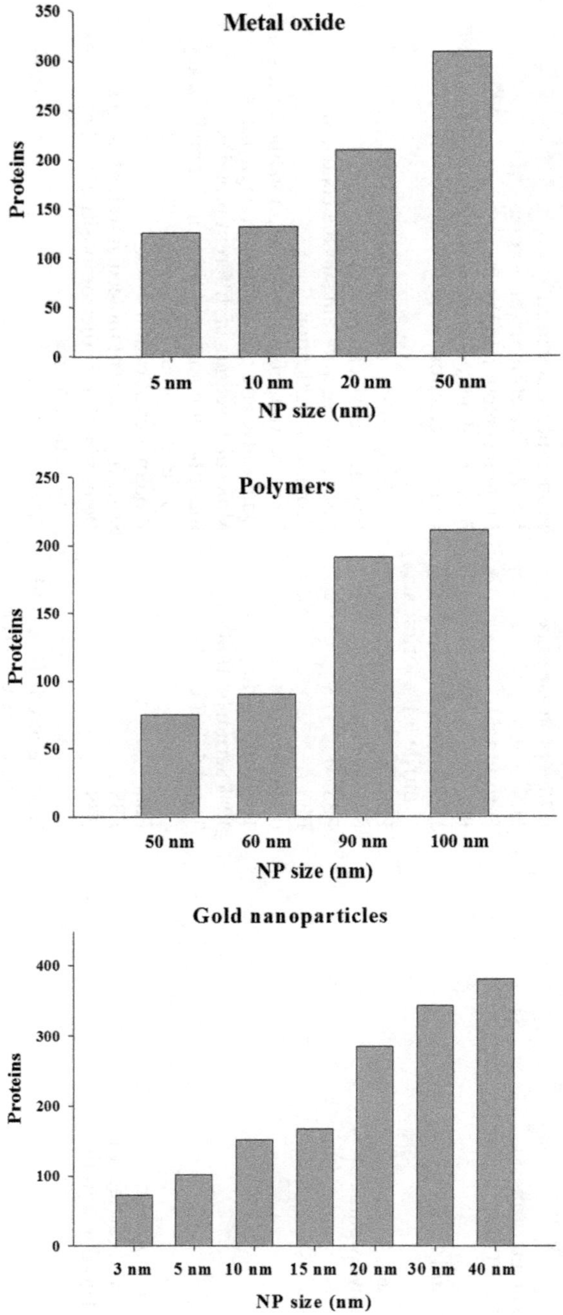

Figure 4.12 Summary of size-dependent protein adsorption by metals, metal oxides and polymeric nanoparticles suggesting an increase in the number of adsorbed proteins (y-axis) with an increase in the sizes of nanoparticles (x-axis) irrespective of the type of material. Adapted from ref. 6, 7, 53 and 54 with permission from the Royal Society of Chemistry.

Table 4.9 Summary of nanoparticles studied using ICP-MS.

S. No	Nanoparticle	Medium	Study	Reference
1	Au	Not specified	Molar concentration of the atoms of dissolved NPs is determined	77
2	AuNP	Aqua regia	Qualitative and quantitative characterization of the PC	41
3	SPIONs	PBS	Cellular iron content	78
4	Ag	FBS	Trace element speciation	79
5	Ag	PBS	Elemental analysis	80
6	Ag	TRIS-acetate buffer	Identification of ions	81
7	TiO_2	RPMI	Degree of cellular uptake and concentration	82
8	Ag	Not specified	Detection of ions	83
9	Au	Not specified	Composition of the hard corona	84
10	Ag	PBS	Protein identification	85
11	Ag	FBS	Identification of NP-bound proteins	86
12	Au	PBS	Studying the behavior of gold nanoparticles of different geometry, size and surface functionalization	87
13	ZnS	Sodium borate buffer (SBB)	Concentration of nanoparticles	88

Table 4.10 A summary of various studies of protein corona identification using other MS techniques.

S. No	NP	MS used	Medium	Study	Reference
1	DC-Chol-DOPE NPs	Nano liquid tandem mass spectrometry (NanoLC-MS-MS)	PBS	Protein identification	89
2	Lipid nanoparticles		0.1% HCOOH	Protein identification	90
3	Protein interactions with small molecule ligands	Electrospray ionization mass spectrometry (ESI-MS)	Ammonium acetate (AmAc) buffer	Quantitative determination of binding strengths	91
4	Carbon NTs		Not specified	Protein identification	20
5	Maghemite nanoparticles	LTQ-Orbitrap XL mass spectrometry	DMEM	Protein identification	92
6	AuNPs		0.05% PBS	Protein corona analysis	93
7	AuNPs from cancer and non-cancer cells	Tandem mass spectrometry	FBS	Protein identification	94
8	Colloidal NPs		Not specified	Protein estimation	95
9	Silica nanoparticles	Proteomics-mass spectrometry	PBS	Protein identification	96
10	CdSe/ZnS quantum dots	Matrix-assisted laser desorption ionization time-of-flight mass spectrometry (MALDI ToF MS)	PBS	Protein identification	97
11	Super magnetic iron oxide NPs		0.1% trifluoroacetic acid	Qualitative and quantitative information of protein present in the corona	98
12	Silica NPs		HEPES	Protein corona detection	99

Other than LCMS and ICP-MS, various MS techniques were used on the basis of the sample and the results required. Complex NPs like DC-Chol-DOPE NPs and lipid NPs were studied using nano liquid tandem MS, which helps to get the exact constitution of the hard corona and soft corona. A variety of mediums were used in studying various MS techniques such as AmAc, DMEM, FBS, HEPES to suspend NPs and to form the PC. Qualitative and quantitative studies with the help of MALDI ToF MS were also reported. In MALDI, sample preparation differs in the ionization of samples compared to other MS techniques.

Overall, going through various studies and protein analysis, it can be said that the composition of a corona can be thoroughly studied with the help of MS and its variants.

Mass spectrometry measures the m/z proportion of particles to distinguish and evaluate molecules in basic and complex structures. MS has turned out to be valuable over a wide range of fields and applications, including proteomics. The improvement of high-throughput and quantitative MS proteomics processes over the past two decades has increased the extent of what we think about protein structure, capacity and variants, and in addition worldwide protein dynamics.

4.6 Summary

It has been well established that whenever NPs come into contact with any biological medium, the proteins present in the medium compete to bind to the NPs' surface. This competition results in the formation of a protein layer around the nanoparticles, which changes over time. This protein layer modifies the nanoparticle surface, altering its synthetic identity and forming a new biological identity that affects its cellular internalization and hence controls the physiological response. A correlation between synthetic identity, biological identity and biological response can allow the prediction of the toxicity and efficacy of a particular nanoparticle, giving the flexibility to modify the system as desired. This protein binding is influenced by specific physical and chemical parameters of the nanoparticles including their size, shape, hydrophobicity (or hydrophilicity), surface charge as well as surface modification. These factors govern the formation of protein coronas and determine their thickness as well as composition.

Proteins present in the medium have different affinities for a nanoparticle surface, which is why the protein corona is considered to be dynamic in nature. The proteins that are tightly bound to the nanoparticle surface, due to their high affinity for the nanoparticle surface, form a hard corona, while the loosely bound proteins with a low affinity will form the soft corona irrespective of the protein abundance and mobility in the medium. The kinetics of protein coronas is well understood by using techniques such as FCS, SEC and SPR-based techniques in combination with ITC, which provides the binding affinities of proteins dictating their exchange rates. The protein concentration in the surrounding medium dramatically influences the thickness and identity of the protein corona. DLS measurements estimate changes in the sizes of nanoparticles following their incubation with media. The thickness of the corona increases with increasing protein concentration in the environment. The higher charge density of the nanoparticles also contributes towards increasing the thickness of the corona. Surface charges of nanoparticles as well as the NP–PC can be analyzed well on the basis of Z-potential measurements. Regardless of the original electric charges on nanoparticles, the trends show that surface charges of nanoparticles often yield, on average, negative values post-incubation with media.

The composition of a protein corona can be analytically examined with variants of MS and the results show different types of proteins and their abundance in the corona or the total number of proteins present on the corona. Coronas formed around polymeric nanoparticles are generally found to be abundant in apolipoproteins, while proteins like albumin, fibrinogen and a few complement proteins are often found on coronas formed around metal- and metal-oxide-nanoparticles.

The thermodynamics of a system, such as the interactions between nanoparticles and protein coronas, proceed with the absorption or release of energy in the form of heat. The total heat energy released or absorbed shows us the enthalpies, entropies, heat capacities, stoichiometries and the rate of binding affinities in a system. In most cases, it has been noticed that the sum total of all the binding energy doesn't add up to the total enthalpy of a system or reaction, hinting that there are other reactions, maybe enzymatic, that happen simultaneously with the reactions that work with entropy-driven forces and can be studied effectively by ITC.

Proteins like HSA bind with NPs with big enthalpy changes, whereas proteins like fibrinogen and lysosomes *etc.* bind with NPs with low to no enthalpy changes. MS-based techniques are useful in the determination of the composition of the proteins present in protein coronas. MS techniques demonstrated that the number of proteins on a NP surface increase with an increase in the size of the NPs. In addition, it has also been shown that different proteins can bind to different nanoparticle surfaces based on the size, shape and chemistry of the NPs and the proteins present in the dispersing medium.

The orientation and arrangement of proteins on a nanoparticle surface need to be studied more clearly as the proteins interacting with the nanoparticles may exhibit conformational changes. These conformational changes often affect the biological response owing to the changed protein structure and can partly be understood using techniques such as NMR. Similarly, nanoparticle surface modifications should be given equal importance as surface coatings greatly influence the nanoparticles' biodistribution, toxicity, cellular internalization, as well as immunological recognition. Other methods have also been used to characterize the protein corona evolution and composition including X-ray based techniques such as SAXS, optical techniques such as SERS, IR and CD; these have not been included in this chapter and the reader is encouraged to make themselves aware of these techniques to know the full spectrum of characterization techniques for NP–PCs.

References

1. C. D. Walkey and W. C. Chan, Understanding and controlling the interaction of nanomaterials with proteins in a physiological environment, *Chem. Soc. Rev.*, 2012, **41**, 2780–2799.
2. C. Gunawan, M. Lim, C. P. Marquis and R. Amal, Nanoparticle-protein corona complexes govern the biological fates and functions of nanoparticles,*J. Mater. Chem. B*, 2014, **2**, 2060–2083.
3. X. Zhang, J. Zhang, F. Zhang and S. Yu, Probing the binding affinity of plasma proteins adsorbed on Au nanoparticles, *Nanoscale*, 2017, **9**, 4787–4792.
4. K. C. Tam and E. Wyn-Jones, Insights on polymer surfactant complex structures during the binding of surfactants to polymers as measured by equilibrium and structural techniques, *Chem. Soc. Rev.*, 2006, **35**, 693–709.
5. L. Baranauskienė, V. Petrikaitė, J. Matulienė and D. Matulis, Titration Calorimetry Standards and the Precision of Isothermal Titration Calorimetry Data, *Int. J. Mol. Sci.*, 2009, **10**, 2752–2762.
6. J. H. Shannahan, X. Lai, P. C. Ke, R. Podila, J. M. Brown and F. A. Witzmann, Silver nanoparticle protein corona composition in cell culture media, *PLoS One*, 2013, **8**, e74001.
7. J. Warneke, T. Dulcks, C. Knapp and D. Gabel, Collision-induced gas-phase reactions of perhalogenated closo-dodecaborate clusters–a comparative study, *Phys. Chem. Chem. Phys.*, 2011, **13**, 5712–5721.
8. E. Casals, T. Pfaller, A. Duschl, G. J. Oostingh and V. Puntes, Time Evolution of the Nanoparticle Protein Corona, *ACS Nano*, 2010, **4**, 3623–3632.
9. A. L. Barrán-Berdón, D. Pozzi, G. Caracciolo, A. L. Capriotti, G. Caruso, C. Cavaliere, A. Riccioli, S. Palchetti and A. Laganà, Time Evolution of Nanoparticle–Protein Corona in Human Plasma: Relevance for Targeted Drug Delivery, *Langmuir*, 2013, **29**, 6485–6494.
10. A. Jedlovszky-Hajdú, F. B. Bombelli, M. P. Monopoli, E. Tombácz and K. A. Dawson, Surface Coatings Shape the Protein Corona of SPIONs with Relevance to Their Application *in Vivo*, *Langmuir*, 2012, **28**, 14983–14991.
11. N. P. Mortensen, G. B. Hurst, W. Wang, C. M. Foster, P. D. Nallathamby and S. T. Retterer, Dynamic development of the protein corona on silica nanoparticles: composition and role in toxicity, *Nanoscale*, 2013, **5**, 6372–6380.

12. S. Khan, A. Gupta and C. K. Nandi, Controlling the Fate of Protein Corona by Tuning Surface Properties of Nanoparticles, *J. Phys. Chem. Lett.*, 2013, **4**, 3747–3752.
13. G. Maiorano, S. Sabella, B. Sorce, V. Brunetti, M. A. Malvindi, R. Cingolani and P. P. Pompa, Effects of Cell Culture Media on the Dynamic Formation of Protein–Nanoparticle Complexes and Influence on the Cellular Response, *ACS Nano*, 2010, **4**, 7481–7491.
14. V. Mirshafiee, M. Mahmoudi, K. Lou, J. Cheng and M. L. Kraft, Protein corona significantly reduces active targeting yield, *Chem. Commun.*, 2013, **49**, 2557–2559.
15. M. Lundqvist, J. Stigler, G. Elia, I. Lynch, T. Cedervall and K. A. Dawson Nanoparticle size and surface properties determine the protein corona with possible implications for biological impacts, *Proc. Natl. Acad. Sci. U. S. A.*, 2008, **105**, 14265–14270.
16. A. Lesniak, F. Fenaroli, M. P. Monopoli, C. Aberg, K. A. Dawson and A. Salvati, Effects of the presence or absence of a protein corona on silica nanoparticle uptake and impact on cells, *ACS Nano*, 2012, **6**, 5845–5857.
17. J. C. Y. Kah, J. Chen, A. Zubieta and K. Hamad-Schifferli, Exploiting the Protein Corona around Gold Nanorods for Loading and Triggered Release, *ACS Nano*, 2012, **6**, 6730–6740.
18. L. Treuel, S. Brandholt, P. Maffre, S. Wiegele, L. Shang and G. U. Nienhaus, Impact of Protein Modification on the Protein Corona on Nanoparticles and Nanoparticle–Cell Interactions, *ACS Nano*, 2014, **8**, 503–513.
19. M. P. Monopoli, D. Walczyk, A. Campbell, G. Elia, I. Lynch, F. Baldelli Bombelli and K. A. Dawson, Physical–Chemical Aspects of Protein Corona: Relevance to *in Vitro* and *in Vivo* Biological Impacts of Nanoparticles, *J. Am. Chem. Soc.*, 2011, **133**, 2525–2534.
20. P. Aggarwal, J. B. Hall, C. B. McLeland, M. A. Dobrovolskaia and S. E. McNeil, Nanoparticle interaction with plasma proteins as it relates to particle biodistribution, biocompatibility and therapeutic efficacy, *Adv. Drug Delivery Rev.*, 2009, **61**, 428–437.
21. M. Safi, J. Courtois, M. Seigneuret, H. Conjeaud and J. F. Berret, The effects of aggregation and protein corona on the cellular internalization of iron oxide nanoparticles, *Biomaterials*, 2011, **32**, 9353–9363.
22. Q. Dai, C. Walkey and W. C. W. Chan, Polyethylene Glycol Backfilling Mitigates the Negative Impact of the Protein Corona on Nanoparticle Cell Targeting, *Angew. Chem., Int. Ed.*, 2014, **53**, 5093–5096.
23. S. Ritz, S. Schöttler, N. Kotman, G. Baier, A. Musyanovych, J. Kuharev, K. Landfester, H. Schild, O. Jahn, S. Tenzer and V. Mailänder, Protein Corona of Nanoparticles: Distinct Proteins Regulate the Cellular Uptake, *Biomacromolecules*, 2015, **16**, 1311–1321.

24. S. Winzen, S. Schoettler, G. Baier, C. Rosenauer, V. Mailaender, K. Landfester and K. Mohr, Complementary analysis of the hard and soft protein corona: sample preparation critically effects corona composition, *Nanoscale*, 2015, **7**, 2992–3001.
25. S. Martin, S.-B. Manuela, S. Hakan, T. Shinji, W. Alexander, S. Carsten, M. H. Stefanie, D. J. Blair and G. K. Wolfgang, Serum protein identification and quantification of the corona of 5, 15 and 80 nm gold nanoparticles, *Nanotechnology*, 2013, **24**, 265103.
26. S. Tenzer, D. Docter, J. Kuharev, A. Musyanovych, V. Fetz, R. Hecht, F. Schlenk, D. Fischer, K. Kiouptsi, C. Reinhardt, K. Landfester, H. Schild, M. Maskos, S. K. Knauer and R. H. Stauber, Rapid formation of plasma protein corona critically affects nanoparticle pathophysiology, *Nat. Nanotechnol.*, 2013, **8**, 772.
27. T. Cedervall, I. Lynch, S. Lindman, T. Berggård, E. Thulin, H. Nilsson, K. A. Dawson and S. Linse, Understanding the nanoparticle–protein corona using methods to quantify exchange rates and affinities of proteins for nanoparticles, *Proc. Natl. Acad. Sci. U. S. A.*, 2007, **104**, 2050–2055.
28. C. Röcker, M. Pötzl, F. Zhang, W. J. Parak and G. U. Nienhaus, A quantitative fluorescence study of protein monolayer formation on colloidal nanoparticles, *Nat. Nanotechnol.*, 2009, **4**, 577.
29. E. Hellstrand, I. Lynch, A. Andersson, T. Drakenberg, B. Dahlback, K. A. Dawson, S. Linse and T. Cedervall, Complete high-density lipoproteins in nanoparticle corona, *FEBS J.*, 2009, **276**, 3372–3381.
30. L. Dong, X. Zhou, L. Hu, Y. Yin and J. Liu, Simultaneous size characterization and mass quantification of the *in vivo* core-biocorona structure and dissolved species of silver nanoparticles, *J. Environ. Sci.*, 2018, **63**, 227–235.
31. I. Lynch, T. Cedervall, M. Lundqvist, C. Cabaleiro-Lago, S. Linse and K. A. Dawson, The nanoparticle-protein complex as a biological entity; a complex fluids and surface science challenge for the 21st century, *Adv. Colloid Interface Sci.*, 2007, **134–135**, 167–174.
32. N. Welsch, Y. Lu, J. Dzubiella and M. Ballauff, Adsorption of proteins to functional polymeric nanoparticles, *Polymer*, 2013, **54**, 2835–2849.
33. D. J. Winzor and C. M. Jackson, Interpretation of the temperature dependence of equilibrium and rate constants, *J. Mol. Recognit.*, 2006, **19**, 389–407.
34. L. Treuel, D. Docter, M. Maskos and R. H. Stauber, Protein corona - from molecular adsorption to physiological complexity, *Beilstein J. Nanotechnol.*, 2015, **6**, 857–873.
35. C. C. Fleischer and C. K. Payne, Secondary Structure of Corona Proteins Determines the Cell Surface Receptors Used by Nanoparticles, *J. Phys. Chem. B*, 2014, **118**, 14017–14026.

36. L.-Y. Yang, S.-Y. Hua, Z.-Q. Zhou, G.-C. Wang, F.-L. Jiang and Y. Liu, Characterization of fullerenol-protein interactions and an extended investigation on cytotoxicity, *Colloids Surf., B*, 2017, **157**, 261–267.
37. S. Mandal, M. Hossain, P. S. Devi, G. S. Kumar and K. Chaudhuri, Interaction of carbon nanoparticles to serum albumin: elucidation of the extent of perturbation of serum albumin conformations and thermodynamical parameters, *J. Hazard. Mater.*, 2013, **248–249**, 238–245.
38. R. Huang, R. P. Carney, K. Ikuma, F. Stellacci and B. L. T. Lau, Effects of Surface Compositional and Structural Heterogeneity on Nanoparticle–Protein Interactions: Different Protein Configurations, *ACS Nano*, 2014, **8**, 5402–5412.
39. H. Liu, S. Chen, H. Cui, J. Hu, H. Cai and W. Deng, Fabrication of triple responsive polymer brushes and their catalytic performance after loading palladium, *RSC Adv.*, 2015, **5**, 72444–72452.
40. M. A. Dobrovolskaia, A. K. Patri, J. Zheng, J. D. Clogston, N. Ayub, P. Aggarwal, B. W. Neun, J. B. Hall and S. E. McNeil, Interaction of colloidal gold nanoparticles with human blood: effects on particle size and analysis of plasma protein binding profiles, *Nanomedicine: Nanotechnol,. Biol. Med.*, 2009, **5**, 106–117.
41. R. Garcia-Alvarez, M. Hadjidemetriou, A. Sanchez-Iglesias, L. M. Liz-Marzan and K. Kostarelos, *In vivo* formation of protein corona on gold nanoparticles. The effect of their size and shape, *Nanoscale*, 2018, **10**, 1256–1264.
42. S. H. D. P. Lacerda, J. J. Park, C. Meuse, D. Pristinski, M. L. Becker, A. Karim and J. F. Douglas, Interaction of Gold Nanoparticles with Common Human Blood Proteins, *ACS Nano*, 2010, **4**, 365–379.
43. A. Sasidharan, J. E. Riviere and N. A. Monteiro-Riviere, Gold and silver nanoparticle interactions with human proteins: impact and implications in biocorona formation, *J. Mater. Chem. B*, 2015, **3**, 2075–2082.
44. J. E. Diederichs, Plasma protein adsorption patterns on liposomes: establishment of analytical procedure, *Electrophoresis*, 1996, **17**, 607–611.
45. T. M. Goppert and R. H. Muller, Plasma protein adsorption of Tween 80- and poloxamer 188-stabilized solid lipid nanoparticles, *J. Drug Targeting*, 2003, **11**, 225–231.
46. K. Thode, M. Luck, W. Semmler, R. H. Muller and M. Kresse, Determination of plasma protein adsorption on magnetic iron oxides: sample preparation, *Pharm. Res.*, 1997, **14**, 905–910.
47. C. Fang, B. Shi, Y. Y. Pei, M. H. Hong, J. Wu and H. Z. Chen, *In vivo* tumor targeting of tumor necrosis factor-alpha-loaded stealth nanoparticles: effect of MePEG molecular weight and particle size, *Eur. J. Pharm. Sci.*, 2006, **27**, 27–36.
48. L. Calzolai, F. Franchini, D. Gilliland and F. Rossi, Protein–nanoparticle interaction: identification of the ubiquitin–gold nanoparticle interaction site, *Nano Lett.*, 2010, **10**, 3101–3105.

49. M. Lundqvist, I. Sethson and B. H. Jonsson, Protein adsorption onto silica nanoparticles: conformational changes depend on the particles' curvature and the protein stability, *Langmuir*, 2004, **20**, 10639–10647.
50. A. Wang, K. Vangala, T. Vo, D. Zhang and N. C. Fitzkee, A Three-Step Model for Protein–Gold Nanoparticle Adsorption, *J. Phys. Chem. C*, 2014, **118**, 8134–8142.
51. E. Casals, T. Pfaller, A. Duschl, G. J. Oostingh and V. Puntes, Time evolution of the nanoparticle protein corona, *ACS Nano*, 2010, **4**, 3623–3632.
52. K. H. Shim, J. Hulme, E. H. Maeng, M. K. Kim and S. S. An, Analysis of zinc oxide nanoparticles binding proteins in rat blood and brain homogenate, *Int. J. Nanomed.*, 2014, **9**(Suppl 2), 217–224.
53. C. D. Walkey and W. C. Chan, Understanding and controlling the interaction of nanomaterials with proteins in a physiological environment, *Chem. Soc. Rev.*, 2012, **41**, 2780–2799.
54. C. Gunawan, M. Lim, C. P. Marquis and R. Amal, Nanoparticle–protein corona complexes govern the biological fates and functions of nanoparticles,*J. Mater. Chem. B*, 2014, **2**, 2060–2083.
55. C. Jecklin Matthias, S. Schauer, E. Dumelin Christoph and R. Zenobi, Label-free determination of protein–ligand binding constants using mass spectrometry and validation using surface plasmon resonance and isothermal titration calorimetry, *J. Mol. Recognit.*, 2009, **22**, 319–329.
56. M. Lundqvist, J. Stigler, G. Elia, I. Lynch, T. Cedervall and K. A. Dawson, Nanoparticle size and surface properties determine the protein corona with possible implications for biological impacts, *Proc. Natl. Acad. Sci. U. S. A.*, 2008, **105**, 14265–14270.
57. T. Cedervall, I. Lynch, S. Lindman, T. Berggard, E. Thulin, H. Nilsson, K. A. Dawson and S. Linse, Understanding the nanoparticle-protein corona using methods to quantify exchange rates and affinities of proteins for nanoparticles, *Proc. Natl. Acad. Sci. U. S. A.*, 2007, **104**, 2050–2055.
58. Y. K. Lee, E. J. Choi, T. J. Webster, S. H. Kim and D. Khang, Effect of the protein corona on nanoparticles for modulating cytotoxicity and immunotoxicity, *Int. J. Nanomed.*, 2015, **10**, 97–113.
59. J. Ashby, S. Pan and W. Zhong, Size and surface functionalization of iron oxide nanoparticles influence the composition and dynamic nature of their protein corona, *ACS Appl. Mater. Interfaces*, 2014, **6**, 15412–15419.
60. R. R. Arvizo, K. Giri, D. Moyano, O. R. Miranda, B. Madden, D. J. McCormick, R. Bhattacharya, V. M. Rotello, J. P. Kocher and P. Mukherjee, Identifying new therapeutic targets via modulation of protein corona formation by engineered nanoparticles, *PLoS One*, 2012, **7**, e33650.
61. A. S. Pitek, A. M. Wen, S. Shukla and N. F. Steinmetz, The Protein Corona of Plant Virus Nanoparticles Influences their Dispersion Properties, Cellular Interactions, and *In Vivo* Fates, *Small*, 2016, **12**, 1758–1769.

62. M. A. Dobrovolskaia, B. W. Neun, S. Man, X. Ye, M. Hansen, A. K. Patri, R. M. Crist and S. E. McNeil, Protein corona composition does not accurately predict hematocompatibility of colloidal gold nanoparticles, *Nanomedicine: Nanotechnol,. Biol. Med.*, 2014, **10**, 1453–1463.
63. N. V. Konduru, R. M. Molina, A. Swami, F. Damiani, G. Pyrgiotakis, P. Lin, P. Andreozzi, T. C. Donaghey, P. Demokritou, S. Krol, W. Kreyling and J. D. Brain, Protein corona: implications for nanoparticle interactions with pulmonary cells, *Part. Fibre Toxicol.*, 2017, **14**, 42.
64. M. P. Monopoli, D. Walczyk, A. Campbell, G. Elia, I. Lynch, F. B. Bombelli and K. A. Dawson, Physical-chemical aspects of protein corona: relevance to *in vitro* and *in vivo* biological impacts of nanoparticles, *J. Am. Chem. Soc.*, 2011, **133**, 2525–2534.
65. P. Foroozandeh and A. A. Aziz, Merging worlds of nanomaterials and biological environment: factors governing protein corona formation on nanoparticles and its biological consequences, *Nanoscale Res. Lett.*, 2015, **10**, 221.
66. U. Sakulkhu, M. Mahmoudi, L. Maurizi, J. Salaklang and H. Hofmann, Protein corona composition of superparamagnetic iron oxide nanoparticles with various physico-chemical properties and coatings, *Sci. Rep.*, 2014, **4**, 5020.
67. A. Amici, G. Caracciolo, L. Digiacomo, V. Gambini, C. Marchini, M. Tilio, A. L. Capriotti, V. Colapicchioni, R. Matassa, G. Familiari, S. Palchetti, D. Pozzi, M. Mahmoudi and A. Lagana, *In vivo* protein corona patterns of lipid nanoparticles, *RSC Adv.*, 2017, **7**, 1137–1145.
68. D. Docter, U. Distler, W. Storck, J. Kuharev, D. Wunsch, A. Hahlbrock, S. K. Knauer, S. Tenzer and R. H. Stauber, Quantitative profiling of the protein coronas that form around nanoparticles, *Nat. Protoc.*, 2014, **9**, 2030–2044.
69. S. Tenzer, D. Docter, J. Kuharev, A. Musyanovych, V. Fetz, R. Hecht, F. Schlenk, D. Fischer, K. Kiouptsi, C. Reinhardt, K. Landfester, H. Schild, M. Maskos, S. K. Knauer and R. H. Stauber, Rapid formation of plasma protein corona critically affects nanoparticle pathophysiology, *Nat. Nanotechnol.*, 2013, **8**, 772–781.
70. D. J. Sobczynski and O. Eniola-Adefeso, Effect of anticoagulants on the protein corona-induced reduced drug carrier adhesion efficiency in human blood flow, *Acta Biomater.*, 2017, **48**, 186–194.
71. M. Das, D. K. Yi and S. S. An, Analyses of protein corona on bare and silica-coated gold nanorods against four mammalian cells, *Int. J. Nanomed.*, 2015, **10**, 1521–1545.
72. A. Jedlovszky-Hajdu, F. B. Bombelli, M. P. Monopoli, E. Tombacz and K. A. Dawson, Surface coatings shape the protein corona of SPIONs with relevance to their application *in vivo*, *Langmuir*, 2012, **28**, 14983–14991.

73. A. Kurtz-Chalot, C. Villiers, J. Pourchez, D. Boudard, M. Martini, P. N. Marche, M. Cottier and V. Forest, Impact of silica nanoparticle surface chemistry on protein corona formation and consequential interactions with biological cells, *Mater. Sci. Eng., C*, 2017, **75**, 16–24.
74. M. Magro, L. Fasolato, E. Bonaiuto, N. A. Andreani, D. Baratella, V. Corraducci, G. Miotto, B. Cardazzo and F. Vianello, Enlightening mineral iron sensing in Pseudomonas fluorescens by surface active maghemite nanoparticles: Involvement of the OprF porin, *Biochim. Biophys. Acta*, 2016, **1860**, 2202–2210.
75. O. K. Kari, T. Rojalin, S. Salmaso, M. Barattin, H. Jarva, S. Meri, M. Yliperttula, T. Viitala and A. Urtti, Multi-parametric surface plasmon resonance platform for studying liposome-serum interactions and protein corona formation, *Drug Delivery Transl. Res.*, 2017, **7**, 228–240.
76. K. Strojan, A. Leonardi, V. B. Bregar, I. Krizaj, J. Svete and M. Pavlin, Dispersion of Nanoparticles in Different Media Importantly Determines the Composition of Their Protein Corona, *PLoS One*, 2017, **12**, e0169552.
77. P. d. Pino, B. Pelaz, Q. Zhang, P. Maffre, G. U. Nienhaus and W. J. Parak, Protein corona formation around nanoparticles - from the past to the future, *Mater. Horiz.*, 2014, **1**, 301–313.
78. C. Vogt, M. Pernemalm, P. Kohonen, S. Laurent, K. Hultenby, M. Vahter, J. Lehtio, M. S. Toprak and B. Fadeel, Proteomics Analysis Reveals Distinct Corona Composition on Magnetic Nanoparticles with Different Surface Coatings: Implications for Interactions with Primary Human Macrophages, *PLoS One*, 2015, **10**, e0129008.
79. M. P. Chantada-Vázquez, A. Moreda-Piñeiro, M. C. Barciela-Alonso and P. Bermejo-Barrera, Spectrometric-based techniques for metal-binding protein assessment in clinical, environmental, and food samples, *Appl. Spectrosc. Rev.*, 2017, **52**, 145–174.
80. T. K. Mudalige, H. Qu and S. W. Linder, Asymmetric Flow-Field Flow Fractionation Hyphenated ICP-MS as an Alternative to Cloud Point Extraction for Quantification of Silver Nanoparticles and Silver Speciation: Application for Nanoparticles with a Protein Corona, *Anal. Chem.*, 2015, **87**, 7395–7401.
81. A. Kakinen, F. Ding, P. Chen, M. Mortimer, A. Kahru and P. C. Ke, Interaction of firefly luciferase and silver nanoparticles and its impact on enzyme activity, *Nanotechnology*, 2013, **24**, 345101.
82. Z. E. Allouni, N. R. Gjerdet, M. R. Cimpan and P. J. Hol, The effect of blood protein adsorption on cellular uptake of anatase TiO_2 nanoparticles, *Int. J. Nanomed.*, 2015, **10**, 687–695.
83. Y. Wen, N. K. Geitner, R. Chen, F. Ding, P. Chen, R. E. Andorfer, P. N. Govindan and P. C. Ke, Binding of cytoskeletal proteins with silver nanoparticles, *RSC Adv.*, 2013, **3**, 22002–22007.

84. M. Matczuk, J. Legat, S. N. Shtykov, M. Jarosz and A. R. Timerbaev, Characterization of the protein corona of gold nanoparticles by an advanced treatment of CE-ICP-MS data, *Electrophoresis*, 2016, **37**, 2257–2259.
85. W. Liu, I. A. M. Worms, N. Herlin-Boime, D. Truffier-Boutry, I. Michaud-Soret, E. Mintz, C. Vidaud and F. Rollin-Genetet, Interaction of silver nanoparticles with metallothionein and ceruloplasmin: impact on metal substitution by Ag(i), corona formation and enzymatic activity, *Nanoscale*, 2017, **9**, 6581–6594.
86. S. Argentiere, C. Cella, M. Cesaria, P. Milani and C. Lenardi, Silver nanoparticles in complex biological media: assessment of colloidal stability and protein corona formation, *J. Nanopart. Res.*, 2016, **18**, 253.
87. J. Legat, M. Matczuk, A. Timerbaev and M. Jarosz, CE Separation and ICP-MS Detection of Gold Nanoparticles and Their Protein Conjugates, *Chromatographia*, 2017, **80**, 1695–1700.
88. M. Garcia-Cortes, E. Sotelo Gonzalez, M. T. Fernandez-Arguelles, J. R. Encinar, J. M. Costa-Fernandez and A. Sanz-Medel, Capping of Mn-Doped ZnS Quantum Dots with DHLA for Their Stabilization in Aqueous Media: Determination of the Nanoparticle Number Concentration and Surface Ligand Density, *Langmuir*, 2017, **33**, 6333–6341.
89. A. L. Barran-Berdon, D. Pozzi, G. Caracciolo, A. L. Capriotti, G. Caruso, C. Cavaliere, A. Riccioli, S. Palchetti and A. Lagana, Time evolution of nanoparticle-protein corona in human plasma: relevance for targeted drug delivery, *Langmuir*, 2013, **29**, 6485–6494.
90. L. Digiacomo, F. Cardarelli, D. Pozzi, S. Palchetti, M. A. Digman, E. Gratton, A. L. Capriotti, M. Mahmoudi and G. Caracciolo, An apolipoprotein-enriched biomolecular corona switches the cellular uptake mechanism and trafficking pathway of lipid nanoparticles, *Nanoscale*, 2017, **9**, 17254–17262.
91. M. C. Jecklin, S. Schauer, C. E. Dumelin and R. Zenobi, Label-free determination of protein-ligand binding constants using mass spectrometry and validation using surface plasmon resonance and isothermal titration calorimetry, *J. Mol. Recognit.*, 2009, **22**, 319–329.
92. R. Venerando, G. Miotto, M. Magro, M. Dallan, D. Baratella, E. Bonaiuto, R. Zboril and F. Vianello, Magnetic Nanoparticles with Covalently Bound Self-Assembled Protein Corona for Advanced Biomedical Applications, *J. Phys. Chem. C*, 2013, **117**, 20320–20331.
93. M. S. Boyles, T. Kristl, A. Andosch, M. Zimmermann, N. Tran, E. Casals, M. Himly, V. Puntes, C. G. Huber, U. Lutz-Meindl and A. Duschl, Chitosan functionalisation of gold nanoparticles encourages particle uptake and induces cytotoxicity and pro-inflammatory conditions in phagocytic cells, as well as enhancing particle interactions with serum components, *J. Nanobiotechnol.*, 2015, **13**, 84.
94. K. Giri, K. Shameer, M. T. Zimmermann, S. Saha, P. K. Chakraborty, A. Sharma, R. R. Arvizo, B. J. Madden, D. J. McCormick, J. P. Kocher, R.

Bhattacharya and P. Mukherjee, Understanding protein-nanoparticle interaction: a new gateway to disease therapeutics, *Bioconjugate Chem.*, 2014, **25**, 1078–1090.

95. C. Carrillo-Carrion, M. Carril and W. J. Parak, Techniques for the experimental investigation of the protein corona, *Curr. Opin. Biotechnol.*, 2017, **46**, 106–113.
96. S. Shrivastava, S. A. McCallum, J. H. Nuffer, X. Qian, R. W. Siegel and J. S. Dordick, Identifying specific protein residues that guide surface interactions and orientation on silica nanoparticles, *Langmuir*, 2013, **29**, 10841–10849.
97. H. Wang, L. Shang, P. Maffre, S. Hohmann, F. Kirschhofer, G. Brenner-Weiss and G. U. Nienhaus, The Nature of a Hard Protein Corona Forming on Quantum Dots Exposed to Human Blood Serum, *Small*, 2016, 5836–5844.
98. D. A. Mbeh, T. Javanbakht, L. Tabet, Y. Merhi, K. Maghni, E. Sacher and L. H. Yahia, Protein Corona Formation on Magnetite Nanoparticles: Effects of Culture Medium Composition, and Its Consequences on Superparamagnetic Nanoparticle Cytotoxicity, *J. Biomed. Nanotechnol.*, 2015, **11**, 828–840.
99. M. Lundqvist, C. Augustsson, M. Lilja, K. Lundkvist, B. Dahlbäck, S. Linse and T. Cedervall, The nanoparticle protein corona formed in human blood or human blood fractions, *PLoS One*, 2017, **12**, e0175871.
100. https://upload.wikimedia.org/wikipedia/commons/d/d8/13C_NMR_ethanol.GIF.
101. http://www.rsc.org/learn-chemistry/resource/res00001304/mass-spectrometry?cmpid=CMP00002746.

CHAPTER 5

An Analytical Approach to Investigate Nanoparticle–Protein Corona Complexes

PAULA RAIJIWALA[a], ALOK PANDYA[b] AND RITESH K. SHUKLA*[a]

[a]Biological and Life Sciences, School of Arts and Sciences, Ahmedabad University, Central Campus, Navrangpura, Ahmedabad-380009, Gujarat, India; [b]Physical Sciences, Institute of Advanced Research, Gandhinagar-382426, Gujarat, India
*E-mail: ritesh.shukla@ahduni.edu.in, shuklakritesh@gmail.com

5.1 Introduction

Over the last two decades, nanomaterials (NMs) have been increasingly used for biomedical applications such as sensing, imaging and drug delivery.[1] A deep understanding of the fundamental interactions of nanomaterials with living matter plays a central role in such various applications of nanoscience.[2] Nowadays, the interactions of NMs with biomolecules (such as proteins) inside the biological environment are increasing tremendously.[3,4] When these NMs come into contact with biomolecules like proteins, the proteins start to adsorb on the surface of the NMs and form a layer called the "protein corona".[5–7] Corona formation depends on the binding affinities of different types of proteins, its composition and respective NM surfaces.[8] Uptake and fate (additional stability, biocompatibility,

Issues in Toxicology No. 40
Nanoparticle–Protein Corona: Biophysics to Biology
Edited by Ashutosh Kumar and Alok Dhawan

Published by the Royal Society of Chemistry, www.rsc.org

and retention time) of NMs in the biological system can be affected by the nature and composition of the corona.[9–11] In biological media, the nanoparticle surface is "dressed" by a corona of biological macromolecules, such as proteins, that loosely divide into a "soft" component due to dynamical exchange of the biomolecules between the medium and the nanoparticles and a "hard" corona due to the high affinity of biological macromolecules for the nanoparticle surface.[4,7,12]

In the last two decades, protein coronas have been used extensively in pharmaceutical research for diagnostics, therapeutics and specific drug-like nanomedicines.[6,7] In the case of nanomedicine, when a drug enters a biological system, proteins from the bloodstream immediately cover the drug, forming a protein corona (PC).[13,14] When proteins interact with NMs, this alters the size of the NMs and forms a new biological entity that can be measured by physiological responses including agglomeration, cellular uptake, circulation lifetime, signaling, kinetics, transport, accumulation, and toxicity.[15] So, we can say that when a PC forms on NMs, it could govern the fate and successes/failures of nanomedicines in terms of efficacy, targeting, toxicity, cellular interaction, cellular uptake, and biodistribution.[16–20] In addition, the structure and composition of a protein corona is also very complex and it depends on the various physico-chemical properties of the NMs (size, shape, composition, surface functional groups, and surface charges), the nature of the biological environment (blood, interstitial fluid, cell cytoplasm, cell organelles *etc.*) and time of interaction.[15] Thus, it is difficult to characterize the protein corona on the basis of one parameter and one analytical method. This chapter is aimed at commenting on recent advances and the most relevant possibilities in terms of experimental methodologies to more completely characterize these protein coronas and their efficiency. Various analytical methods have been categorized on the basis of separation/isolation of proteins as well as on the basis of different parameters. This chapter will begin with the classification of the corona.

5.2 Classification of the Corona

Corona formation is mainly the adsorption of proteins on the surface of NMs, which is governed by protein–NM binding affinities as well as their exchange of time.[15] Hence, on the basis of *binding affinity* and the *exchange of time*, protein coronas are classified into: (1) the hard corona and (2) soft corona; while on the basis of binding affinity when proteins adsorb with a high affinity on the surface of nanoparticles (NPs), this is known as the hard corona (Figure 5.1). It consists of firmly bound proteins that do not easily desorb. While proteins that adsorb with a low affinity on the surface of NPs is known as the soft corona. It consists of loosely bound proteins that can easily desorb (Figure 5.1). However, on the basis of exchange of time, when a protein corona shows a larger exchange time (several hours), it is called the hard corona and when the exchange of time is shorter, it is called the soft corona.[21]

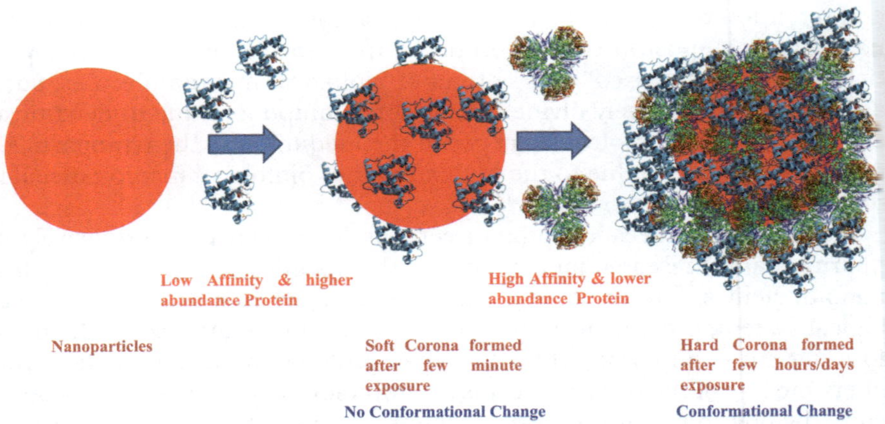

Figure 5.1 Schematic representation of protein corona–NP complex formation.

5.2.1 Hard Corona Formation

The hard corona is formed when proteins adsorb with a high affinity on the surface of nanoparticles (NPs) and consists of firmly bound proteins that do not easily desorb. Aggarwal *et al.* have reported that highly abundant proteins such as albumin, immunoglobulin G (IgG), fibrinogen and apolipoproteins found in blood plasma are present in the coronas of NPs.[13] Over time, these proteins might be substituted by lower concentrations of proteins with a higher affinity to the surface of NPs. Lundqvist *et al.* have studied the hard corona that formed around various types of NPs including copolymer and polystyrene NPs with different size and surface properties.[4] Lundqvist *et al.* studied the incubation of the NPs with plasma and, thereafter, transferred the nanoparticle–plasma complexes to their corresponding hard protein coronas into cytosolic fluid.[4] In a subsequent incubation, the hard protein corona was determined and compared to the incubated fluids (plasma and cytosolic fluid) separately. In this study, three different NPs (9 nm silica, 50 nm polystyrene, and 50 nm carboxyl-modified polystyrene particles) were incubated with either human plasma, cytosolic fluid, or in plasma followed by cytosolic fluid and hard corona.[4] The results depicted that the hard corona developed significantly in the second biological fluids but the final corona holds a "fingerprint" of its history.[15] With the help of this report, a transport pathways map could be developed, which is exploited by NPs and ultimately helps to foresee the fate and behavior of NPs in the body.

"Ex-particle" and "on particle" are two approaches that can be used to analyze the composition of a hard corona.[3,22] But it is debatable which one is the best choice of approach for analysis. According to some of the literature, in the case of an ex-particle approach where proteins are identified by 2D-PAGE, the chances of sample contamination by albumin proteins present in plasma are high. Moreover, in the 2D-PAGE method, proteins are identified on the basis of a spot position.[22] In this method, spot positions are compared with a standard reference map, which may lead to misleading results if two spots overlap.[3] Apart from this limitation, the ex-particle approach provides better sample fractionation as well as multiple techniques – electrophoresis and tandem-mass spectroscopy analysis – which may help to identify various proteins.[3,23] Therefore, the type of approach used to analyze a hard corona is not a limitation, but obtaining a concrete result independently from the selected approach is the major limitation.[3] This limitation could be resolved using multiple combinations such as electrophoresis combined either with MS or a chromatography technique.[3,22,23]

5.2.2 Soft Corona Formation

When proteins have a weak interaction with the corona or are loosely bound to the surface of NPs, this is known as the soft corona. For example, with PEGylated NPs in which the PEG is weakly bound to the NP surface, no hard corona is observed.[24] A challenge in protein corona formation is unravelling how certain proteins competitively adsorb on the NPs' surface. Though there are many clues that exhibit that it is a collective process and thus it is difficult to explain protein-binding studies on an individual basis.[15]

5.3 Analytical Methods for Protein Corona Assessment

5.3.1 On the Basis of Protein Isolation/Separation

The protein–NP interaction is a dynamic process[25,26] during the formation of a protein corona, which is facilitated by the binding affinities and protein equilibrium constants to the respective NP surface, and composition variations in the protein of a biological matrix.[3,5] Also, the nature of the corona can affect the fate, uptake, and performance of NPs in a biological context.[9–11] Indeed, it may lead to stable NPs with less toxicity, an efficient cellular uptake of NPs or a better *in vivo* response.[27–31] Hence, the analysis of protein corona formation is a subject of interest.[32–35] Analysis of the protein–NP interaction is also essential for nano-based therapeutic drugs that are being used in biomedical applications.[36] Proteins that are part of a nanoparticle protein corona are being isolated and identified in order to understand the bio-reactivity of specific NPs.[37] Concerning analysis, the adsorbed proteins can be analyzed by various techniques. Numerous methods are available that can be used to study NP–protein interactions, but for experimental analysis, these methods are classified as (1) direct methods and (2) indirect methods.[5] Amalgamation of these methods provides a more complete picture of protein–nanoparticle interactions, and the proteins' structure and conformation upon their interactions with the nanoparticles.[37,38]

5.3.1.1 Direct Methods

In the direct method, various techniques, such as transmission electron microscopy (TEM),[32] inductively coupled plasma mass spectroscopy (ICP-MS),[39] gel electrophoresis (GE), liquid chromatography–mass spectrometry (LC-MS),[40,41] circular dichroism (CD),[42] nuclear magnetic resonance (NMR) spectroscopy,[43] Fourier-transform infrared spectrometry (FT-IR)[44] and surface enhanced Raman scattering (SERS),[45,46] can be applied (Figure 5.2).

Methods directly involved in the analysis of proteins adsorbed on the NPs'/NMs' surface are known as direct methods.[5] The presence of free proteins along with the proteins adsorbed on the NPs'/NMs' surface are the major drawback in the analysis of direct methods where free proteins may interfere with the read-out. Therefore, free proteins have to be removed prior to analysis.[33] These free proteins, which may be lost during the purification step, are referred to as soft coronas.

5.3.1.1.1 Transmission Electron Microscopy. This is a qualitative method of analysis in which adsorbed proteins can be directly detected and visualized after negative staining. For quantitative analysis of proteins, standard biochemical assays like the Bradford or bicinchoninic acid assay (BCA) can be applied.[27,47] However, this technique requires sample preparation to understand the effect on the morphology of NP–PC complexes. In addition, counterstaining is also required, since the small size of the NPs and the thin protein layer may provide insufficient contrast.

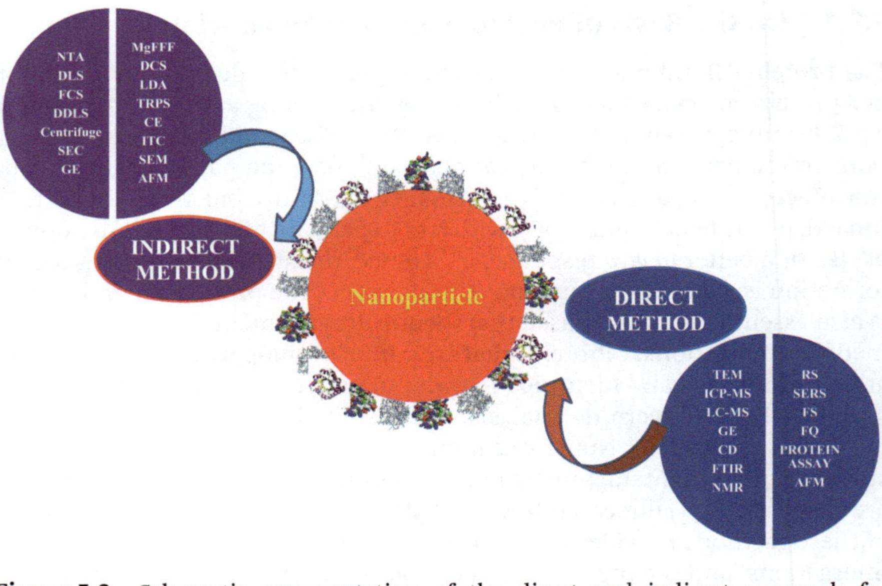

Figure 5.2 Schematic representation of the direct and indirect approach for protein corona analysis.

5.3.1.1.2 Inductively Coupled Plasma Mass Spectrometry (ICP-MS). This technique has to be used when proteins are adsorbed onto the surface of inorganic NPs. This technique allows the calculation of the total sulfur content (comes from the cysteine residue) present in the proteins as well as calculate the total metal content coming from the inorganic NPs.[39] In this fashion, ICP-MS not only quantitatively assesses the NPs but also, with the protein quantification assays, can provide the number of proteins/NPs.[48]

5.3.1.1.3 Gel Electrophoresis (GE). Gel electrophoresis (GE) is a direct method to identify the composition of protein coronas using protein molecular weight (Mw) as a standard marker. Sodium dodecyl sulfate polyacrylamide gel electrophoresis (SDS-PAGE) is the most commonly used gel electrophoresis system in which proteins are denatured and disulfide bonds are partially digested prior to the electrophoresis step.[49–52] The protein corona is a complex mixture of proteins similar to the proteins present in serum or plasma and GE can provide only qualitative information. Thus, this technique is frequently coupled with a MS-based technique where proteins get quantified.[53,54] Selected bands are recovered from gel and can be analyzed by MS-based techniques like tandem mass spectrometry (MS/MS), matrix-assisted laser desorption/ionization time-of-flight mass spectrometry (MALDI-TOF MS), or electrospray ionization mass spectrometry (ESI-MS). The composition of the protein corona can be achieved by comparing the mass spectral peak of the recorded protein fragments with existing databases.[55–57] But, for absolute compositional analysis of proteins, the liquid chromatography–mass spectrometry (LC-MS) technique is indispensable.[40,41] With all these methods, protein purification is a necessary step before analysis, which makes these methods destructive in nature. TEM and GE techniques are capable of identifying the adsorbed proteins but quantification is only possible when MS-based techniques (LC-MS and ICP-MS) are coupled. In addition, MS-based techniques are also not perfect because they cannot analyze the structural changes of adsorbed proteins. However, some other methods are available that can detect structural/conformational changes of proteins when adsorbed onto NPs' surfaces. Circular dichroism (CD) and Fourier-transform infrared spectrometry (FT-IR) are the most common techniques.[5]

5.3.1.1.4 Circular Dichroism (CD) Spectroscopy. CD spectroscopy encompasses absorption in the far (180–250 nm) and near (250 nm–visible) UV range of the spectrum. CD is a technique that provides information on the secondary structure of proteins.[42] So, with the help of this technique, structural/conformational changes in adsorbed proteins can be detected.

5.3.1.1.5 Fourier-transform Infrared Spectrometry (FT-IR). FT-IR offers useful information on protein–nanoparticle interactions. FT-IR is a sensitive and non-destructive technique in which shifts and shape changes of adsorbed proteins in vibrational bands can be easily compared with amide bands present in it. FT-IR has the ability to provide a comprehensive depiction of bonds, their lengths, strengths, angles and conformations, chemical and physical structures, redox states, hydrogen bonding, electric fields, conformational freedom and dynamic motions.[56] FT-IR has also been used to monitor the binding of cysteine-containing proteins from the corona to NPs.[44]

5.3.1.1.6 Nuclear Magnetic Resonance (NMR) Spectroscopy. NMR spectroscopy offers expedient information on protein structure and it is also used for studying NP–protein interactions. NMR spectroscopy provides information with the possibility to distinguish the exact protein domain that is adsorbed upon the NP surface.[38] In general, NMR is a powerful technique, which directly analyses the nanoparticle's formation and its morphology, in both the solid and the solution phase at the molecular scale level.[45] The presence of a complex mixture of proteins in the corona and its high molecular weight make the interpretation of NMR spectra very complex, therefore, this technique is rarely used for protein corona analysis.[43] In NMR, free proteins present in the sample contribute to the signal, which impedes the result, hence, purification of the free proteins is required.

5.3.1.1.7 Raman Spectroscopy. Raman spectroscopy is a very useful technique for studying protein–nanoparticle interactions. It comprises an interaction between light from the laser and a molecule, which results in scattered light. When light from the laser interacts with a sample, the light is scattered and the frequency also shifts as per the molecular vibrational frequency of the molecules present in the sample.[57]

5.3.1.1.8 Surface Enhanced Raman Scattering (SERS). In the case of protein coronas, the proteins are close to the surface of metal NPs so that when SERS is applied, the Raman signal gets amplified by numerous orders of extent. Thus, only adsorbed proteins contribute to the signal, which is contrary to NMR.[45,46] This shows that CD, FT-IR, NMR and SERS are techniques used for protein corona analysis without a purification step.[5,45,46] In other words, these are known as non-destructive methods where there is no need to digest the protein–NP complex prior to protein corona analysis. In addition, these techniques mainly provide information related to structural and conformational changes of the adsorbed proteins. In this order, atomic force microscopy (AFM) is another technique that can unfold the proteins by applying mechanical force and in this way, structural changes of elements can be assessed.[5] But, at present, no protein corona study has been reported using AFM.

5.3.1.1.9 Fluorescence Spectroscopy. Fluorescence spectroscopy is a technique that can be applied to study the size, shape, structure and dynamics of proteins. It is demonstrated to be very useful for the study of the structure and conformation of proteins.[56] In this method, definite wavelengths of light interact with the sample and excite the electrons present in the molecules of the sample, which results in an emission of a certain wavelength of light.[57,58]

Fluorescence quenching (FQ) is a key factor that regulates the accuracy of fluorescence spectroscopy.[32] This factor is widely used because of its sensitivity, reproducibility and appropriateness. In general, quenching is a decrease in the intensity of fluorescence when the excited state of a fluorophore interacts with its surroundings. So, we can say that fluorescence quenching (FQ) could play an important role in protein corona formation.[59] In general, it is also well known that proteins have a fluorescence property so it could be another parameter for protein analysis. When proteins are adsorbed on a NP's surface, it leads to quenching of the native fluorescence of the proteins (*i.e.*, proteins containing tryptophan, tyrosine and phenylalanine). In this case, when proteins are adsorbed upon the NP's surface, tryptophan, tyrosine and phenylalanine residues easily come into contact with the metallic surface of the NPs and get quenched.[60,61] Smaller NPs are more efficient in quenching compared to larger NPs. In principle, changes in wavelength intensity (λ_{max}) in the continuous fluorescence emission intensity spectrum can help to assess the conformational changes in proteins.[56]

From here, we postulate that fluorescence quenching (FQ) could be another technique through which adsorbed proteins are analyzed.[60,61] For adequate quenching, the proteins (fluorescent parts) have to come close to the vicinity of the NP's surface (inorganic part) and thus polymer shells around the NP cores may significantly reduce quenching. However, the FQ technique has its own limitations like overestimation, which is caused by light absorption or scattering.[5]

5.3.1.2 *Indirect Methods*

There are multiple analytical techniques available that are frequently used for NP characterization and are now being extended to indirectly scrutinize the protein coronas.[5] With the help of these indirect techniques, protein coronas can be analyzed *via* measuring changes in their size, charge, density, mass, absorbance and fluorescence.

There are numerous techniques available that can be used to measure the sizes of NPs upon adsorption of proteins. Techniques to measure the sizes of NPs are TEM,[32] scanning electron microscopy (SEM),[62] AFM,[62] nanoparticle tracking analysis (NTA),[63] dynamic light scattering (DLS),[64] fluorescence correlation spectroscopy (FCS),[65] NMR,[43] differential centrifugal sedimentation (DCS)[66] and GE[32] (Figure 5.2).

These techniques provide information on the thickness of the protein layer by comparing the hydrodynamic diameter of NPs before and after corona formation. The only assumption is that the NPs are not agglomerated. When proteins are adsorbed onto the NPs' surfaces, the hydrodynamic size of the NPs increases, which can be quantified by TEM after negative staining,[32] SEM, or AFM.[62]

5.3.1.2.1 Nanoparticle Tracking Analysis (NTA). When proteins are adsorbed onto NPs' surfaces, their mass is increased and because of this, the Brownian motion of the NPs slows down. This change in Brownian motion can be quantified by the NTA technique, which is based on the movement of individual NPs traced by light scattering.[63]

5.3.1.2.2 Dynamic Light Scattering (DLS). DLS is a frequently used technique for the study of NP–protein interactions.[10,64,67] However, DLS is not appropriate when NPs are of a size below 10 nm, as unbound proteins may interfere with the measurements. Unbound proteins can be eliminated by separation methods prior to DLS analysis, which results in the desorption of bound proteins to solutions towards the formation of a new equilibrium.[33]

5.3.1.2.3 Fluorescence Correlation Spectroscopy (FCS). FCS is a technique based on fluorescence labeling of the NP in which the signal originates only from the fluorescent NPs. In FCS, protein corona analysis can be implemented *in situ* without eliminating unbound proteins.[65] Thus, this technique also allows kinetic studies.[68]

For high accuracy in NPs' size measurement, dual focus FCS (2fFCS) is an upgraded alternative technique that includes absolute calibrated standards.[69] FCS can be applied to wide range of proteins and NP surfaces.[32,69,70] FCS can measure the dissociation coefficient of a protein corona as a quantitative parameter.[33] For example, in a recent study, FCS had shown a very low affinity of the adsorption of plasma proteins (HSA, ApoE3, and C3) to lipid-coated quantum dots, specifically weaker than for quantum dots coated with dihydrolipoic acid.[71] Interestingly, human blood serum, which contains various types of proteins, also formed a corona from it and, remarkably, this corona consisted of a mono-layer as a result of the FCS being applied to blood serum.[72]

To achieve absolute data on the hydrodynamic diameters of NP–protein complexes, DLS and FCS are the most appropriate techniques. In both of these techniques, the diffusion coefficients of the NP–protein complexes can be derived from the diffusion of the NPs in/out and fluctuation of the readout signal that is fitted with an autocorrelation function.[73,74] This diffusion coefficient then helps to determine the hydrodynamic diameter of the NPs using the Stokes–Einstein equation.[73]

Regarding the principle of analysis, DLS and FCS are closely related to each other. DLS measures fluctuations in scattered light, whereas FCS measures fluctuations in fluorescence emitted from the NPs. The major drawback of both techniques is that they are not appropriate for larger NPs (>100 nm), because when proteins are adsorbed onto larger NP surfaces, their relative size is increased so much that it will fall within the experimental error of measurement.

5.3.1.2.4 Depolarized DLS (DDLS). This is an alternative DLS technique that is applied to the study of protein coronas.[75] This technique depends on the optical anisotropy of NPs that are inhomogeneous polycrystalline structures and are non-spherical in shape.[75] With this optical anisotropy coupled with localized surface plasmon resonance (LSPR), upon scattering, a depolarized dot pattern is formed whose sequential fluctuations provide accurate information on the NPs' size. On the contrary, depolarized scattering from the biological matrix is almost undetectable.

In addition, a non-optical technique like NMR may also be applied to determine diffusion rate and thus hydrodynamic diameter for protein corona studies.[76,77] Changes in the hydrodynamic diameters of NPs can also be implied by the movement of NPs in matrices.

5.3.1.2.5 Centrifugation. Centrifugation is the most commonly used method for separation. It is based on the gradient of density. In case of protein coronas, this method separates out nanomaterials according to their density as compared to free proteins.[4,21,25,78,79] It is a simple, fast and efficient method that helps to retrieve sufficient proteins for their identification through mass spectrometry where the quality of identification is dependent on the amount of samples.

5.3.1.2.6 Size Exclusion Chromatography (SEC). SEC is an alternative method to centrifugation in which a protein corona is separated in a less disturbing manner as compared to centrifugation. In this method, unbound proteins are separated from NP–PC complexes *via* a column containing a porous stationary phase. In this method, when NP–PC complexes are bigger than the pore size of the stationary phase, they do not infiltrate into the pores and easily come out from the column before the unbound or free proteins, which can enter the pore and persist for longer in the column before being released. In this separation method, unbound or weakly bound proteins may also be retrieved.[6,7,12]

5.3.1.2.7 Gel Electrophoresis (GE). GE can perform qualitative measurements of adsorbed proteins on NP surfaces with the help of size exclusion chromatography (SEC) in which sorting of protein–NP complexes can be done on the basis of size.[80] Normally, in GE, charged particles can migrate through the pores of the gel when an electric field is applied. When proteins are adsorbed on a NP's surface, the complex size increases and migration through pores is driven by gravity. This means that smaller complexes move faster on the gel compared to larger ones that can be sorted by SEC.

5.3.1.2.8 Magnetic Separation/Magnetic Flow Field Fractionation (MgFFF). This is a chromatography-like method in which elution of magnetic NPs is carried out in a single liquid phase. In this method, a magnetic field is applied perpendicular to the direction of the sample flow, which drifts *via* a thin and empty channel.[81] In this method, proteins are screened through their distinct exchange kinetics in the corona around NPs. This method also provides separation in non-equilibrium conditions in which protein–NP complexes are continuously dissociated inside the column. In this way, dissociated proteins are frequently carried away from the complex *via* a protein-free mobile phase.[82]

5.3.1.2.9 Differential Centrifugal Sedimentation (DCS). DCS is a widely used technique to measure the quantity of adsorbed proteins on a NP's surface. But simultaneously, it usually misjudges the thickness of the protein corona. In principal, DCS estimates the sedimentation time of NPs using a density gradient exposed to a centrifugal force.[66,83] The sedimentation time is a function of the NPs' size and density and, in this way, the mean density of NPs and their hydrodynamic diameter can be determined.

5.3.1.2.10 Laser Doppler Anemometry (LDA). LDA is a technique based on changes in the surface charges of the NPs. In LDA, when NPs are exposed to a laser beam, they move towards the oppositely charged electrode of an applied electric field and this drift in velocity helps us to measure the Z-potential of NPs.[84,85] In the case of protein coronas, protein adsorption may make the Z-potential vary, which is based on the initial charge of the NPs, the characteristics of the proteins, their isoelectric point and the pH of the medium. LDA may provide a quantitative assessment of adsorbed proteins.[10,49]

5.3.1.2.11 Tunable Resistive Pulse Sensing (TRPS). TRPS is another technique that measures the Z-potential.[86,87] In this technique, the surface charges of NPs are measured by mapping the "blockade" signals in electric current when the NPs pass through the pore of a membrane. The amplitude of the blockade signal, that is the surface charge of a NP, is directly proportional to the amount of protein adsorbed on the NP's surface. This technique has been effectively used to visualize the formation of protein coronas on NPs placed in serum/plasma.[87]

5.3.1.2.12 Capillary Electrophoresis (CE). CE is a technique that operates in an open column without any packing materials.[32,88] It measures the electrophoretic mobility of the NPs, which is correlated with the charge of the particles. This technique helps to quantify affinities between individual proteins and NPs.

As a function of the protein concentration, comparative changes in the retention of the NP–protein complex can be a parameter for protein binding to NPs. It can be fitted with appropriate mathematical Hill models[33] for finding KD (the dissociation constant).[89] Conventional CE with UV detection is not a very sensitive technique but the sensitivity is increased significantly with laser induced fluorescence (LIF) or MS detection.

5.3.1.2.13 Isothermal Titration Calorimetry (ITC). ITC is basically used to determine various thermodynamic parameters such as binding affinity constant, stoichiometry, and enthalpy changes, which play a key role in protein corona formation.[90,91] ITC helps to measure the heat of a NP solution upon injections of protein aliquots in a time-dependent manner and, over a period of time, provides incremental heat as a function of the molar ratio between proteins and NPs. However, these thermal exchanges may occur due to either NP–protein complex formation or protein denaturation events. Moreover, the major limitation of ITC is that it requires high sample concentrations for analysis.

5.3.2 On the Basis of Characterization Parameters

Protein coronas can be analyzed on the basis of different characterization parameters. To elucidate these parameters, different analytical methods can be used (Figure 5.3). These parameters through which protein coronas are characterized/analyzed can be classified as follows:

(i) Structure (thickness)
(ii) Protein quantification
(iii) Protein affinity and stoichiometry
(iv) Protein conformation
(v) Composition

5.3.2.1 Structure Analysis of Protein Coronas

5.3.2.1.1 Dynamic Light Scattering (DLS). DLS is a technique that allows the measurement of the hydrodynamic diameters of colloidal particles. Hence, this technique is useful to measure the diameters of NPs after they have been incubated in a biological environment (liquid phase).[25] Many studies have reported that the DLS technique is useful to assess the degree of protein corona formation by correlating it with the increase in the hydrodynamic diameters of NPs exposed to serum or plasma. In this case, when NPs are exposed to serum or plasma, the proteins in the serum or plasma can have binding affinities with the NPs' surface and form a corona around the NPs' surface and, hence, the effective hydrodynamic diameter of the NPs has increased.[21,79,92] DLS can be used in both *in situ* and *ex situ* approaches in protein corona analysis, which is a major advantage of this technique. However, to obtain reliable results, DLS requires a mono-disperse sample solution of the NP–PC complexes, which are spherical in shape as these could affect the hydrodynamic diameter measurement.

5.3.2.1.2 Differential Centrifugal Sedimentation (DCS). DCS is a separation technique based on the size and density of the component present in a mixture. In this technique, large and dense objects sediment easily at a low centrifugal force. This technique applies gradient centrifugation in which samples are repeatedly centrifuged followed by discarding of the pellet and this is repeated with an increased centrifugal force.[21,79]

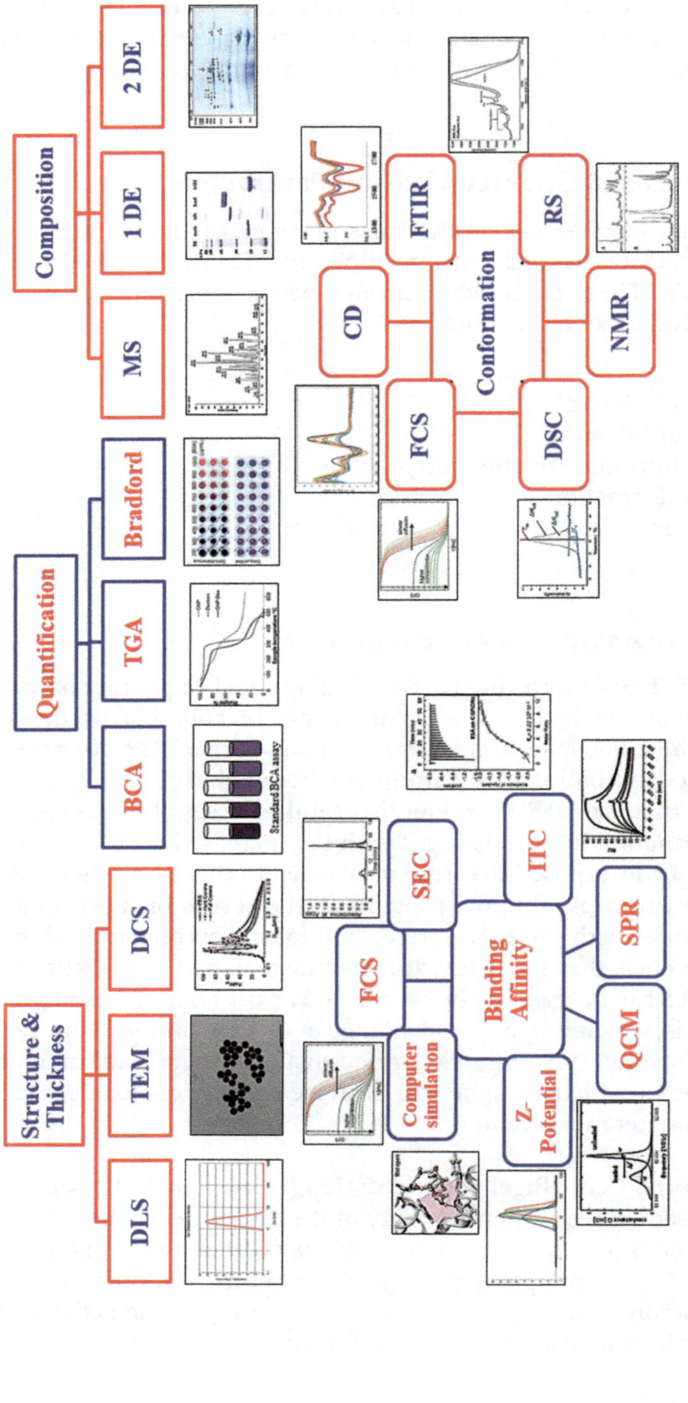

Figure 5.3 Schematic representation of the analytical methods used for the characterization of protein coronas.

5.3.2.1.3 Transmission Electron Microscopy (TEM). In the case of protein coronas, TEM is a technique that can help to determine the thickness of a protein corona present upon NPs. Sample preparation is a crucial step in this technique, which may affect the morphology of the NP–PC complex.[79] In addition, since the NPs and the protein corona layer may not be distinguished easily, counterstaining is needed.[93]

5.3.2.2 Protein Quantification

5.3.2.2.1 Bicinchoninic Acid (BCA) Assay. In this assay, peptide bonds present in proteins reduce Cu^{2+} to Cu^{1+} under alkaline conditions and form a BCA–Cu^{1+} purple complex (λ_{max} = 562 nm).[94] In the case of protein coronas, the BCA assay is applied to quantify the total adsorbed proteins on the NPs' surface.[95–97] The major advantage of this assay is that a small amount of sample is required and its compatibility with various reagents and buffers. However, this assay is time consuming and expensive compared to other protein quantification methods like the Bradford assay.[94]

5.3.2.2.2 Bradford Assay. This is also a colorimetric assay to detect proteins in samples. In this assay, Coomassie brilliant dye binds with proteins and forms a complex with a colorimetric change in the solution from red (λ_{max} = 465 nm) to blue (λ_{max} = 595 nm), due to a shift in the absorption peak.[94] This method is fast, sensitive and requires a small amount of sample for analysis. In addition, it is a less expensive method for protein quantification compared to others.

5.3.2.2.3 Thermogravimetric Analysis (TGA). In general, due to the thermal decomposition reaction, weight variations occur in organic or semi-organic materials that can be measured by this technique. With this principle, the weight of a protein adsorbed onto a NP's surface can be measured by this technique after the decomposition reaction.[98]

5.3.2.3 Binding Affinity/Stoichiometry and Protein Interaction

5.3.2.3.1 Fluorescence Correlation Spectroscopy (FCS). FCS is a technique based on fluorescence labeling of the NP in which the signal originates only from the fluorescent NPs. It measures the fluctuations in fluorescence emitted from the NPs. In FCS, protein corona analysis can be implemented *in situ* without eliminating unbound proteins.[66] Thus, this technique also allows kinetic studies.[68] This technique provides information on the adsorption affinities of protein molecules on NP surfaces by measuring binding curves between exposed NPs and different concentrations of protein.[99]

5.3.2.3.2 Size Exclusion Chromatography (SEC). This technique can also assist in measuring the affinity of proteins to bind to NP surfaces. In general, this technique can separate the components on the basis of their size.

Size-based separation depends mainly on the size and geometry of the pores, which restrict the access of the molecules in accordance with their Stokes radii.[100] The large molecules elute easily because they are excluded from the pores. In the case of protein coronas, proteins with a high binding affinity to the NPs' surface elute frequently with NPs; however, proteins with a low affinity will elute without the NPs.[6]

5.3.2.3.3 Isothermal Titration Calorimetry (ITC). ITC is an appropriate method for stoichiometric analysis in which the affinity and enthalpy of protein–NP interactions can be measured. In this method, changes in enthalpy are measured at a constant temperature between sample and reference cells Differences in enthalpy occur due to the addition of proteins to a NP suspension in the sample cell. This method also provides information about the number of bound proteins per NP, if both the concentrations of NPs and added proteins are known.[7]

5.3.2.3.4 Surface Plasmon Resonance (SPR). This method is used to measure adsorption kinetics. In this method, NPs are attached to the gold surface of a sensor chip and proteins are introduced to flow over the modified surface. When the additional molecules are adsorbed onto the NPs' surface, the oscillation of the surface plasmon waves changes, which is measured by SPR.[7,101,102]

5.3.2.3.5 Quartz Crystal Microbalance (QCM). This method can be used to measure changes in mass during protein corona formation. In this method, it is essential that the proteins or NPs are static on top of a gold surface located on a quartz crystal. A binding partner is introduced into the sample chamber, which passes through the quartz surface and, concurrently, frequency is measured. In this way, under real-time conditions, protein–NP binding profiles are quantitatively measured. In addition, association and dissociation constants can also be measured using the Langmuir adsorption isotherm.[103]

5.3.2.3.6 Z-potential Measurement. Zeta potential is the charge that develops at the interface between two different media. In other words, zeta potential is a function of the surface charge of the particle. In the case of protein corona formation, when proteins adsorb onto the surface of NPs, the surface property of the NPs changes or we can say that the zeta potential of the protein–NP complex is different from the zeta potential of the NPs alone. So, it is interpreted that changes in zeta potential could be a good approach to screen protein–NP interactions.[25]

5.3.2.3.7 Computer Simulation. *In silico*, or computer simulation, is another possible approach to analyze protein–NP interactions. It is a predictive approach in which, without lab work, a protein corona can be characterized using different software. In addition, it also provides information about protein orientation and conformation in three-dimensional view as well as sequential determination. In the case of protein corona formation, this method applies surface ligand structure, and surface curvature as a function to study the adsorption of protein onto a NP's surface.[104]

5.3.2.4 Protein Conformation

5.3.2.4.1 Circular Dichroism (CD) Spectroscopy. CD spectroscopy encompasses absorption in the far (180–250 nm) and near (250 nm–visible) UV range of the spectrum. CD is a technique that provides information about the secondary structure of proteins.[42] So, with the help of this technique, structural/conformational changes in adsorbed proteins can be detected. The drawback with this technique is that it requires high concentrations of the sample as well as its inability to be applied to complex protein mixtures.[101]

5.3.2.4.2 Fourier-transform Infrared (FTIR) Spectroscopy. Similar to CD spectroscopy, FT-IR also provides information on how to measure the conformational changes of proteins. FT-IR is a sensitive and non-destructive technique in which shifts and shape changes of adsorbed proteins in vibrational bands can be easily compared with amide bonds present in it. Among all the amide bonds, amide I is the most sensitive, and it is used to measure protein conformation.[101] FT-IR has the ability to provide a comprehensive depiction of bonds, their lengths, strengths, angles and conformation, chemical and physical structures, redox states, hydrogen bonding, electric fields, conformational freedom and dynamic motions.[57] FT-IR has also been used to monitor the binding of cysteine-containing proteins from the corona to NPs.[44]

5.3.2.4.3 Raman Spectroscopy (RS). Raman spectroscopy is a very useful technique for studying protein–nanoparticle interactions in aqueous solutions. It comprises the interaction of light from the laser with a molecule that results in scattered light. When light that comes from the laser interacts with a sample, the light is scattered and the frequency also shifts as per the molecular vibrational frequency of the molecules present in the sample. Spectral interpretation of RS is easier as compared to IR spectra. In RS spectra, electron-rich groups or vibrations of double/triple bonds exhibit intense bands as compared to electron-poor groups or vibration of a single bond.[101]

5.3.2.4.4 Nuclear Magnetic Resonance (NMR) Spectroscopy. NMR spectroscopy can provide expedient information on a protein's structure, dynamics, reaction rate and its chemical composition. NMR spectroscopy provides information about the possibility to distinguish the exact protein domain (residue specific structural information) that is adsorbed upon the NP surface.[38] In general, NMR is a powerful technique, which directly analyses the nanoparticle formation and its morphology, in both the solid and the solution phase at the molecular scale level.[45] The presence of a complex mixture of proteins in the corona and its high molecular weight make the interpretation of NMR spectra very complex, therefore, this technique is rarely used for protein corona analysis.[105] In NMR, free proteins present in a sample contribute to the signal, which impedes the result, hence, purification of the free protein is required.

5.3.2.4.5 Differential Scanning Calorimetry (DSC) Spectroscopy. This technique is applied to measure the changes in enthalpy when it is associated with thermal denaturation of a molecule at a constant rate. Thus, with the help of this technique, protein stability can be checked after its absorption onto a NP's surface.[106]

5.3.2.4.6 Fluorescence Correlation Spectrometry (FCS). FCS can provide information about protein conformation. In this method, information about a protein's conformation is obtained by measuring the changes in intensity of the fluorescence emission spectra.[56]

5.3.2.5 Composition

To identify the protein composition, various techniques like gel electrophoresis (sodium dodecyl sulfate–polyacrylamide gel electrophoresis [SDS–PAGE] and two-dimensional gel electrophoresis [2-DE]) and mass spectrometry (MS) can be used only *ex situ*, after isolation of protein–NP complexes from biological fluids.[3]

5.3.2.5.1 One-dimensional Gel Electrophoresis (1-DE or SDS–PAGE). Gel electrophoresis (GE) is a direct method to identify the composition of protein coronas using protein molecular weight (Mw) as a standard marker. Sodium dodecyl sulfate–polyacrylamide gel electrophoresis (SDS-PAGE) is the most commonly used gel electrophoresis system in which proteins are denatured and disulfide bonds are partially digested prior to the electrophoresis step.[49–52] A protein corona is a complex mixture of proteins similar to the proteins present in serum or plasma and GE can provide only qualitative information. Thus, this technique is frequently coupled with a MS-based technique where proteins are quantified.[53,54] Selected bands are recovered from the gel and can be analyzed by MS-based techniques like tandem mass spectrometry (MS/MS), matrix-assisted laser desorption/ionization time-of-flight mass spectrometry (MALDI-TOF MS), or electrospray ionization mass spectrometry (ESI-MS). The composition of the protein corona can be achieved by comparing the mass spectral peak of the recorded protein fragments with existing databases.[55,58,72] But, for absolute compositional analysis of proteins, the liquid chromatography–mass spectrometry (LC-MS) technique is indispensable.[40,41]

5.3.2.5.2 Two-dimensional Gel Electrophoresis (2-DE). In this technique, the protein separates in two steps. Isoelectric focusing (IEF) is the first step of this technique in which proteins are separated according to their isoelectric point. In the second step, proteins are fractionated according to their molecular weights. Thereafter, protein bands are visualized using a staining method and quantified using imaging software.[107] Protein identification is also possible using this technique, when a 2D gel is compared to a 2D map of proteins.[95,96,107]

5.3.2.5.3 Mass Spectrometry (MS). MS is the most commonly used technique for protein identification.[9,108] In this technique, proteins are required to be broken down into smaller peptides using a proteolytic enzyme, to decrease the size of the analytes to make them more appropriate in terms of mass range for the instrument. Thereafter, the small peptides are ionized using the ion source and brought into the high vacuum region of the instrument. The peptide ions are separated according to their mass-to-charge ratio under the strong electromagnetic field in a long drift tube. As a result, the mass spectra depict the primary sequence of the peptide present in the sample mixture. These data are then compared with a reference database to identify the protein.[101] In addition, with the help of this protocol, MS can be useful to identify proteins in coronas using gel and non-gel-based methods.

Gel-based Method. This method is most commonly used to measure the protein arrangement located in the coronas around NPs.[4,24] In this method, the protein is first separated by SDS-PAGE, then the gel is cut with the bands of protein and incubated with a proteolytic enzyme. As a result, the proteins are broken down into small peptides, which are then analyzed by MS.

Non-gel-based Method. In this method, proteins are first incubated with a proteolytic enzyme and, as a result, small peptides are formed that can be analyzed directly by MS. Before the action of the proteolytic enzyme, protein denaturation is an essential step that makes the protein domain open for enzyme activity.[101]

In summary, prior to MS analysis, formation of peptides and their separation are crucial steps in both methods.

5.4 Other Recent Techniques

The chemical nose is the latest promising method that has been developed to quantify proteins and their mixtures within human serum using AuNPs.[109] Principal component analysis (PCA) is a method that allows quantification of various proteins simultaneously and provides unique fingerprints of proteins after their interaction with NPs. When plasmonic NPs are incubated with proteins, it leads to a red shift and the plasmon spectra become wider, therefore, in this case, UV–vis absorption spectroscopy is extensively used.

In the case of non-plasmonic NPs, adsorption of proteins can be detected using a surface plasmon resonance (SPR) device, as changes occur in the angular positions of the SPR spectra. Various proteins can also be detected using a multiplexed SPR method in which microchips with multiple microfluidic channels have been applied for analysis.[110] In this order, second harmonic light scattering (SHLS) is another powerful technique to quantify weakly interacting proteins, which cannot be simply sensed by conventional techniques.[111] The efficiency of SHLS improved significantly when the fundamental or harmonic wavelengths were pitched near their confined surface plasmon resonance. In SHLS, thermodynamic parameters like free energy adsorption and number of protein molecules adsorbed on the NP surface are measured from the changes in SHLS signal using a modified Langmuir model. In addition, during protein corona formation, changes in mass also occur that can be measured by quartz crystal microbalance (QCM) and suspended microchannel resonator (SMR) methods. In QCM, it is essential that the NPs are static on top of an oscillating quartz surface, which further measures changes in mass (increase) as proteins adsorb.[62] Likewise, SMR measures the mass of a NP suspension flushed through a microchannel inside the resonator.[62] Sensitivity is a major advantage of these techniques, which ranges from femtogram to attogram. Changes in frequency and, therefore, alteration in mass are allied with total loading mass (including water and ions) as well as viscoelastic contributions from the adsorbing protein, which make the quantitative analysis difficult. In addition, Carril *et al.* (2017) have presented a methodology for determining protein corona formation in complex media.[112] Nanoparticles were labeled with ^{19}F and their diffusion coefficient measured using ^{19}F diffusion-ordered nuclear magnetic resonance (NMR) spectroscopy. ^{19}F diffusion NMR measurements of hydrodynamic radii allow for *in situ* characterization of NPs in complex environments by quantification of protein adsorption to the surface of NPs, as determined by an increase in hydrodynamic radius.

5.5 Protein-corona-enabled Biosensors

Some cancers, like pancreatic cancer, are often diagnosed at an advanced stage. Thus, developing new methods for the early detection of pancreatic cancer is an urgent task. Caputo *et al.* (2017) developed a new pancreatic cancer diagnostic technology based on nano-bio-interactions by taking blood samples from 20 pancreatic cancer patients, and five patients without malignancy were left to interact with designed lipid nanoparticles leading to the formation of a hard "protein corona" at the nanoparticle surface.[113] The isolated protein patterns were characterized by sodium dodecyl sulfate–polyacrylamide gel electrophoresis (SDS–PAGE). Caputo *et al.* (2017) found that the protein coronas of the pancreatic cancer patients were much more enriched than those of the healthy individuals. Statistical analysis of the SDS-PAGE results shows a clear distinction between healthy and pancreatic cancer patients with a total discriminate correctness rate of 88%.[113]

This work would further validate the potential of protein coronas for preventative diagnostics, which open up new avenues in terms of treatment and outcomes.

5.6 Future Perspective

Nowadays, colloidal nanoparticles (NPs) are used as a multipurpose platform for *in vivo* nanomedicine.[112] When NPs enter the blood circulation, they undergo radical changes, such as the formation of a protein corona.[24] The protein corona represents one of the most promising tools for drug-targeting and drug-delivery.[3] However, a deeper understanding of the complex dynamics that happen after their *in vivo* administration is required.[3] Particularly, plasma proteins tend to form a new surface with NPs. The high exposure of this surface to NPs have a strong impact on NP bio-distribution, targeting efficacy and also toxicity.[36] The protein corona consists of two poorly delimited layers, known as the hard corona and soft corona, which are affected by the complexity of the environment and the protein–surface equilibrium formed during *in vivo* blood circulation.[3] The hard corona is formed by proteins strongly associated to the NPs, while the soft corona is an outer layer consisting of loosely bound proteins. As has been seen, several studies attempted to investigate the hard corona, which can be easily isolated, but shows poor reproducibility, owing to variation in experimental conditions. As a result, full mapping of the hard corona for different NPs is still lacking. Moreover, the current knowledge on the soft corona, which plays a significant role in the initial interaction of NPs once *in vivo*, is also very limited, mainly due to the difficulties in preserving it after purification. Therefore, multi-disciplinary approaches leading to the obtainment of a major amount of information about the protein corona and its properties are strongly needed for a full understanding.

Nowadays, nanoparticles are not only formed chemically, but also are associated with biological molecules such as proteins, which are stably or weakly adsorbed onto their surfaces.[3] Nanoparticles and their PCs are new "biological entities". These interactions also strongly impact on their safety and functionality performances more than the composition of the nanoparticles, size, and shapes. Since this concept is relatively new, a number of issues need to be addressed:

1. *Physiological response of a nanomedicine.*
2. *Technologies' viability and reliability to discriminate and describe the HC, SC, or both.*
3. *Skills to understand the impact of the PC on the destiny of nanomedicines.*

To date, many of these issues are almost completely unsolved, but some indications and future directions could be hypothesized. Firstly, it is reliable that the whole protein corona influences the bio-distribution of protein corona–NP complexes, but it could be hypothesized that some specific NP accumulation could be possible due to a selective interaction of a subset of the associated PC with specific cells or receptors.[21] Another important lack in protein corona research is connected with technologies. PC–NP complexes could be characterized by integrating information on morphology such as imaging-spectroscopy-scattering-based techniques and on structure/composition of the PC such as cryo-electron microscopy and protein crystallography. Thus, multi-disciplinary or trans-disciplinary approaches are needed in order to obtain much more information about the protein corona and its properties to fully understand the real impact of the protein corona on nanomedicines, and, therefore, to better support a more safety and conscious application of nanotechnology in medicine. Similarly, the methodology needs to be developed in a turbid environment such as in the presence of cells. While several obstacles need to be solved, the presented methodology, in principle, may provide an opportunity to measure the hydrodynamic radii of NPs *in vivo*. Realizing this issue, it would be a challenge and opportunity toward understanding PC–NP complexes and what is going on *in vivo* after NPs have been administered.

The protein corona–NP complexes also have the potential for preventative diagnostics research, which will open new avenues in terms of treatment and outcomes.

5.7 Conclusions

The protein corona has a complex composition as well as arrangement (composition – whether it forms a soft corona or a hard corona; arrangement – whether it adsorbs in a monolayer or in multilayers). Therefore, using only one specific technique is not sufficient for analysis, but a combination of multiple techniques can be used for this purpose. For a better understanding of how to analyze the various aspects of protein coronas, the use of complementary characterization techniques plays a vital role. Because all the characterization techniques are based on different types of principles, noticeable differences can be observed when they are applied. Thus, only multi-disciplinary approaches will provide the maximum amount of information about the protein corona.

Acknowledgements

RKS gratefully acknowledges the Science and Engineering Research Board (SERB), Department of Science and Technology, Government of India for providing a grant under Early Career Research Award (ECR/2016/001310) scheme. RKS also acknowledges the Financial Assistance received from the Department of Biotechnology, Government of India under the project "NanoToF: Toxicological evaluation and risk assessment on Nanomaterials in Food" (grant number BT/PR10414/PFN/20/961/2014). AP gratefully acknowledges the funding from SERB, India under Young Scientist Start-up Grant (YSS/2015/000258).

References

1. M. Elsabahy and K. L. Wooley, Design of polymeric nanoparticles for biomedical delivery applications, *Chem. Soc. Rev.*, 2012, **41**, 2545–2561.
2. S. Harper, C. Usenko, J. E. Hutchison, B. L. S. Maddux and R. L. Tanguay, *In vivo* biodistribution and toxicity depends on nanomaterial composition, size, surface functionalisation and route of exposure, *J. Exp. Nanosci.*, 2008, **3**, 195–206.
3. F. Pederzoli, G. Tosi, M. A. Vandelli, D. Belletti, F. Forni and B. Ruozi, Protein corona and nanoparticles: how can we investigate on? *Wiley Interdiscip. Rev.: Nanomed. Nanobiotechnol.*, 2017, **9**, e1467.
4. M. Lundqvist, J. Stigler, G. Elia, I. Lynch, T. Cedervall and K. A. Dawson, Nanoparticle size and surface properties determine the protein corona with possible implications for biological impacts, *Proc. Natl. Acad. Sci. U. S. A.*, 2008, **105**, 14265–14270.
5. C. Carrillo-Carrion, M. Carril and W. J. Parak, Techniques for the experimental investigation of the protein corona, *Curr. Opin. Biotechnol.*, 2017, **46**, 106–113.
6. I. Lynch, T. Cedervall, M. Lundqvist, C. Cabaleiro-Lago, S. Linse and K. A. Dawson, The nanoparticle-protein complex as a biological entity; a complex fluids and surface science challenge for the 21st century, *Adv. Colloid Interface Sci.*, 2007, **134–135**, 167–174.
7. T. Cedervall, I. Lynch, S. Lindman, T. Berggard, E. Thulin, H. Nilsson, K. A. Dawson and S. Linse, Understanding the nanoparticle-protein corona using methods to quantify exchange rates and affinities of proteins for nanoparticles, *Proc. Natl. Acad. Sci. U. S. A.*, 2007, **104**, 2050–2055.
8. C. Ge, J. Tian, Y. Zhao, C. Chen, R. Zhou and Z. Chai, Towards understanding of nanoparticle-protein corona, *Arch. Toxicol.*, 2015, **89**, 519–539.
9. D. Docter, S. Strieth, D. Westmeier, O. Hayden, M. Gao, S. K. Knauer and R. H. Stauber, No king without a crown–impact of the nanomaterial-protein corona on nanobiomedicine, *Nanomedicine*, 2015, **10**, 503–519.
10. A. Sasidharan, J. E. Riviere and N. A. Monteiro-Riviere, Gold and silver nanoparticle interactions with human proteins: impact and implications in biocorona formation, *J. Mater. Chem. B*, 2015, **3**, 2075–2082.
11. C. Corbo, R. Molinaro, A. Parodi, N. E. Toledano Furman, F. Salvatore and E. Tasciotti, The impact of nanoparticle protein corona on cytotoxicity, immunotoxicity and target drug delivery, *Nanomedicine*, 2016, **11**, 81–100.
12. T. Cedervall, I. Lynch, M. Foy, T. Berggard, S. C. Donnelly, G. Cagney, S. Linse and K. A. Dawson, Detailed identification of plasma proteins adsorbed on copolymer nanoparticles, *Angew. Chem., Int. Ed.*, 2007, **46**, 5754–5756.

13. P. Aggarwal, J. B. Hall, C. B. McLeland, M. A. Dobrovolskaia and S. E. McNeil, Nanoparticle interaction with plasma proteins as it relates to particle biodistribution, biocompatibility and therapeutic efficacy, *Adv. Drug Delivery Rev.*, 2009, **61**, 428–437.
14. G. Caracciolo, Liposome-protein corona in a physiological environment: challenges and opportunities for targeted delivery of nanomedicines, *Nanomedicine*, 2015, **11**, 543–557.
15. M. Rahman, S. Laurent, N. Tawil, L. Yahia and M. Mahmoudi, *Protein-nanoparticle Interactions*, Springer Nature Switzerland AG, Cham, 2013.
16. P. Foroozandeh and A. A. Aziz, Merging worlds of nanomaterials and biological environment: factors governing protein corona formation on nanoparticles and its biological consequences, *Nanoscale Res. Lett.*, 2015, **10**, 221.
17. C. Gunawan, M. Lim, C. P. Marquis and R. Amal, Nanoparticle–protein corona complexes govern the biological fates and functions of nanoparticles, *J. Mater. Chem. B*, 2014, **2**, 2060–2083.
18. P. P. Karmali and D. Simberg, Interactions of nanoparticles with plasma proteins: implication on clearance and toxicity of drug delivery systems, *Expert Opin. Drug Delivery*, 2011, **8**, 343–357.
19. A. Kumar, E. M. Bicer, A. B. Morgan, P. E. Pfeffer, M. Monopoli, K. A. Dawson, J. Eriksson, K. Edwards, S. Lynham, M. Arno, A. F. Behndig, A. Blomberg, G. Somers, D. Hassall, L. A. Dailey, B. Forbes and I. S. Mudway, Enrichment of immunoregulatory proteins in the biomolecular corona of nanoparticles within human respiratory tract lining fluid, *Nanomedicine*, 2016, **12**, 1033–1043.
20. J. Wolfram, K. Suri, Y. Yang, J. Shen, C. Celia, M. Fresta, Y. Zhao, H. Shen and M. Ferrari, Shrinkage of pegylated and non-pegylated liposomes in serum, *Colloids Surf., B*, 2014, **114**, 294–300.
21. M. P. Monopoli, D. Walczyk, A. Campbell, G. Elia, I. Lynch, F. B. Bombelli and K. A. Dawson, Physical-chemical aspects of protein corona: relevance to *in vitro* and *in vivo* biological impacts of nanoparticles, *J. Am. Chem. Soc.*, 2011, **133**, 2525–2534.
22. K. Sempf, T. Arrey, S. Gelperina, T. Schorge, B. Meyer, M. Karas and J. Kreuter, Adsorption of plasma proteins on uncoated PLGA nanoparticles, *Eur. J. Pharm. Biopharm.*, 2013, **85**, 53–60.
23. H. R. Kim, K. Andrieux, C. Delomenie, H. Chacun, M. Appel, D. Desmaële, F. Taran, D. Georgin, P. Couvreur and M. Taverna, Analysis of plasma protein adsorption onto PEGylated nanoparticles by complementary methods: 2-DE, CE and Protein Lab-on-chip® system, *Electrophoresis*, 2007, **28**, 2252–2261.
24. M. Lundqvist, J. Stigler, T. Cedervall, T. Berggard, M. B. Flanagan, I. Lynch, G. Elia and K. Dawson, The evolution of the protein corona around nanoparticles: a test study, *ACS Nano*, 2011, **5**, 7503–7509.
25. E. Casals, T. Pfaller, A. Duschl, G. J. Oostingh and V. Puntes, Time evolution of the nanoparticle protein corona, *ACS Nano*, 2010, **4**, 3623–3632.

26. S. Tenzer, D. Docter, J. Kuharev, A. Musyanovych, V. Fetz, R. Hecht, F. Schlenk, D. Fischer, K. Kiouptsi, C. Reinhardt, K. Landfester, H. Schild, M. Maskos, S. K. Knauer and R. H. Stauber, Rapid formation of plasma protein corona critically affects nanoparticle pathophysiology, *Nat. Nanotechnol.*, 2013, **8**, 772–781.
27. A. O. Luby, E. K. Breitner and K. K. Comfort, Preliminary protein corona formation stabilizes gold nanoparticles and improves deposition efficiency, *Appl. Nanosci.*, 2016, **6**, 827–836.
28. S. R. Saptarshi, A. Duschl and A. L. Lopata, Interaction of nanoparticles with proteins: relation to bio-reactivity of the nanoparticle, *J. Nanobiotechnol.*, 2013, **11**, 26.
29. S. Ashraf, J. Park, M. A. Bichelberger, K. Kantner, R. Hartmann, P. Maffre, A. H. Said, N. Feliu, J. Lee and D. Lee, Zwitterionic surface coating of quantum dots reduces protein adsorption and cellular uptake, *Nanoscale*, 2016, **8**, 17794–17800.
30. S. Ritz, S. Schottler, N. Kotman, G. Baier, A. Musyanovych, J. Kuharev, K. Landfester, H. Schild, O. Jahn, S. Tenzer and V. Mailander, Protein corona of nanoparticles: distinct proteins regulate the cellular uptake, *Biomacromolecules*, 2015, **16**, 1311–1321.
31. A. Salvati, A. S. Pitek, M. P. Monopoli, K. Prapainop, F. B. Bombelli, D. R. Hristov, P. M. Kelly, C. Aberg, E. Mahon and K. A. Dawson, Transferrin-functionalized nanoparticles lose their targeting capabilities when a biomolecule corona adsorbs on the surface, *Nat. Nanotechnol.*, 2013, **8**, 137–143.
32. J. Huhn, C. Fedeli, Q. Zhang, A. Masood, P. Del Pino, N. M. Khashab, E. Papini and W. J. Parak, Dissociation coefficients of protein adsorption to nanoparticles as quantitative metrics for description of the protein corona: A comparison of experimental techniques and methodological relevance, *Int. J. Biochem. Cell Biol.*, 2016, **75**, 148–161.
33. P. del Pino, B. Pelaz, Q. Zhang, P. Maffre, G. U. Nienhaus and W. J. Parak, Protein corona formation around nanoparticles – from the past to the future, *Mater. Horiz.*, 2014, **1**, 301–313.
34. S. Zanganeh, R. Spitler, M. Erfanzadeh, A. M. Alkilany and M. Mahmoudi, Protein corona: Opportunities and challenges, *Int. J. Biochem. Cell Biol.*, 2016, **75**, 143–147.
35. B. Kharazian, N. L. Hadipour and M. R. Ejtehadi, Understanding the nanoparticle-protein corona complexes using computational and experimental methods, *Int. J. Biochem. Cell Biol.*, 2016, **75**, 162–174.
36. V. H. Nguyen and B. J. Lee, Protein corona: a new approach for nanomedicine design, *Int. J. Nanomed.*, 2017, **12**, 3137–3151.
37. J. J. Sutkovic, A review on Nanoparticle and Protein interaction in biomedical applications, *Periodicals Eng. Nat. Sci. (PEN)*, 2016, **4**, 34–40.
38. S. R. Saptarshi, A. Duschl and A. L. Lopata, Interaction of nanoparticles with proteins: relation to bio-reactivity of the nanoparticle, *J. Nanobiotechnol.*, 2013, **11**, 26.

39. M. Matczuk, J. Legat, S. N. Shtykov, M. Jarosz and A. R. Timerbaev, Characterization of the protein corona of gold nanoparticles by an advanced treatment of CE-ICP-MS data, *Electrophoresis*, 2016, **37**, 2257–2259.
40. D. Pozzi, G. Caracciolo, L. Digiacomo, V. Colapicchioni, S. Palchetti, A. L. Capriotti, C. Cavaliere, R. Zenezini Chiozzi, A. Puglisi and A. Lagana, The biomolecular corona of nanoparticles in circulating biological media, *Nanoscale*, 2015, **7**, 13958–13966.
41. G. Miotto, M. Magro, M. Terzo, M. Zaccarin, L. Da Dalt, E. Bonaiuto, D. Baratella, G. Gabai and F. Vianello, Protein corona as a proteome fingerprint: The example of hidden biomarkers for cow mastitis, *Colloids Surf., B*, 2016, **140**, 40–49.
42. S. Li, Z. Peng and R. M. Leblanc, Method To Determine Protein Concentration in the Protein-Nanoparticle Conjugates Aqueous Solution Using Circular Dichroism Spectroscopy, *Anal. Chem.*, 2015, **87**, 6455–6459.
43. M. Assfalg, L. Ragona, K. Pagano, M. D'Onofrio, S. Zanzoni, S. Tomaselli and H. Molinari, The study of transient protein-nanoparticle interactions by solution NMR spectroscopy, *Biochim. Biophys. Acta*, 2016, **1864**, 102–114.
44. M. Wang, C. Fu, X. Liu, Z. Lin, N. Yang and S. Yu, Probing the mechanism of plasma protein adsorption on Au and Ag nanoparticles with FT-IR spectroscopy, *Nanoscale*, 2015, **7**, 15191–15196.
45. L. E. Marbella and J. E. Millstone, NMR techniques for noble metal nanoparticles, *Chem. Mater.*, 2015, **27**, 2721–2739.
46. R. A. Alvarez-Puebla, A. Agarwal, P. Manna, B. P. Khanal, P. Aldeanueva-Potel, E. Carbo-Argibay, N. Pazos-Perez, L. Vigderman, E. R. Zubarev, N. A. Kotov and L. M. Liz-Marzan, Gold nanorods 3D-supercrystals as surface enhanced Raman scattering spectroscopy substrates for the rapid detection of scrambled prions, *Proc. Natl. Acad. Sci. U. S. A.*, 2011, **108**, 8157–8161.
47. A. Bigdeli, S. Palchetti, D. Pozzi, M. R. Hormozi-Nezhad, F. Baldelli Bombelli, G. Caracciolo and M. Mahmoudi, Exploring Cellular Interactions of Liposomes Using Protein Corona Fingerprints and Physicochemical Properties, *ACS Nano*, 2016, **10**, 3723–3737.
48. A. Jedlovszky-Hajdu, F. B. Bombelli, M. P. Monopoli, E. Tombacz and K. A. Dawson, Surface coatings shape the protein corona of SPIONs with relevance to their application *in vivo*, *Langmuir*, 2012, **28**, 14983–14991.
49. C. Grafe, A. Weidner, M. V. Luhe, C. Bergemann, F. H. Schacher, J. H. Clement and S. Dutz, Intentional formation of a protein corona on nanoparticles: Serum concentration affects protein corona mass, surface charge, and nanoparticle-cell interaction, *Int. J. Biochem. Cell Biol.*, 2016, **75**, 196–202.
50. N. J. Braun, M. C. DeBrosse, S. M. Hussain and K. K. Comfort, Modification of the protein corona-nanoparticle complex by physiological factors, *Mater. Sci. Eng., C*, 2016, **64**, 34–42.

51. R. Safavi-Sohi, S. Maghari, M. Raoufi, S. A. Jalali, M. J. Hajipour, A. Ghassempour and M. Mahmoudi, Bypassing protein corona issue on active targeting: zwitterionic coatings dictate specific interactions of targeting moieties and cell receptors, *ACS Appl. Mater. Interfaces*, 2016, **8**, 22808–22818.
52. V. Colapicchioni, M. Tilio, L. Digiacomo, V. Gambini, S. Palchetti, C. Marchini, D. Pozzi, S. Occhipinti, A. Amici and G. Caracciolo, Personalized liposome-protein corona in the blood of breast, gastric and pancreatic cancer patients, *Int. J. Biochem. Cell Biol.*, 2016, **75**, 180–187.
53. N. A. Hani and W. Hui-Fen, Proteomics analysis of the mode of antibacterial action of nanoparticles and their interactions with proteins, *TrAC, Trends Anal. Chem.*, 2015, **65**, 30–46.
54. D. J. O'Connell, F. B. Bombelli, A. S. Pitek, M. P. Monopoli, D. J. Cahill and K. A. Dawson, Characterization of the bionano interface and mapping extrinsic interactions of the corona of nanomaterials, *Nanoscale*, 2015, 7, 15268–15276.
55. D. Mbeh, T. Javanbakht, L. Tabet, Y. Merhi, K. Maghni, E. Sacher and L. H. Yahia, Protein corona formation on magnetite nanoparticles: effects of culture medium composition, and its consequences on superparamagnetic nanoparticle cytotoxicity, *J. Biomed. Nanotechnol.*, 2015, **11**, 828–840.
56. S. H. Lacerda, J. J. Park, C. Meuse, D. Pristinski, M. L. Becker, A. Karim and J. F. Douglas, Interaction of gold nanoparticles with common human blood proteins, *ACS Nano*, 2010, **4**, 365–379.
57. T. J. Kamerzell, R. Esfandiary, S. B. Joshi, C. R. Middaugh and D. B. Volkin, Protein-excipient interactions: mechanisms and biophysical characterization applied to protein formulation development, *Adv. Drug Delivery Rev.*, 2011, **63**, 1118–1159.
58. N. Fernandez-Iglesias and J. Bettmer, Complementary mass spectrometric techniques for the quantification of the protein corona: a case study on gold nanoparticles and human serum proteins, *Nanoscale*, 2015, 7, 14324–14331.
59. A. A. Sousa, A Note on the use of Steady-State Fluorescence Quenching to Quantify Nanoparticle-Protein Interactions, *J. Fluoresc.*, 2015, **25**, 1567–1575.
60. C. C. Fleischer and C. K. Payne, Secondary structure of corona proteins determines the cell surface receptors used by nanoparticles, *J. Phys. Chem. B*, 2014, **118**, 14017–14026.
61. F. Ahmad, Y. Zhou, Z. Ling, Q. Xiang and X. Zhou, Systematic elucidation of interactive unfolding and corona formation of bovine serum albumin with cobalt ferrite nanoparticles, *RSC Adv.*, 2016, **6**, 35719–35730.
62. X. Wang, R. Berger, J. I. Ramos, T. Wang, K. Koynov, G. Liu, H.-J. Butt and S. Wu, Nanopatterns of polymer brushes for understanding protein adsorption on the nanoscale, *RSC Adv.*, 2014, **4**, 45059–45064.

63. A. Jedlovszky-Hajdú, F. B. Bombelli, M. P. Monopoli, E. Tombácz and K. A. Dawson, Surface coatings shape the protein corona of SPIONs with relevance to their application *in vivo*, *Langmuir*, 2012, **28**, 14983–14991.
64. R. Capomaccio, I. O. Jimenez, P. Colpo, D. Gilliland, G. Ceccone, F. Rossi and L. Calzolai, Determination of the structure and morphology of gold nanoparticle-HSA protein complexes, *Nanoscale*, 2015, **7**, 17653–17657.
65. C. Rocker, M. Potzl, F. Zhang, W. J. Parak and G. U. Nienhaus, A quantitative fluorescence study of protein monolayer formation on colloidal nanoparticles, *Nat. Nanotechnol.*, 2009, **4**, 577–580.
66. D. Maiolo, P. Bergese, E. Mahon, K. A. Dawson and M. P. Monopoli, Surfactant titration of nanoparticle–protein corona, *Anal. Chem.*, 2014, **86**, 12055–12063.
67. K. Fischer and M. Schmidt, Pitfalls and novel applications of particle sizing by dynamic light scattering, *Biomaterials*, 2016, **98**, 79–91.
68. S. Milani, F. B. Bombelli, A. S. Pitek, K. A. Dawson and J. Radler, Reversible versus irreversible binding of transferrin to polystyrene nanoparticles: soft and hard corona, *ACS Nano*, 2012, **6**, 2532–2541.
69. P. Maffre, K. Nienhaus, F. Amin, W. J. Parak and G. U. Nienhaus, Characterization of protein adsorption onto FePt nanoparticles using dual-focus fluorescence correlation spectroscopy, *Beilstein J. Nanotechnol.*, 2011, **2**, 374–383.
70. B. Pelaz, P. del Pino, P. Maffre, R. Hartmann, M. Gallego, S. Rivera-Fernandez, J. M. de la Fuente, G. U. Nienhaus and W. J. Parak, Surface Functionalization of Nanoparticles with Polyethylene Glycol: Effects on Protein Adsorption and Cellular Uptake, *ACS Nano*, 2015, **9**, 6996–7008.
71. Y. Klapper, P. Maffre, L. Shang, K. N. Ekdahl, B. Nilsson, S. Hettler, M. Dries, D. Gerthsen and G. U. Nienhaus, Low affinity binding of plasma proteins to lipid-coated quantum dots as observed by *in situ* fluorescence correlation spectroscopy, *Nanoscale*, 2015, **7**, 9980–9984.
72. H. Wang, L. Shang, P. Maffre, S. Hohmann, F. Kirschhofer, G. Brenner-Weiss and G. U. Nienhaus, The Nature of a Hard Protein Corona Forming on Quantum Dots Exposed to Human Blood Serum, *Small*, 2016, **12**, 5836–5844.
73. S. Dominguez-Medina, S. Chen, J. Blankenburg, P. Swanglap, C. F. Landes and S. Link, Measuring the Hydrodynamic Size of Nanoparticles Using Fluctuation Correlation Spectroscopy, *Annu. Rev. Phys. Chem.*, 2016, **67**, 489–514.
74. T. Liedl, S. Keller, F. C. Simmel, J. O. Radler and W. J. Parak, Fluorescent nanocrystals as colloidal probes in complex fluids measured by fluorescence correlation spectroscopy, *Small*, 2005, **1**, 997–1003.

75. S. Balog, L. Rodriguez-Lorenzo, C. A. Monnier, M. Obiols-Rabasa, B. Rothen-Rutishauser, P. Schurtenberger and A. Petri-Fink, Characterizing nanoparticles in complex biological media and physiological fluids with depolarized dynamic light scattering, *Nanoscale*, 2015, **7**, 5991–5997.
76. K. E. Sapsford, K. M. Tyner, B. J. Dair, J. R. Deschamps and I. L. Medintz, Analyzing Nanomaterial Bioconjugates: A Review of Current and Emerging Purification and Characterization Techniques, *Anal. Chem.*, 2011, **83**, 4453–4488.
77. C. Porsch, Y. Zhang, Å. Östlund, P. Damberg, C. Ducani, E. Malmström and A. M. Nyström, In Vitro Evaluation of Non-Protein Adsorbing Breast Cancer Theranostics Based on 19F-Polymer Containing Nanoparticles, *Part. Part. Syst. Charact.*, 2013, **30**, 381–390.
78. J. Sund, H. Alenius, M. Vippola, K. Savolainen and A. Puustinen, Proteomic characterization of engineered nanomaterial-protein interactions in relation to surface reactivity, *ACS Nano*, 2011, **5**, 4300–4309.
79. D. Walczyk, F. B. Bombelli, M. P. Monopoli, I. Lynch and K. A. Dawson, What the Cell "Sees" in Bionanoscience, *J. Am. Chem. Soc.*, 2010, **132**, 5761–5768.
80. R. A. Sperling, T. Liedl, S. Duhr, S. Kudera, M. Zanella, C. A. J. Lin, W. H. Chang, D. Braun and W. J. Parak, Size Determination of (Bio)conjugated Water-Soluble Colloidal Nanoparticles: A Comparison of Different Techniques, *J. Phys. Chem. C*, 2007, **111**, 11552–11559.
81. F. A. Messaud, R. D. Sanderson, J. R. Runyon, T. Otte, H. Pasch and S. K. R. Williams, An overview on field-flow fractionation techniques and their applications in the separation and characterization of polymers, *Prog. Polym. Sci.*, 2009, **34**, 351–368.
82. J. Ashby, S. Pan and W. Zhong, Size and Surface Functionalization of Iron Oxide Nanoparticles Influence the Composition and Dynamic Nature of Their Protein Corona, *ACS Appl. Mater. Interfaces*, 2014, **6**, 15412–15419.
83. P. M. Kelly, C. Aberg, E. Polo, A. O'Connell, J. Cookman, J. Fallon, Z. Krpetic and K. A. Dawson, Mapping protein binding sites on the biomolecular corona of nanoparticles, *Nat. Nanotechnol.*, 2015, **10**, 472–479.
84. C. Y. Tay, M. I. Setyawati, J. Xie, W. J. Parak and D. T. Leong, Back to basics: exploiting the innate physico-chemical characteristics of nanomaterials for biomedical applications, *Adv. Funct. Mater.*, 2014, **24**, 5936–5955.
85. T. L. Doane, C.-H. Chuang, R. J. Hill and C. Burda, Nanoparticle ζ-Potentials, *Acc. Chem. Res.*, 2012, **45**, 317–326.
86. A. Sikora, A. G. Shard and C. Minelli, Size and zeta-Potential Measurement of Silica Nanoparticles in Serum Using Tunable Resistive Pulse Sensing, *Langmuir*, 2016, **32**, 2216–2224.

87. E. L. Blundell, M. J. Healey, E. Holton, M. Sivakumaran, S. Manstana and M. Platt, Characterisation of the protein corona using tunable resistive pulse sensing: determining the change and distribution of a particle's surface charge, *Anal. Bioanal. Chem.*, 2016, **408**, 5757–5768.
88. L. Trapiella-Alfonso, G. Ramirez-Garcia, F. d'Orlye and A. Varenne, Electromigration separation methodologies for the characterization of nanoparticles and the evaluation of their behaviour in biological systems, *TrAC, Trends Anal. Chem.*, 2016, **84**, 121–130.
89. J. Wang, J. Li, J. Li, Y. Qin, C. Wang, L. Qiu and P. Jiang, In-capillary self-assembly study of quantum dots and protein using fluorescence coupled capillary electrophoresis, *Electrophoresis*, 2015, **36**, 1523–1528.
90. K. Srinivasan, S. Parimal, M. M. Lopez, S. A. McCallum and S. M. Cramer, Investigation into the Molecular and Thermodynamic Basis of Protein Interactions in Multimodal Chromatography Using Functionalized Nanoparticles, *Langmuir*, 2014, **30**, 13205–13216.
91. R. Huang and B. L. T. Lau, Biomolecule-nanoparticle interactions: Elucidation of the thermodynamics by isothermal titration calorimetry, *Biochim. Biophys. Acta*, 2016, **1860**, 945–956.
92. G. Maiorano, S. Sabella, B. Sorce, V. Brunetti, M. A. Malvindi, R. Cingolani and P. P. Pompa, Effects of cell culture media on the dynamic formation of protein-nanoparticle complexes and influence on the cellular response, *ACS Nano*, 2010, **4**, 7481–7491.
93. M. Mahmoudi, A. M. Abdelmonem, S. Behzadi, J. H. Clement, S. Dutz, M. R. Ejtehadi, R. Hartmann, K. Kantner, U. Linne and P. Maffre, Temperature: the "ignored" factor at the nanoBio interface, *ACS Nano*, 2013, **7**, 6555–6562.
94. S. Chutipongtanate, K. Watcharatanyatip, T. Homvises, K. Jaturongkakul and V. Thongboonkerd, Systematic comparisons of various spectrophotometric and colorimetric methods to measure concentrations of protein, peptide and amino acid: detectable limits, linear dynamic ranges, interferences, practicality and unit costs, *Talanta*, 2012, **98**, 123–129.
95. A. Gessner, A. Lieske, B. R. Paulke and R. H. Muller, Functional groups on polystyrene model nanoparticles: influence on protein adsorption, *J. Biomed. Mater. Res., Part A*, 2003, **65**, 319–326.
96. T. M. Goppert and R. H. Muller, Protein adsorption patterns on poloxamer- and poloxamine-stabilized solid lipid nanoparticles (SLN), *Eur. J. Pharm. Biopharm.*, 2005, **60**, 361–372.
97. S. K. Sohaebuddin, P. T. Thevenot, D. Baker, J. W. Eaton and L. Tang, Nanomaterial cytotoxicity is composition, size, and cell type dependent, *Part. Fibre Toxicol.*, 2010, **7**, 22.
98. A. M. Clemments, P. Botella and C. C. Landry, Protein Adsorption From Biofluids on Silica Nanoparticles: Corona Analysis as a Function of Particle Diameter and Porosity, *ACS Appl. Mater. Interfaces*, 2015, **7**, 21682–21689.
99. G. U. Nienhaus, P. Maffre and K. Nienhaus, Studying the protein corona on nanoparticles by FCS, *Methods Enzymol.*, 2013, **519**, 115–137.

100. P. Hong, S. Koza and E. S. Bouvier, Size-Exclusion Chromatography for the Analysis of Protein Biotherapeutics and their Aggregates, *J. Liq. Chromatogr. Relat. Technol.*, 2012, **35**, 2923–2950.
101. M. Mahmoudi, I. Lynch, M. R. Ejtehadi, M. P. Monopoli, F. B. Bombelli and S. Laurent, Protein-nanoparticle interactions: opportunities and challenges, *Chem. Rev.*, 2011, **111**, 5610–5637.
102. J. H. Teichroeb, J. A. Forrest and L. W. Jones, Size-dependent denaturing kinetics of bovine serum albumin adsorbed onto gold nanospheres, *Eur. Phys. J. E*, 2008, **26**, 411–415.
103. E. D. Kaufman, J. Belyea, M. C. Johnson, Z. M. Nicholson, J. L. Ricks, P. K. Shah, M. Bayless, T. Pettersson, Z. Feldotö, E. Blomberg, P. Claesson and S. Franzen, Probing Protein Adsorption onto Mercaptoundecanoic Acid Stabilized Gold Nanoparticles and Surfaces by Quartz Crystal Microbalance and ζ-Potential Measurements, *Langmuir*, 2007, **23**, 6053–6062.
104. C. D. Walkey and W. C. Chan, Understanding and controlling the interaction of nanomaterials with proteins in a physiological environment, *Chem. Soc. Rev.*, 2012, **41**, 2780–2799.
105. M. Mahmoudi, M. P. Monopoli, M. Rezaei, I. Lynch, F. Bertoli, J. J. McManus and K. A. Dawson, The protein corona mediates the impact of nanomaterials and slows amyloid beta fibrillation, *ChemBioChem*, 2013, **14**, 568–572.
106. N. Brandes, P. B. Welzel, C. Werner and L. W. Kroh, Adsorption-induced conformational changes of proteins onto ceramic particles: differential scanning calorimetry and FTIR analysis, *J. Colloid Interface Sci.*, 2006, **299**, 56–69.
107. M. Lück, K.-F. Pistel, Y.-X. Li, T. Blunk, R. H. Müller and T. Kissel, Plasma protein adsorption on biodegradable microspheres consisting of poly (D,L-lactide-co-glycolide), poly (L-lactide) or ABA triblock copolymers containing poly (oxyethylene): influence of production method and polymer composition, *J. Controlled Release*, 1998, **55**, 107–120.
108. S. Tenzer, D. Docter, S. Rosfa, A. Wlodarski, J. Kuharev, A. Rekik, S. K. Knauer, C. Bantz, T. Nawroth, C. Bier, J. Sirirattanapan, W. Mann, L. Treuel, R. Zellner, M. Maskos, H. Schild and R. H. Stauber, Nanoparticle Size Is a Critical Physicochemical Determinant of the Human Blood Plasma Corona: A Comprehensive Quantitative Proteomic Analysis, *ACS Nano*, 2011, **5**, 7155–7167.
109. J. L. Rogowski, M. S. Verma, P. Z. Chen and F. X. Gu, A "chemical nose" biosensor for detecting proteins in complex mixtures, *Analyst*, 2016, **141**, 5627–5636.
110. A. Patra, T. Ding, G. Engudar, Y. Wang, M. M. Dykas, B. Liedberg, J. C. Kah, T. Venkatesan and C. L. Drum, Component-Specific Analysis of Plasma Protein Corona Formation on Gold Nanoparticles Using Multiplexed Surface Plasmon Resonance, *Small*, 2016, **12**, 1174–1182.

111. A. Das, A. Chakrabarti and P. K. Das, Probing protein adsorption on a nanoparticle surface using second harmonic light scattering, *Phys. Chem. Chem. Phys.*, 2016, **18**, 24325–24331.
112. M. Carril, D. Padro, P. del Pino, C. Carrillo-Carrion, M. Gallego and W. J. Parak, In situ detection of the protein corona in complex environments, *Nat. Commun.*, 2017, **8**, 1542.
113. D. Caputo, M. Papi, R. Coppola, S. Palchetti, L. Digiacomo, G. Caracciolo and D. Pozzi, A protein corona-enabled blood test for early cancer detection, *Nanoscale*, 2017, **9**, 349–354.

CHAPTER 6

Impact of Nanoparticle–Protein Interactions on Biological Assays

KAVITA DUBEY, ONILA LUGUN AND ALOK KUMAR PANDEY*

CSIR- Indian Institute of Toxicology Research, Vishvigyan Bhawan, 31, Mahatma Gandhi Marg, Lucknow-226001, Uttar Pradesh, India
*E-mail: alokpandey@iitr.res.in, pandeyalok2006@gmail.com

6.1 Introduction

Nanotechnology is the rapidly developing branch of science that deals with nanometer-sized objects. Developing devices and systems are also included in the core potential of nanotechnology. About a decade ago, the study of nanomaterials began because of their size-dependent physical and chemical properties that, nowadays, have entered a completely different realm regarding commercial applications. The unique properties of nanoparticles are becoming useful in a wide range of applications, one of which is bio-medical applications where they are used as nano-vaccines and nano-drugs where the only problem is the lack of knowledge about the biocompatibility and the risks of exposure to the nanoparticles.

Issues in Toxicology No. 40
Nanoparticle–Protein Corona: Biophysics to Biology
Edited by Ashutosh Kumar and Alok Dhawan

Published by the Royal Society of Chemistry, www.rsc.org

6.2 Factors Affecting Nanoparticles and Protein Interactions

Various factors contribute to the interactions of nanoparticles with protein coronas in a biological system, such as the size, shape, surface area, chemical composition, surface charge, hydrophobicity, the presence of various functional groups, pH and the temperature, which in turn affect protein adsorption on the surface of the nanoparticles.

6.2.1 Nanoparticle Size and Shape

6.2.1.1 Size of Nanoparticles

The size of the nanoparticles plays a vital role in modulating the protein's structure and thereby its function. Lysosome protein lost its structure and function significantly because of 100 nm silica nanoparticles as compared to 4 nm-sized particles.[1] Less perturbation of the protein structure is observed when the protein is adsorbed onto smaller nanoparticles. Further, there is a retention of less native-like enzyme activity for adsorption on larger nanoparticles probably because of their smaller surface curvature when compared with the smaller nanoparticles.[2] Negatively charged polyacrylic acid gold nanoparticles (12 nm) had a higher binding affinity with fibrinogen when compared with 7 nm nanoparticles, which shows that nanoparticle–fibrinogen interactions can have distinct binding configurations and affinities depending on the size of the nanoparticles, which will ultimately affect the nanoparticle–protein interaction and aggregation.[3]

6.2.1.2 The Shape of Nanoparticles

The shape of the nanoparticles also plays a very important role in the formation of protein coronas on nanoparticles. Deng *et al.* reported in a study that titanium dioxide (TiO_2) nanorods and nanotubes show differential plasma protein adsorptions.[4] Fibrinogen bound mainly to TiO_2 nanotubes whereas IgM and IgG proteins were found to be bound to TiO_2 nanorods. In terms of quantitative assessment, proteins were found to be more effectively bound to TiO_2 nanospheres than to TiO_2 nanorods and nanotubes, thus proving that the binding efficiency of a protein is dependent on the shape of the nanoparticles.

6.2.2 Surface Properties of Nanoparticles

6.2.2.1 Surface Charge of Nanoparticles

Some common proteins like fibrinogen bind with a high affinity to both positively charged and negatively charged nanoparticles as compared to neutral ones, with different orientations leading to varying binding kinetics and digestion patterns of the fibrinogen by plasmin and henceforth different biological impacts.[5] Surface modification can lead to the increase or decrease of the net surface charge as was reported in the case of Au nanoparticles where their surface was modified with electrolytes such as citrate and hexadecyltrimethyl ammonium bromide to increase their dispersibility. This increase in the net surface charge of the Au nanoparticles produced an electrostatic attraction to the oppositely charged functional groups in the proteins. Hence, more serum proteins were adsorbed on the surface.[6]

6.2.2.2 Hydrophilicity and Hydrophobicity

The hydrophilicity or the hydrophobicity of any nanoparticle surface is a temperature-dependent phenomenon. As in the case of silicon nanoparticles, the hydrophilicity decreased upon heating at temperatures up to 800 °C. Therefore, the amounts of different proteins (bovine serum albumin: BSA; hen egg lysozyme: HEL; ribonuclease: RNase; bovine lactoperoxidase: LPO) adsorbed on the surface of two nanosized silica nanopowders (one is the silica in pristine form and other is the silica nanoparticles heated up to 800 °C, thereby changing its hydrophilicity) were different as there was a considerable variation in the hydrophilic character and surface charge.[7]

6.2.2.3 Chemical Modification of the Nanoparticle Surface

There are methods by which the nanoparticle surface is chemically fabricated to avoid the adsorption of proteins, for example using polyethylene glycol (PEG), also referred to as "PEGylation". This gives the nanoparticle surface a stealthy character, so it is shielded by the immune cells. By controlling the density of PEG on the surface of the nanoparticles, they can stay in circulation for longer durations.[8] In some other studies, siliconate-treated zinc oxide (ZnO) nanoparticles showed no detectable binding of plasma proteins as compared to normal ZnO nanoparticles.

6.2.3 Other Physical and Chemical Aspects of Protein Coronas

6.2.3.1 Incubation Temperature of Nanoparticles

The corona composition is indeed affected by the temperature at which the nanoparticles and protein are maintained in the solution. It has been shown that the composition of the protein corona formed upon the exposure of the nanoparticles to heat-inactivated proteins (by heating at 56 °C) and non-heat-inactivated proteins are quite different.[9] Polymer-coated iron platinum (FePt) nanoparticles were incubated with human serum albumin (HSA) and apotransferrin (apo-Tf) at different concentrations and temperatures ranging from 5° to 45 °C, which revealed that varying the incubation temperature could cause a significant change in the protein corona formation and composition and the effect was especially pronounced at the physiologically relevant temperature window of 37–41 °C. A higher affinity of apo-transferrin towards the FePt nanoparticles was seen at 43 °C as compared to 9 °C and 22 °C.[10]

6.2.3.2 Concentration of Nanoparticles and Proteins

The ratio of nanoparticles to proteins determines the coating efficiency of serum proteins. Here, the concentrations of different nanoparticles determine the types and extent of coronas formed. The proteins bound to these nanoparticles vary with the nanoparticle concentrations and express no change in their respective zeta-potentials. At higher nanoparticle concentrations, the agglomeration sizes are increased significantly. Low protein plasma concentrations always form a thin protein corona and *vice versa*.[11]

6.2.3.3 Types of Nanoparticles

In a study done on TiO_2, SiO_2 and ZnO_2 nanoparticles, it was clearly shown that protein binding to these nanoparticles was similar for the former two nanoparticles but was notably different for the latter one, which proves that different types of nanoparticles show different patterns of adsorption of protein. This differential adsorption indicates that the surface of the nanoparticles is highly influenced by protein binding.

6.3 Biological Influence

Nanoparticles hold a key role in the field of nanomedicine. The same nanoparticles that possess unique properties like their quantum properties or their capacity to carry or adsorb other compounds that make them suitable for a wide range of applications, such as in drug delivery and nano-diagnostics, also make them quite harmful and toxic. The field of nanomedicine has various shortcomings and challenges to work upon like formulation stability, shelf life, biocompatibility, bio-distribution, and targeting. The progress that has been made in this field is not being used on a wide scale because of some of the fundamental challenges that should be looked at. There are various factors at the nano–bio interface that have been overlooked for ages. Here, we focus on protein coronas on nanoparticle surfaces.

6.3.1 Modification of *In Vitro* Assays

The approaches we have been using for the toxicity evaluation of nanoparticles are toxicological protocols that were developed before the arrival of the first nano-based applications.[12] This means that essential modifications need to be done to evaluate the toxicity of nanoparticles. It was suggested by Berridge *et al.* that the presence of MTT (3-(4,5-dimethylthiazol-2-yl)-2,5-diphenyltetrazolium bromide) dye can cause interference; its interaction with nanoparticles can modify the structure of the dye into formazan crystals, which will give false positive results.[13] This assay could give negative outcomes in terms of the interactions of the nanoparticles along with the protein coronas. When nanoparticles are resuspended in a cell culture medium, the high surface area of the nanoparticles readily interacts with the biomolecules of the medium and forms a protein corona layer that tends to change the nutritional balance of the medium and thereby its pH. This increase in the acidity of the medium could itself be toxic for the cells. This was observed when polyvinyl-alcohol-coated superparamagnetic iron oxide nanoparticles (SPIONs) were resuspended in a cell culture medium; after spinning and removing the particles, there was a considerable change in the pH and the structure/content of the protein in the medium. When this medium was incubated with mouse fibroblast cells, it was revealed that the medium was responsible for inducing almost 20% toxicity. This problem could be minimized by the use of corona-coated nanoparticles for toxicity evaluations.[14] It was also seen that if factors like plasma gradient, the temperature of the target tissue, interfacial effects and personalized protein coronas are considered, then a consistency in toxicity results could be obtained. So, if all these factors are kept in mind while performing *in vitro* assays, then they could be a better and more reliable predictor of *in vivo* behavior.

6.3.1.1 Agglomeration, Cellular Uptake, Cellular Response

The physico–chemical properties of nanoparticles, including size, shape, surface charge and surface chemistry, are known to strongly govern the cellular uptake efficiency, transport, and fate of nanoparticles. Not only this, but the biological factors are also crucial. Various studies have revealed that the same nanoparticles can behave in very different ways and induce varying biological outcomes in the presence or absence of a protein corona. In the normal scenario, uptake of the nanoparticles inside the cells is a two-step process that begins with the initial adhesion of the nanoparticles to the cell surface, their interactions with the different components of the cell membrane and then their subsequent internalization through energy-dependent pathways. The possible interactions between the biomolecules of the corona and the receptors on the membrane could lead to recognition and a corresponding uptake of the nanoparticles. In most of the studies in the literature, it has been observed that cellular uptake under serum-free conditions is higher as compared to the same nanoparticles along with the serum. In the case of Si nanoparticles, they internalized very differently (in terms of quantity and processes) when exposed to cells in a serum-free medium and in a complete medium. Thus, Si nanoparticles without serum proteins showed a high degree of adhesion to the cellular membrane and, therefore, greater internalization.[15]

In the same way, bare carboxylated polystyrene nanoparticles showed higher attachment to the cell membrane when compared to the serum conditions.[16] Thus, nanoparticles with high energy on the bare surface typically adsorb strongly to the cell membrane and this strong adhesion at least in part contributes to a higher uptake efficiency. However, in the presence of serum, the formation of the corona lowers the surface energy and, thereby, decreases the unspecific interaction and so the interaction between the nanoparticle–corona complex and the cell membrane is greatly reduced.

The internalization of dihydrolipoic-acid-coated quantum dots (DHLA-QDs) with native (HSA) and a modified protein corona (by succinic anhydride) inside human cervix adenocarcinoma epithelial (HeLa) cells was studied to investigate the effect on cellular uptake. This study suggested that the presence of an engineered corona is not the sole reason for the differential uptake level; it also has a marked effect on the intracellular location of the nanoparticles. The overall uptake yield is controlled by the availability of nanoparticles on the plasma membrane, which ultimately depends on the membrane's affinity to different nanoparticles.[17] There are studies that took advantage of the magnetic properties of silica-coated magnetite nanoparticles to recover sub-cellular fractions of nanoparticles as the analysis of the proteins revealed that much of the serum-derived corona was preserved during uptake.[18] The presence and absence of the protein corona attenuate the overall interactions of nanoparticles with red blood cells (RBCs) and thereby regulates the hemolytic activity of the nanoparticles. This shows the protective role of the protein corona towards improved blood compatibility and can give a broader window for the design of therapeutic parameters with limited toxicity.[19]

6.3.1.2 *Cytotoxicity*

Once taken up by the cells, nanoparticles can accumulate in different organelles, especially in the lysosomes, intracellular vacuoles as shown in the case of HeLa cells, or in the cytoplasm or the mitochondria, as demonstrated with copolymer nanoparticles.[20,21] The two important repercussions of nanoparticle uptake by cells are cytotoxicity and immune modulation. Adsorption of the protein on the nanoparticle surface occurs almost instantly. So, mostly it is assumed that there is always an indirect interaction between the nanoparticles and the cell surface through the nanoparticle–protein corona. The chemical and physical properties of the nanoparticles play very important roles in the potentially toxic or non-toxic effects put forth by the protein corona. For instance, the positive charge present on the nanoparticles can lead to their increased interactions with the negatively charged cell surface, which can lead to a disintegrated membrane and ultimately cell death.[22] So, the presence of a coating (both inorganic and biological) can mask these properties, as in the case of ZnO nanoparticles where their toxicity was significantly reduced in the presence of a coating of a protein corona and silica in human hepatocarcinoma and skin dermal fibroblast cells, respectively.[23,24] In the case of carbon nanotubes, when they were enveloped by blood proteins, a reduction in their cytotoxic effects was observed in THP-1 cells and umbilical vein endothelial cells.[25,26] Nevertheless, the nanoparticles and the protein coronas can also reduce systemic toxicity. Studies have demonstrated that freshly isolated human thrombocytes are activated and have a hemolytic effect on red blood cells when kept with bare silica nanoparticles, whereas this cell phenotype was not seen in the case of nanoparticle–protein corona complexes and the red blood cells were protected from the respective damage.[27] The protein corona also increases the stability of the nanoparticles if the secondary degradation product in the biological fluids poses a potential risk. The formation of a protein corona on ZnO and Ag nanoparticles restricted the dissolution of these nanoparticles and so the release of Zn and Ag ions were inhibited, which is one of the causative factors of toxicity.[23,28] The protein corona also holds back the generation of reactive oxygen species (ROS) from nanoparticles, especially when it is composed of materials with semiconductor features.[29]

Proteins adsorbed on the surfaces of nanoparticles can also cause thermal instability, thereby, the conformational changes in the proteins might lead to protein denaturation. TiO_2 nanoparticles restricted tubulin polymerization, which then induced conformational changes in tubulin resulting in protein function alteration.[30] In the same way, gold nanoparticles induced conformational changes in serum albumin.[31]

6.3.1.3 Immunotoxicity

The protein corona brings about differential recognition of nanoparticles by immune cells. The protein corona acts as an interface between the immune cells and the nanoparticles, which can significantly affect the particle–cell interactions and pathway activation. For example, disulfide-stabilized poly(methacrylic acid) nanoporous polymer particles demonstrated differential recognition for monocytes (THP-1) and differentiated macrophage-like cells (dTHP-1) cells, which were assessed by the levels of internalization of the nanoparticles. When compared to the levels of uptake by pristine nanoparticles, the uptake was strongly inhibited in the presence of BSA, *i.e.*, the protein corona.[32] The complement cascade can be switched on/off by modifying the surface properties of the nanoparticles, of which the protein corona is an important part. Yu *et al.* revealed that the different conformational states of the glycopolymer chain present on the surface of the nanoparticles act as the "molecular switch" for the complement system's activation and amplification.[33] The nanoparticle-protein corona complex exhibits a "risk" signal; then the adaptive immune processes can be stimulated. The release of inflammatory cytokines was seen due to the protein corona, which activates receptors (Mac-1) that in turn activate a signaling pathway like NF-κB.[34]

6.3.1.4 Genotoxicity

Genotoxicity could be induced due to the interactions of nanoparticles with different proteins (cell cycle checkpoint proteins, nuclear proteins, and mitotic spindles) and their residues.[35] Gelatin nanoparticles disrupt and disorganize the F-actin and h-tubulin of the cytoskeleton.[36] Nanoparticles restrict the polymerization of microtubule assembly, which in turn could lead to micronucleus formation. A study demonstrated no significant difference in the number of micronuclei in human granulosa cells incubated with SPIONs coated with BSA and lauric acid as compared to normal cells. This means that protein coronas can also be used to enhance the biocompatibility of nanoparticles and in the mitigation of genotoxicity.[37]

6.3.2 Effect on Drug Release

It was believed that nano drug delivery systems were limited by the presence of proteins around them, which shielded the nanocarrier and altered the drug release profile. Among the various synthetic nano-carriers examined the release profile of drug-loaded SPIONs was remarkably attenuated by the corona that formed.[38] A reduced burst effect in the drug release profile was observed in both protein–drug conjugates (abraxane) and drug-loaded PEGF-coated SPIONs. Conversely, the protein corona plays a crucial role in controlling drug payloads and reduces the effects of anti-cancer drugs and their therapeutic efficacy.[39] A multilayered protein corona around nanoparticles is also used effectively to improve carrier properties for drug release by holding up high amounts of therapeutic payloads. In a recent study, it was demonstrated that gold nanorods can be loaded with DNA and doxorubicin at a capacity 5–10-fold higher than that achieved previously.[40] In addition to this, the protein corona was formed from an endogenous protein, so undesirable immune responses to the therapeutic carrier could be eliminated or minimized. The nanoparticle-protein corona is a promising tool for improving drug delivery. The different natures of the parts of a protein corona, *i.e.*, the hard corona and the soft corona, can be used to manipulate and optimize the properties of therapeutic delivery. It is expected that if nanoparticles with a protein corona have an abundance of soft corona protein in the environment, the exchange will be limited, leading to a low payload release and *vice versa*.[41] Thus, the use of different classes of proteins can influence the exchange and, consequently, the release profile.

6.3.3 Effect on Protein Fibrillation

From various studies, it was confirmed that protein–NP interactions lead to an alteration in the naïve conformation of proteins. According to the paradigm, proteins only work in their fixed three-dimensional structure, and an alteration in the structure can have a major influence on various functional aspects of proteins and cells as well. It was found that in many cases, nanoparticles have been associated with diseases due to protein misfolding like amyloidosis, various neurodegenerative conditions (Alzheimer's, Creutzfeldt–Jakob disease), and dialysis-related amyloidosis. A similar process is amyloid fibril formation in which proteins aggregate to form a folded structure that sticks among them to form an insoluble fibrillar form. Amyloid fibrillation is a nucleation-dependent polymerization process in which oligomeric proteins act as nuclei, but recently some nanoparticles have been tested as nucleation centers for protein binding.[42] Many nanoparticles have been reported to assist amyloid fibrillation.[43] Their small size and high surface-to-volume ratios help nanoparticles to interact with proteins, but other factors, like polymeric surface charge and salt concentration, also act as modulators for protein nucleation kinetics for protein fibrillation. Experimentally, it was observed that TiO_2 nanoparticles promote β-amyloid fibrillation *in vitro*, which is the main cause of pathogenesis associated with Alzheimer's disease (AD).[44] Fibrillation of synthetic Ab42 was monitored in the presence of TiO_2 nanoparticles and was found to reduce the lag time required to carry out the nucleation process and showed enhanced protein binding. A study on nanoparticles, such as copolymer particles, quantum dots, cerium oxide nanoparticles, and carbon nanotubes, showed enhanced nucleation of human 2-microglobulin on the nanoparticle surfaces leading to a multilayered oligomer formation.[45] Though much research has been directed towards the field of nanoparticles that form amyloids, there is evidence that particles also play a role in suppressing protein fibrillation. It was hypothesized that charged particles interact with charged residues on denaturing proteins to facilitate refolding of misfolded proteins and prevent their aggregation. The structures and functions of thermally denatured chymotrypsin, lysozyme and papain were restored after treatment with 2-(10-mercaptodecyl)-malonic-acid-functionalized 2 nm core gold nanoparticles (AuDA).[46] It was found that CdTe nanoparticles showed a phenomenon of preferential binding to oligomers rather than monomers, thereby preventing the formation of higher order aggregates such as protofibrils and fibrils, resulting in significantly reduced fibril formation.[47] This understanding of protein fibrillation has the potential to enhance our knowledge about various amyloid diseases as well as their therapeutic opportunities.

Most *in vitro* aggregation assays are based on the readouts of a UV/Vis spectrophotometer or fluorescence emission measurements, being label-free detection methods based on light transmittance. Another alternative assay is measuring the alterations in fluorescence signals of different fluorophores that report different protein aggregation states. Among them, the most popular is thioflavin T, which specifically binds with β-sheet structures of amyloids and emits signals upon binding.[48] In this direction, one more advanced assay is based on the use of fluorophores exhibiting a dual emission state. The best example is the assay based on fluorophores presenting excited state intramolecular proton transfer.[49]

6.3.4 Effect on Targeting Capabilities

In the field of nanomedicine, due to their targeting ability, nanoparticles have been used in practice for years. Proteins adsorbed on their surfaces have a direct effect on the modulation of the pharmacological and toxicological profiles of nanoparticles in the biological milieu. They also have a role in the potential therapeutic and diagnostic functionalities perceived by biological systems on exposure to nanoparticles. While designing nanoparticle surfaces for desired targeting and therapeutic delivery, the role of adsorption of multiple proteins in the form of protein coronas should also be considered as a very important aspect.

Due to their small size, nanoparticles act as suitable vehicles for targeted delivery approaches, but to improve the targeting efficiency, very often ligands are used to functionalize the surfaces of the particles. However, when placed in a biological environment, a nanoparticle loses its specificity in targeting due to the formation of a protein corona. It may be due to proteins covering the ligands and hindering targeted delivery. There are cases of a loss of selectivity in receptor-mediated endocytosis and targeting capabilities of transferrin-functionalized nanoparticles due to adsorption of proteins on their surfaces.[50] Fluorescence microscopy imaging also confirmed that the targeting ability of bicyclononyne (BCN)-functionalized silica nanoparticles to an azide-modified silicon substrate was reduced in the presence of a serum corona.[51] Protein binding was found to induce loss of specificity in targeting capacity due to conferring a new identity to the particle. It was found that the presence of sera reduced the targeted uptake of single-domain antibody-conjugated fluorescent silica nanoparticles in cells.[52] Plasma protein adsorption can also lead to ligand modifications following altered targeted delivery. A similar study was conducted using transferrin and RGD peptide on Au nanoparticles as targeting ligands, and a reduction in the cellular uptake was observed when incubated with plasma proteins.[53] Corona composition also has a direct effect on the specific targeting and localization of particles. Using a fluorescence molecular tomography system (FMT), localization of a polysorbate-80-stabilized nanoparticle was analyzed in BALB/c mice.[54] It was observed that functionalized nanoparticles facilitate the adsorption of apolipoprotein E4 on the particle surface and they drive the specific targeted movement of the particles towards the brain from the liver and intestinal region. Apolipoproteins like apo-E and apoA-I are covalently attached to the nanoparticle surface, which helps to enhance the passage of nanoparticles across the blood–brain barrier (BBB).[55] In a study to target cancer cells, protein coronas rich in vitronectin were used to target MDA-MB-435S that over-expresses αvβ3 and αvβ5 integrins, two major vitronectin receptors. The findings suggested that the cellular uptake of the vitronectin-rich protein layer on lipoplexes was two-fold greater than that of bare supports.

6.3.5 Optical Detection Applications

Studies have revealed that protein coronas affect the protein relaxivity and MRI contrast efficiency of SPIONs. The main advantage of SPIONs as MRI contrast agents is their higher sensitivity resulting in the contrast enhancement of magnetic resonance images. In some cases, protein coronas can reduce the ability of the contrast agents. The different physico–chemical properties that occur when coating SPIONs with dextran can lead to the formation of a protein corona with a different composition. It was also observed that in the case of negatively charged SPIONs, the relaxivity was increased slightly, which is the measure of the increase in contrast efficiency.[56]

6.3.6 *In Vivo* Biodistribution

To date, there are only a few examples in the literature of protein coronas formed on nanoparticles *in vivo*. Owing to their ease of operation in an *in vitro* system, various studies have been conducted on protein coronas with respect to nanoparticle distribution. But, factors like blood flow dynamics, the immune response, and interactions with circulating and endothelial lining cells after injection into an animal system are hard to extrapolate from *in vitro* data. Though it is more reflective of the true complex biological environment that is experienced by the nanoparticles, the major limiting factor associated with *in vivo* studies are the recovery of nanoparticles post-administration and the highly dynamic nature of corona proteins. So, the fact that there is a significant difference between corona distributions *ex vivo* and *in vivo* is not unanticipated.

The retention and distribution of nanoparticles in the systemic circulation can be correlated with their recognition by the mononuclear phagocytic system and subsequent clearance or their filtration rates. From a nanoparticle distribution point of view, the size of the particle plays a critical role, as particles >100 nm can easily be recognized and eliminated by the Kupffer cells and macrophages while particles <10 nm are easily filtered by glomerular filtration.[57,58] That is why binding to a protein corona can change the size of a particle, thereby severely altering the fate of nanoparticles in the circulation and their distribution. A major impact of protein binding is the opsonization of nanoparticles, which leads to their easy recognition by the reticuloendothelial system and rapid removal.[59] It was also proposed that serum protein binding with nanoparticles induces a conformational change in the structure thereby altering it from its inactive form to give an active conformation and enhanced opsonisation.[60] Complement activation is an excellent example of this mechanism. Increasing plasma half-life by avoiding protein binding is the most extensively published work to date. In clinical settings, PEGylation leads to less adsorption of protein on a nanoparticle surface and so it plays a critical role in clearance *in vivo*. Analysis of the results indicates that PEG density on the surface of PEG-PLGA nanoparticles is the key determinant for particle clearance in animals.[61]

Not only does the corona formation speed up the clearance process, sometimes binding of some bystander proteins reduces the non-specific binding and provides stealth to nanoparticles. Specific binding of proteins that do not have a receptor binding sequence makes the nanoparticles less prone to recognition as compared to naked particles' surfaces.[62] Protein binding also directs or targets nanoparticles to a particular area of the body. Adsorption of apolipoprotein E (ApoE) onto the surfaces of polysorbate-80-coated nanoparticles helps the transport of the particles through the BBB to the brain's interior.[63]

6.4 Personalized Protein Coronas

The plasma protein present in any patient changes with different types of diseases (*e.g.*, cancer, obesity, diabetes, hemodialysis, hemophilia, rheumatism), especially a change in the protein composition or the conformation. In such cases, the modified plasma protein is measured as a predictive marker for such diseases that can be taken forward for treatment. Now, more precise studies have revealed that changes in plasma protein concentrations and structures widely affect the protein coronas on nanoparticles. For example, some plasma proteins are components of acute/chronic inflammatory responses – increased levels of fibrinogen-alpha, alpha-antitrypsin, and C-reactive protein are linked to diabetes, and C-reactive protein and tumor necrosis factor are connected to cardiovascular diseases.[64,65] In a study, personalized protein coronas formed on silica and polystyrene nanoparticles from plasma obtained from patients with different diseases revealed that the proteins changed in composition and amount.[66] From an application point of view, Zheng *et al.* developed a simple protein-corona-based assay for the early screening of prostate cancer with the help of gold nanoparticles.[67] So, in the near future, we can expect the development of disease-specific custom nanoparticles with this novel concept of personalized protein coronas, along with the recent findings, in mind.

6.5 Analytical Methods Influenced by Protein Coronas

As most of the nanoparticles are ultimately going to end up in biological systems, it is important for us to understand their interactions in the biological milieu. There are batteries of techniques that are affected by nanoparticle–protein interactions and can be utilized for the study of the physiological interplay between the two (Table 6.1).

6.5.1 Spectroscopic Methods

Spectroscopy is a simple and effective analytical technique used to evaluate the absorbance of any particle. Whether it is a protein or nanoparticle, both possess a characteristic absorption spectrum and any change in this can be used to evaluate binding. It is characterized by shifting and an increase in the intensity of the absorption spectra of protein–NP complexes.[68] In a recent work, graphene oxide-silver nanocomposites (GO-AgNCPs) with BSA were found to red shift by approximately 6 nm, which is attributed to the BSA and the nanoparticles interacting. Further, in a similar experiment, there was a less obvious blue shift of about 1.1 nm

Table 6.1 Analytical methods influenced by protein corona and nanoparticle interactions.

S. No.	Analytical method	Parameters analyzed	Advantages and disadvantages	NPs	Protein of interaction	Effect of interaction	Reference
1.	UV/Vis spectroscopy	Absorbance	-Easy, simple, effective, sensitive to variables like pH, temperature, electrolytes *etc.*	i-Graphene oxide silver nanocomposites (GO-AgNCPs) ii-Hydroxyapatite nanoparticle	-Bovine serum albumin (BSA) -BSA	-Redshift of spectra -Blueshift	68 and 69
2.	Hyperspectral micrographs	Multispectral imaging		i-Ag-NP	-Serum	-Distinct shifts in hybridized plasmon energy	70
3.	Fluorescence spectroscopy	-Intrinsic fluorescence	-Easy to operate, less stable, quenching	-Hydroxyapatite nanorods -GO-AgNCPs -AuMBA, AuGSH	-BSA -BSA -CrataBL	Quenching of fluorescence intensity	69, 68 and 72
4.	Fourier-transform infrared spectroscopy	-Conformation information of protein	-Not limited by protein size or the physical state of the samples, sensitive to structure, cheap, can't give fine structural details	-HSA, trypsin, HB and Mb	-Au/AgNPs	-Trypsin-AuNPs showed structural changes	73
5.	Raman spectroscopy	-Vibrational spectroscopy to give structural information	-Can work in aqueous solution, greater spectral simplicity, no sample preparation required, non-destructive, high specificity, high sensitivity is needed for detection of Raman spectrum	-Bare CdS quantum dots (QDs)	-Human adult hemoglobin (Hb)	-Cysteine involved in bond formation on the surface of the CdS QDs	75
6	Circular dichroism	-absorption bands of optically active chiral molecules for structural studies	-Gives average of total population, can't determine structure alteration at amino acid level	-Black carbon nanoparticles	-BSA	-Slight decrease in the α-helix content	78
7.	Nuclear magnetic resonance	-Structural information	-Fine detail analysis, work for wide range of samples, organic, inorganic, solution and solid state, qualitative and quantitative, expensive, expertise needed for result interpretation	-AuNPs	-Glutathione, bovine carbonic anhydrase, BSA *etc.*	-Proteins without cysteine showed reversible binding	80

(continued)

Table 6.1 (*continued*)

S. No.	Analytical method	Parameters analyzed	Advantages and disadvantages	NPs	Protein of interaction	Effect of interaction	Reference
8.	Surface plasmon resonance	-Binding kinetics to determine interaction	-Sensitive to refractive index of medium, restricted sensitivity limit	-AuNPs	-HSA, fibrinogen, ApoA1, and polyclonal IgG	-ApoA1 with lowest dissociation constant binds most strongly	82
9.	Transmission electron microscopy	-Direct visualization	-Only for qualitative analysis, not applicable for quantitative studies	-$MnFe_2O_4$ NPs	-BSA	-Can be visualized with negative staining	83
10.	Dynamic light scattering	-By measuring hydrodynamic diameter	-Simple, fast measurement, hydration shells formation affects hydrodynamic diameter, limited by shape, monodispersed sample required	-AuNPs	-HSA	-Hydrodynamic diameter increases	84
11.	Fluctuation correlation spectroscopy	-Measuring size by fluctuation in the fluorescence intensity emitted		-FePt NPs	-apoA-I, apoE4, HAS, apotransferrin	-Size increase was observed	87
12.	Nanoparticle tracking analysis	-Brownian motion of particles	-Limited sample concentration, should be monodispersed	-SPION	-Human plasma	-Higher hydrodynamic sizes	89
13.	Laser Doppler anemometry	-Surface charge	-Sample should be monodispersed				
14.	Tunable resistive pulse sensing	-Zeta-potential measurement	-Indicates stability of dispersion	-Carboxylated polystyrene particles	-fibrinogen, albumin, γ-globulin, human blood serum	-Changes in particle zeta-potentials	92

15.	Quartz crystal microbalance	-Mass variation	-Label-free, high resolution mass sensing, surface properties like surface charge, roughness may influence results	-Mercaptoundecanoic-acid-stabilized AuNPs	-BSA, Mb and cytochrome *c* (CytC)	-Adsorption isotherms gave idea about binding	93
16.	Differential centrifugation sedimentation	-Differential sedimentation in density gradient		-Surface-carboxylated polystyrene particles, sulfonated polystyrene and SiO_2 nanoparticles	-Human blood plasma	-Different size distribution for different species	95
17.	SDS-PAGE	-Relative mobility	-Qualitative method, suitable for complex protein mixture, denaturation and digestion of proteins are needed	-Polyethylenimine-coated magnetic iron oxide nanoparticles	-FCS	-Serum concentration affects protein corona mass	96
18.	Capillary electrophoresis	-Relative mobility	-No impact of hydrodynamic size				
19.	Flow field flow fractionation	-Fractionation		-SPION -Au-NP	-IgG/albumin-depleted serum -HSA	-Structural information	84 and 99
20.	Isothermal titration calorimetry	-Binding stoichiometry	-Thermodynamic characterization allows further optimization of compound, no labeling or immobilization is required, not affected by size or optical property of sample	-Hydroxyapatite nanorods	-BSA	Exothermic interaction	69

indicating structural changes of the BSA protein while interacting with hydroxyapatite nanoparticles.[69] Similar to this, hyperspectral imaging is an advanced analytical tool that uses the combinatorial approach of spectroscopy and microscopy. It accumulates the reflectance spectrum of each pixel in real time and represents them in hyperspectral micrographs. The cellular uptake of silver nanoparticles with and without protein coronas was recently explored using hyperspectral imaging by analyzing the shift of plasmon energies, and revealed intracellular modifications due to protein corona formation.[70] In fluorescence spectroscopy, to show nanoparticle–protein interactions, the intrinsic emission properties of the protein are exploited. This fluorescence is mainly attributed to the protein due to aromatic amino acids and after absorbing ultraviolet light, particle binding can result in the quenching of these chromospheres. For macromolecules, the degree of fluorescence quenching can provide information about binding mechanism, intermolecular distance, binding mode *etc.*[71] While elucidating the fluorescence intensity of BSA protein, it showed an emission band at 345 nm, while titration of this protein with the hydroxyapatite nanoparticle leads to quenching and a simultaneous decrease in the fluorescence spectra.[69] Similarly, the fluorescence intensity of BSA decreases as compared to its native form, which showed an emission maximum at 345 nm, when present along with GO-AgNCPs.[68] Lira *et al.* have used CrataBL intrinsic fluorescence for fluorescence titration quenching by gold NPs as a function of salt concentration.[72]

Fourier-transform infrared spectroscopy (FT-IR), which is sensitive to vibrational modes, has been used to detect alterations in the vibrational band of the amide–I bond to determine the conformations of proteins.[73] In a recent study, the interactions of Au and Ag nanoparticles with different plasma proteins, like HSA, hemoglobin (Hb), myoglobin (Mb) and trypsin, were monitored using FT-IR spectroscopy.[73] This gives information about the overall structural change in the protein structure, which might be due to their interactions and formation of NP–protein complexes.[74] Raman spectroscopy is another form of vibrational spectroscopy, which gives structural information by using vibrational bands associated with peptide chains, sulfur-containing side chains and aromatic side chains.[75] A slight modification of orientation in the heme group was observed while studying the interaction of hemoglobin with bare CdS quantum dots showing protein–NP interactions.[75] Although it is similar to other vibrational spectroscopy methods like FT-IR, it has two main advantages: it can measure protein–NP complexes in aqueous solutions and it shows greater spectral simplicity due to a more intense band generation.[76] For years, circular dichroism (CD) spectroscopy has been used for the detection of protein secondary structure. Any change in the secondary structures (α-helix, β-sheet) of these biomolecules can be detected by the altered CD spectra in the UV region, which is characteristic of binding interactions with proteins. Due to the non-chiral nature of nanoparticles, they rarely affect protein CD spectra. Native BSA exhibits two negative absorption peaks at 209 nm and 222 nm, which are characteristic of the α-helix present in BSA.[77] When black carbon is allowed to interact with BSA protein, a very slight increase in the negative peak without any shift in wavelength was observed, suggesting a minor decrease in the α-helix without affecting the conformation of the BSA.[78] In a recent study, BSA protein showed a negative peak in the spectrum at 200–290 nm indicating the loss of the α-helical structures and an increase in the β-turns while interacting with nanoparticles.[79] Nuclear magnetic resonance (NMR) enables the study of the interactions of NP–protein complexes by providing information on the secondary structures of proteins. It is best suited to complex mixtures like biological fluids and high molecular weight proteins. As Au nanoparticles are attractive delivery vehicles in the biomedical field, their interactions with proteins are studied intensively with NMR, including the chemical nature of the interactions.[80] For kinetics studies of various protein–nanoparticle interactions, including association and dissociation, surface plasmon resonance (SPR) has been used extensively.[81,82] The unique label-free real-time monitoring ability of their molecular binding events makes them a very useful tool for studying nanoparticle–protein interactions. In a study, multiplexing data acquisition using SPR is used to show a comparative study of the corona formation of various serum proteins with gold nanoparticles.[82]

6.5.2 Methods Based on Size

Proteins adsorbed on the surfaces of nanoparticles can be visualized after negative staining under a transmission electron microscope.[83] It is only a qualitative analysis for the detection of proteins adsorbed on the surfaces of the particles and cannot be used for quantitative analysis. Proteins adsorbed onto the surfaces of nanoparticles can be observed directly by analyzing the change in the hydrodynamic diameter by using dynamic light scattering (DLS). As more and more proteins are adsorbed onto the surfaces of the nanoparticles, their hydrodynamic size increases.[84,85] DLS is best suited to small particles.[85] Fluctuation correlation spectroscopy (FCS) can be used to measure the size of a particle based on the fluctuations in the fluorescence intensity emitted by the particle.[86] It works on the same principle as DLS, by measuring the hydrodynamic size of the particle, which is the readout of diffusion coefficient.[87] Protein adsorption directly affects the Brownian motion of nanoparticles, which can be detected by nanoparticle tracking analysis (NTA). Due to protein binding, the mass of a nanoparticle increases and thereby slows down the Brownian motion of the particle.[88] NTA detects the movement of each nanoparticle by detecting the light scattered by the particles.[89]

6.5.3 Methods Based on Charge

Irrespective of the initial potential measurement, there is always a shift in the potential observed with the evolution of a protein corona.[69] For example, when the protein concentration increases, it imparts a decrease in the zeta-potential of the hydroxyapatite nanoparticle implying binding of negatively charged BSA. This alteration in charge can be detected by measuring the zeta-potential of the particles. The main principle behind laser doppler anemometry (LDA) is the charge alteration on the surfaces of nanoparticles with respect to protein binding. In LDA, nanoparticles are illuminated with a laser beam and are allowed to move in an applied electric field. With respect to their drift velocity, their zeta-potential is determined.[90,91] Further, the zeta-potentials of nanoparticles can also be determined by using tunable resistive pulse sensing (TRPS). When a NP passes through a membrane, it creates a blockade in the electric current passing through. On association with any protein component, this blockade signal changes and this can be correlated with the thickness of the protein layer.[88] Using TRPS, Blundell *et al.* characterized the protein corona formed when human serum and plasma samples were treated with carboxylated polystyrene particles.[92]

In response to an applied external strain, some materials generate an electric charge called the piezoelectric effect and on this phenomenon, quartz crystal microbalance (QCM) sensing technology works. Two techniques have been developed based on wave propagation phenomena for a label-free quantitative assay format for non-rigid biological recognition surface plasmon resonance (SPR) and QCM. QCM uses acoustic waves to measure the frequency of the minimum acoustic impedance, whereas SPR uses optical waves to measure the minimum angle of reflection. From the studies conducted by comparing the pre-washed and post-washed data of QCM with dissipation, it was found that BSA forms a relatively well-defined reversible layer as compared to Mb, and cytochrome c (CytC) on mercaptoundecanoic-acid-stabilized Au nanoparticles and surfaces.[93] Further, for screening protein recognition motifs exposed on nanoparticle surfaces, QCM provided a rapid and label-free method.[94]

6.5.4 Methods Based on Fractionation

On the basis of the nanoparticle size and density variation due to protein binding, variation is observed in differential sedimentation in a density gradient (sucrose gradient) exposed under centrifugal force. Differential centrifugal sedimentation (DCS) was used to study the structures and stabilities of protein–NP complexes in human plasma.[95] The thickness of the protein corona was estimated to be 10 nm for silica and polystyrene nanomaterials. In a complex protein mixture, it can give information on the size distribution of NP–protein complexes in a semi-quantitative manner and can resolve distinct particles with more than 5% size differences.[76] Moreover, the technique could not become popular due to some underestimations observed while showing protein shell thicknesses.[88]

Any alteration in the surface charge and mass of nanoparticles due to protein binding can affect nanoparticle movement in an applied electric field. Due to protein interactions, there is an increase in the mass of protein–nanoparticle conjugates, and, thereby, their movement is comparatively retarded in an applied electric field.[83] One-dimensional gel electrophoresis (SDS-PAGE) can also give a detailed idea of the proteins involved in corona formation after separating them from the nanoparticles and their subsequent denaturation and digestion.[96] As it is a qualitative method, it will only give an idea of the types of proteins involved in corona formation. Capillary electrophoresis (CE) is based on the relative mobility of NPs and protein–NP complexes in an applied electric field with respect to their charge.[97] As it works without any matrix material, the hydrodynamic size has a lot less impact on its separation, which is the main parameter in gel electrophoresis. CE allows the determination of a protein concentration in a protein–NP complex as a function of altered retention time by fitting it with appropriate mathematical models like the Hill model to calculate the dissociation constant.[98]

The flow field flow fractionation (F4) technique can be performed for a wide colloidal size range separation and distinguishes proteins from protein–NP complexes with high resolution. It can be combined with different fields such as gravitational, centrifugal, magnetic, electrical *etc.* to obtain differential fractions. Recently, asymmetric flow field flow fractionation has been used to separate unbound proteins from AuNP–HSA complexes by size in a complex polydisperse sample mixture accurately.[84] Also, combining this technique with centrifugation helped with the separation of protein coronas formed on SPIONs by their relative dissociation rates when incubated in the presence of major serum proteins like IgG and albumin-depleted human serum.[99]

6.5.5 Methods Based on Temperature

Isothermal titration calorimetry (ITC) works by detecting the generated or adsorbed heat formed by protein–nanoparticle complexes and thereby obtaining both affinity and maximal binding capacity to nanoparticles.[81] Cedervall *et al.* estimated the stoichiometry, enthalpy, and affinity of HSA and nanoparticle interactions. From the negative injection signals, they implied it was an exothermic process. While using ITC for exploring the interaction between BSA and hydroxyapatite nanorods, they found it was an exothermic reaction and it decreased with an increase in the HA–BSA ratio.[69]

6.6 Future Perspectives

Although interactions between nanoparticles and protein coronas have been studied for decades, very few important aspects have emerged so far. The complexity of the environmental systems is a barrier to understanding the nanoparticle and protein corona interaction. There is an urgent need for simplified model systems and new analytical tools to study this interaction and its function. To get a comprehensive overview of corona profiles, and to recognize the potential biological impact, a multilevel approach is surely mandatory with high-throughput screening along with a bioinformatics data mining approach (Figure 6.1).

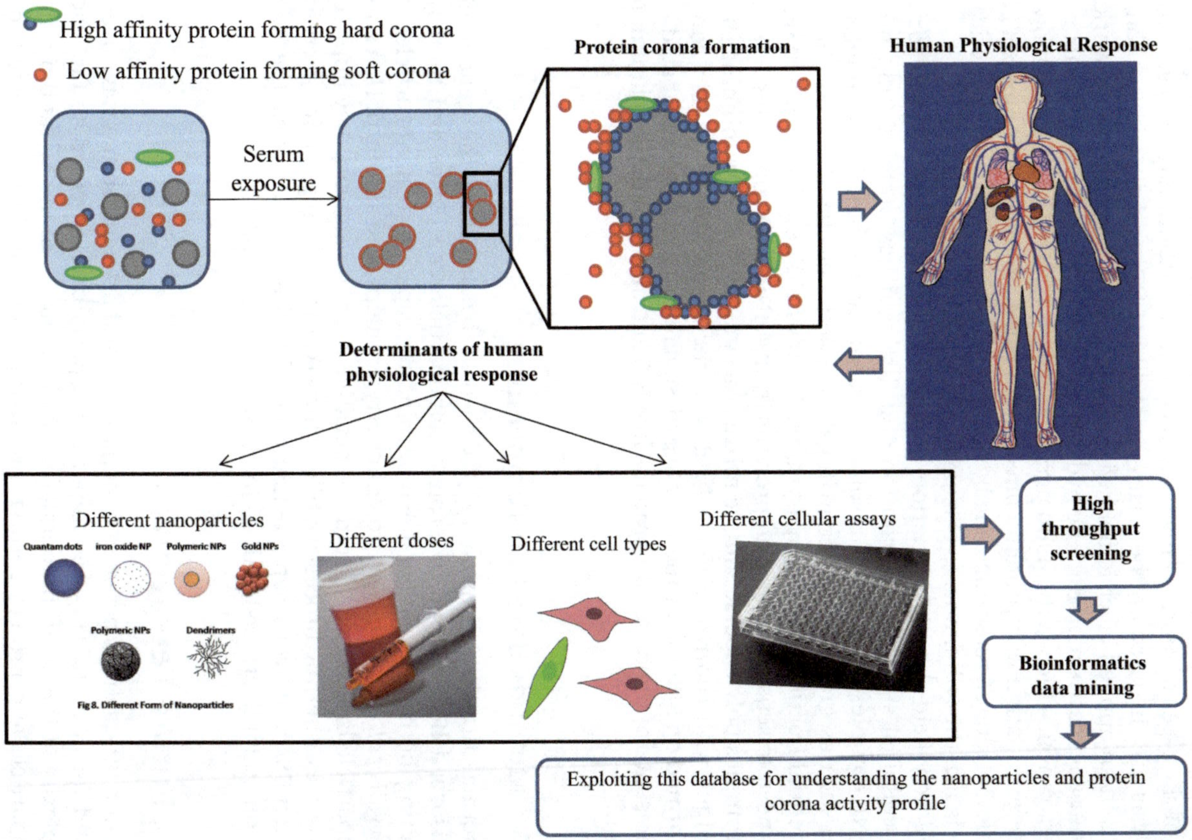

Figure 6.1 Graphical representation of the use of bioinformatics to determine the nanoparticle and protein corona activity profile in humans.

6.7 Conclusion

The rapid development of nano-biotechnology for the enhanced performance of products is leading to an increased risk to living systems. Various technologies are available for studying nanoparticle–protein interactions; however, there is a need for the development of new arrays of techniques for a better understanding of nanoparticles when incubated in biological milieu for the alteration of the physio–chemical and structural properties of both. Also, there is a need to understand the possible binding targets and types of interactions occurring in living systems with implications for nanomedicine and nanotoxicology. Various important questions need to be addressed, such as whether the interaction leads to a monolayer or a multilayer, and, if formed, whether the layer is a soft corona or a hard corona, and what are the influences on biological assays?

References

1. A. A. Vertegel, R. W. Siegel and J. S. Dordick, Silica nanoparticle size influences the structure and enzymatic activity of adsorbed lysozyme, *Langmuir*, 2004, **20**(16), 6800–6807.
2. M. Lundqvist, I. Sethson and B.-H. Jonsson, Protein adsorption onto silica nanoparticles: conformational changes depend on the particles' curvature and the protein stability, *Langmuir*, 2004, **20**(24), 10639–10647.
3. Z. J. Deng, M. Liang, I. Toth, M. J. Monteiro and R. F. Minchin, Molecular interaction of poly (acrylic acid) gold nanoparticles with human fibrinogen, *ACS Nano*, 2012, **6**(10), 8962–8969.
4. Z. J. Deng, M. Liang, I. Toth, M. Monteiro and R. F. Minchin, Plasma protein binding of positively and negatively charged polymer-coated gold nanoparticles elicits different biological responses, *Nanotoxicology*, 2012, **7**(3), 314–322.
5. M. Lundqvist, J. Stigler, G. Elia, I. Lynch, T. Cedervall and K. A. Dawson, Nanoparticle size and surface properties determine the protein corona with possible implications for biological impacts, *Proc. Natl. Acad. Sci. U. S. A.*, 2008, **105**(38), 14265–14270.
6. P. Wang, X. Wang, L. Wang, X. Hou, W. Liu and C. Chen, Interaction of gold nanoparticles with proteins and cells, *Sci. Technol. Adv. Mater.*, 2015, **16**(3), 034610.
7. F. Turci, E. Ghibaudi, M. Colonna, B. Boscolo, I. Fenoglio and B. Fubini, An integrated approach to the study of the interaction between proteins and nanoparticles, *Langmuir*, 2010, **26**(11), 8336–8346.
8. J. L. Perry, K. G. Reuter, M. P. Kai, K. P. Herlihy, S. W. Jones and J. C. Luft, *et al.*, PEGylated PRINT nanoparticles: the impact of PEG density on protein binding, macrophage association, biodistribution, and pharmacokinetics, *Nano Lett.*, 2012, **12**(10), 5304–5310.

9. A. Lesniak, A. Campbell, M. P. Monopoli, I. Lynch, A. Salvati and K. A. Dawson, Serum heat inactivation affects protein corona composition and nanoparticle uptake, *Biomaterials*, 2010, **31**(36), 9511–9518.
10. M. Mahmoudi, A. M. Abdelmonem, S. Behzadi, J. H. Clement, S. Dutz and M. R. Ejtehadi, *et al.*, Temperature: the "ignored" factor at the nano-bio interface, *ACS Nano*, 2013, 7(8), 6555–6562.
11. M. Rahman, S. Laurent, N. Tawil, L. H. Yahia and M. Mahmoudi, Nanoparticle and protein corona, *Protein-nanoparticle Interactions*, Springer, 2013 pp. 21–44.
12. M. Azhdarzadeh, A. A. Saei, S. Sharifi, M. J. Hajipour, A. M. Alkilany and M. Sharifzadeh, *et al.*, Nanotoxicology: advances and pitfalls in research methodology, *Nanomedicine*, 2015, **10**(18), 2931–2952.
13. M. V. Berridge, P. M. Herst and A. S. Tan, Tetrazolium dyes as tools in cell biology: new insights into their cellular reduction, *Biotechnol. Annu. Rev.*, 2005, **11**, 127–152.
14. M. Mahmoudi, A. Simchi, M. Imani, M. A. Shokrgozar, A. S. Milani and U. O. Häfeli, *et al.*, A new approach for the in vitro identification of the cytotoxicity of superparamagnetic iron oxide nanoparticles, *Colloids Surf., B*, 2010, **75**(1), 300–309.
15. A. Lesniak, F. Fenaroli, M. P. Monopoli, C. Åberg, K. A. Dawson and A. Salvati, Effects of the presence or absence of a protein corona on silica nanoparticle uptake and impact on cells, *ACS Nano*, 2012, **6**(7), 5845–5857.
16. A. Lesniak, A. Salvati, M. J. Santos-Martinez, M. W. Radomski, K. A. Dawson and C. Åberg, Nanoparticle adhesion to the cell membrane and its effect on nanoparticle uptake efficiency, *J. Am. Chem. Soc.*, 2013, **135**(4), 1438–1444.
17. F. Bertoli, G. L. Davies, M. P. Monopoli, M. Moloney, Y. K. Gun'ko and A. Salvati, *et al.*, Magnetic Nanoparticles to Recover Cellular Organelles and Study the Time Resolved Nanoparticle-Cell Interactome throughout Uptake, *Small*, 2014, **10**(16), 3307–3315.
18. L. Treuel, S. Brandholt, P. Maffre, S. Wiegele, L. Shang and G. U. Nienhaus, Impact of protein modification on the protein corona on nanoparticles and nanoparticle–cell interactions, *ACS Nano*, 2014, **8**(1), 503–513.
19. K. Saha, D. F. Moyano and V. M. Rotello, Protein coronas suppress the hemolytic activity of hydrophilic and hydrophobic nanoparticles, *Mater. Horiz.*, 2014, **1**(1), 102–105.
20. H. N. Yehia, R. K. Draper, C. Mikoryak, E. K. Walker, P. Bajaj and I. H. Musselman, *et al.*, Single-walled carbon nanotube interactions with HeLa cells, *J. Nanobiotechnol.*, 2007, **5**(1), 8.
21. J. Davda and V. Labhasetwar, Characterization of nanoparticle uptake by endothelial cells, *Int. J. Pharm.*, 2002, **233**(1–2), 51–59.
22. P. Ruenraroengsak, P. Novak, D. Berhanu, A. J. Thorley, E. Valsami-Jones and J. Gorelik, *et al.*, Respiratory epithelial cytotoxicity and membrane damage (holes) caused by amine-modified nanoparticles, *Nanotoxicology*, 2012, **6**(1), 94–108.

23. H. Yin, R. Chen, P. S. Casey, P. C. Ke, T. P. Davis and C. Chen, Reducing the cytotoxicity of ZnO nanoparticles by a pre-formed protein corona in a supplemented cell culture medium, *RSC Adv.*, 2015, **5**(90), 73963–73973.
24. M. Ramasamy, M. Das, S. S. A. An and D. K. Yi, Role of surface modification in zinc oxide nanoparticles and its toxicity assessment toward human dermal fibroblast cells, *Int. J. Nanomed.*, 2014, **9**, 3707.
25. K. Kostarelos, The long and short of carbon nanotube toxicity, *Nat. Biotechnol.*, 2008, **26**(7), 774.
26. Y. Zhao, G. Xing and Z. Chai, Nanotoxicology: are carbon nanotubes safe?, *Nat. Nanotechnol.*, 2008, **3**(4), 191.
27. S. Tenzer, D. Docter, J. Kuharev, A. Musyanovych, V. Fetz and R. Hecht, *et al.*, Rapid formation of plasma protein corona critically affects nanoparticle pathophysiology, *Nat. Nanotechnol.*, 2013, **8**(10), 772.
28. S. Kittler, C. Greulich, J. Diendorf, M. Koller and M. Epple, Toxicity of silver nanoparticles increases during storage because of slow dissolution under release of silver ions, *Chem. Mater.*, 2010, **22**(16), 4548–4554.
29. A. Manke, L. Wang and Y. Rojanasakul, Mechanisms of nanoparticle-induced oxidative stress and toxicity, *BioMed Res. Int.*, 2013, **2013**, 1–15.
30. Z. N. Gheshlaghi, G. H. Riazi, S. Ahmadian, M. Ghafari and R. Mahinpour, Toxicity and interaction of titanium dioxide nanoparticles with microtubule protein, *Acta Biochim. Biophys. Sin.*, 2008, **40**(9), 777–782.
31. N. Wangoo, C. R. Suri and G. Shekhawat, Interaction of gold nanoparticles with protein: a spectroscopic study to monitor protein conformational changes, *Appl. Phys. Lett.*, 2008, **92**(13), 133104.
32. Y. Yan, K. T. Gause, M. M. Kamphuis, C.-S. Ang, N. M. O'Brien-Simpson and J. C. Lenzo, *et al.*, Differential roles of the protein corona in the cellular uptake of nanoporous polymer particles by monocyte and macrophage cell lines, *ACS Nano*, 2013, **7**(12), 10960–10970.
33. K. Yu, B. F. Lai, J. H. Foley, M. J. Krisinger, E. M. Conway and J. N. Kizhakkedathu, Modulation of complement activation and amplification on nanoparticle surfaces by glycopolymer conformation and chemistry, *ACS Nano*, 2014, **8**(8), 7687–7703.
34. Z. J. Deng, M. Liang, M. Monteiro, I. Toth and R. F. Minchin, Nanoparticle-induced unfolding of fibrinogen promotes Mac-1 receptor activation and inflammation, *Nat. Nanotechnol.*, 2011, **6**(1), 39.
35. K. Kansara, P. Patel, D. Shah, R. K. Shukla, S. Singh and A. Kumar, *et al.*, TiO2 nanoparticles induce DNA double strand breaks and cell cycle arrest in human alveolar cells, *Environ. Mol. Mutagen.*, 2015, **56**(2), 204–217.
36. A. K. Gupta, M. Gupta, S. J. Yarwood and A. S. Curtis, Effect of cellular uptake of gelatin nanoparticles on adhesion, morphology and cytoskeleton organisation of human fibroblasts, *J. Controlled Release*, 2004, **95**(2), 197–207.

37. M. Pöttler, A. Staicu, J. Zaloga, H. Unterweger, B. Weigel and E. Schreiber, *et al.*, Genotoxicity of superparamagnetic iron oxide nanoparticles in granulosa cells, *Int. J. Mol. Sci.*, 2015, **16**(11), 26280–26290.
38. S. Behzadi, V. Serpooshan, R. Sakhtianchi, B. Müller, K. Landfester and D. Crespy, *et al.*, Protein corona change the drug release profile of nanocarriers: the "overlooked" factor at the nanobio interface, *Colloids Surf. B*, 2014, **123**, 143–149.
39. M. Mahmoudi, N. Bertrand, H. Zope and O. C. Farokhzad, Emerging understanding of the protein corona at the nano-bio interfaces, *Nano Today*, 2016, **11**(6), 817–832.
40. J. C. Y. Kah, J. Chen, A. Zubieta and K. Hamad-Schifferli, Exploiting the protein corona around gold nanorods for loading and triggered release, *ACS Nano*, 2012, **6**(8), 6730–6740.
41. M. Lundqvist, J. Stigler, T. Cedervall, T. Berggård, M. B. Flanagan and I. Lynch, *et al.*, The evolution of the protein corona around nanoparticles: a test study, *ACS Nano*, 2011, **5**(9), 7503–7509.
42. Y. D. Álvarez, J. V. Pellegrotti and F. D. Stefani, Gold Nanoparticles as Nucleation Centers for Amyloid Fibrillation, *Use of Nanoparticles in Neuroscience*, Springer, 2018, pp. 269–291.
43. M. Zaman, E. Ahmad, A. Qadeer, G. Rabbani and R. H. Khan, Nanoparticles in relation to peptide and protein aggregation, *Int. J. Nanomed.*, 2014, **9**, 899.
44. W.-h. Wu, X. Sun, Y.-p. Yu, J. Hu, L. Zhao and Q. Liu, *et al.*, TiO2 nanoparticles promote β-amyloid fibrillation in vitro, *Biochem. Biophys. Res. Commun.*, 2008, **373**(2), 315–318.
45. S. Linse, C. Cabaleiro-Lago, W.-F. Xue, I. Lynch, S. Lindman and E. Thulin, *et al.*, Nucleation of protein fibrillation by nanoparticles, *Proc. Natl. Acad. Sci. U. S. A.*, 2007, **104**(21), 8691–8696.
46. M. De and V. M. Rotello, Synthetic "chaperones": nanoparticle-mediated refolding of thermally denatured proteins, *Chem. Commun.*, 2008, (30), 3504–3506.
47. S. I. Yoo, M. Yang, J. R. Brender, V. Subramanian, K. Sun and N. E. Joo, *et al.*, Inhibition of amyloid peptide fibrillation by inorganic nanoparticles: functional similarities with proteins, *Angew. Chem., Int. Ed.*, 2011, **50**(22), 5110–5115.
48. H. LeVine III, Quantification of β-sheet amyloid fibril structures with thioflavin T, *Methods in Enzymology*, Elsevier, 1999, vol. 309, pp. 274–284.
49. D. A. Yushchenko, J. A. Fauerbach, S. Thirunavukkuarasu, E. A. Jares-Erijman and T. M. Jovin, Fluorescent ratiometric MFC probe sensitive to early stages of α-synuclein aggregation, *J. Am. Chem. Soc.*, 2010, **132**(23), 7860–7861.
50. A. Salvati, A. S. Pitek, M. P. Monopoli, K. Prapainop, F. B. Bombelli and D. R. Hristov, *et al.*, Transferrin-functionalized nanoparticles lose their targeting capabilities when a biomolecule corona adsorbs on the surface, *Nat. Nanotechnol.*, 2013, **8**(2), 137.

51. V. Mirshafiee, M. Mahmoudi, K. Lou, J. Cheng and M. L. Kraft, Protein corona significantly reduces active targeting yield, *Chem. Commun.*, 2013, **49**(25), 2557–2559.
52. K. Zarschler, K. Prapainop, E. Mahon, L. Rocks, M. Bramini and P. Kelly, *et al.*, Diagnostic nanoparticle targeting of the EGF-receptor in complex biological conditions using single-domain antibodies, *Nanoscale*, 2014, **6**(11), 6046–6056.
53. W. Xiao, J. Xiong, S. Zhang, Y. Xiong, H. Zhang and H. Gao, Influence of ligands property and particle size of gold nanoparticles on the protein adsorption and corresponding targeting ability, *Int. J. Pharm.*, 2018, **538**(1–2), 105–111.
54. R. Dal Magro, B. Albertini, S. Beretta, R. Rigolio, E. Donzelli and A. Chiorazzi, *et al.*, Artificial apolipoprotein corona enables nanoparticle brain targeting, *Nanomedicine: Nanotechnol., Biol. Med.*, 2018, **14**(2), 429–438.
55. G. Caracciolo, D. Pozzi, A. Capriotti, C. Cavaliere, F. Cardarelli and A. Bifone, *et al.*, Cancer cell targeting of lipid gene vectors by protein corona, *Nanotechnology*, 2012, **3**, 354–357.
56. H. Amiri, L. Bordonali, A. Lascialfari, S. Wan, M. P. Monopoli and I. Lynch, *et al.*, Protein corona affects the relaxivity and MRI contrast efficiency of magnetic nanoparticles, *Nanoscale*, 2013, **5**(18), 8656–8665.
57. H. S. Choi, W. Liu, P. Misra, E. Tanaka, J. P. Zimmer and B. I. Ipe, *et al.*, Renal clearance of quantum dots, *Nat. Biotechnol.*, 2007, **25**(10), 1165.
58. E. Sadauskas, H. Wallin, M. Stoltenberg, U. Vogel, P. Doering and A. Larsen, *et al.*, Kupffer cells are central in the removal of nanoparticles from the organism, *Part. Fibre Toxicol.*, 2007, **4**(1), 1.
59. B. Wang, X. He, Z. Zhang, Y. Zhao and W. Feng, Metabolism of nanomaterials *in vivo*: blood circulation and organ clearance, *Acc. Chem. Res.*, 2012, **46**(3), 761–769.
60. D. E. Owens III and N. A. Peppas, Opsonization, biodistribution, and pharmacokinetics of polymeric nanoparticles, *Int. J. Pharm.*, 2006, **307**(1), 93–102.
61. N. Bertrand, P. Grenier, M. Mahmoudi, E. M. Lima, E. A. Appel and F. Dormont, *et al.*, Mechanistic understanding of *in vivo* protein corona formation on polymeric nanoparticles and impact on pharmacokinetics, *Nat. Commun.*, 2017, **8**(1), 777.
62. I. Lynch, A. Salvati and K. A. Dawson, Protein-nanoparticle interactions: what does the cell see?, *Nat. Nanotechnol.*, 2009, **4**(9), 546.
63. A. E. Gulyaev, S. E. Gelperina, I. N. Skidan, A. S. Antropov, G. Y. Kivman and J. Kreuter, Significant transport of doxorubicin into the brain with polysorbate 80-coated nanoparticles, *Pharm. Res.*, 1999, **16**(10), 1564–1569.
64. A. Festa, R. D'Agostino, R. P. Tracy and S. M. Haffner, Elevated levels of acute-phase proteins and plasminogen activator inhibitor-1 predict the development of type 2 diabetes: the insulin resistance atherosclerosis study, *Diabetes*, 2002, **51**(4), 1131–1137.

65. P. M. Ridker, C. H. Hennekens, J. E. Buring and N. Rifai, C-reactive protein and other markers of inflammation in the prediction of cardiovascular disease in women, *N. Engl. J. Med.*, 2000, **342**(12), 836–843.
66. M. J. Hajipour, S. Laurent, A. Aghaie, F. Rezaee and M. Mahmoudi, Personalized protein coronas: a "key" factor at the nanobiointerface, *Biomater. Sci.*, 2014, **2**(9), 1210–1221.
67. T. Zheng, N. Pierre-Pierre, X. Yan, Q. Huo, A. J. Almodovar and F. Valerio, *et al.*, Gold nanoparticle-enabled blood test for early stage cancer detection and risk assessment, *ACS Appl. Mater. Interfaces*, 2015, **7**(12), 6819–6827.
68. X. Xu, X. Mao, Y. Wang, D. Li, Z. Du and W. Wu, *et al.*, Study on the interaction of graphene oxide-silver nanocomposites with bovine serum albumin and the formation of nanoparticle-protein corona, *Int. J. Biol. Macromol.*, 2018, **116**, 492–501.
69. R. Rial, B. Tichnell, B. Latimer, Z. Liu, P. V. Messina and J. M. Ruso, Structural and Kinetic Visualization of the Protein Corona on Bioceramic Nanoparticles, *Langmuir*, 2018, **34**(7), 2471–2480.
70. J. H. Shannahan, R. Podila and J. M. Brown, A hyperspectral and toxicological analysis of protein corona impact on silver nanoparticle properties, intracellular modifications, and macrophage activation, *Int. J. Nanomed.*, 2015, **10**, 6509.
71. D. Sarkar, Probing the interaction of a globular protein with a small fluorescent probe in the presence of silver nanoparticles: spectroscopic characterization of its domain specific association and dissociation, *RSC Adv.*, 2013, **3**(46), 24389–24399.
72. A. L. Lira, R. S. Ferreira, R. J. Torquato, H. Zhao, M. L. V. Oliva and S. A. Hassan, *et al.*, Binding kinetics of ultrasmall gold nanoparticles with proteins, *Nanoscale*, 2018, **10**(7), 3235–3244.
73. M. Wang, C. Fu, X. Liu, Z. Lin, N. Yang and S. Yu, Probing the mechanism of plasma protein adsorption on Au and Ag nanoparticles with FT-IR spectroscopy, *Nanoscale*, 2015, **7**(37), 15191–15196.
74. H. S. Mandal and H.-B. Kraatz, Effect of the surface curvature on the secondary structure of peptides adsorbed on nanoparticles, *J. Am. Chem. Soc.*, 2007, **129**(20), 6356–6357.
75. X.-C. Shen, X.-Y. Liou, L.-P. Ye, H. Liang and Z.-Y. Wang, Spectroscopic studies on the interaction between human hemoglobin and CdS quantum dots, *J. Colloid Interface Sci.*, 2007, **311**(2), 400–406.
76. M. Mahmoudi, I. Lynch, M. R. Ejtehadi, M. P. Monopoli, F. B. Bombelli and S. Laurent, Protein–nanoparticle interactions: opportunities and challenges, *Chem. Rev.*, 2011, **111**(9), 5610–5637.
77. Z. Liao, X. Yu, Q. Yao and P. Yi, Interaction between pirenoxine and bovine serum albumin in aqueous solution, *Spectrochim. Acta, Part A*, 2014, **129**, 314–319.
78. H. Wu, M. Chen, M. Shang, X. Li, K. Mu and S. Fan, *et al.*, Insights into the binding behavior of bovine serum albumin to black carbon nanoparticles and induced cytotoxicity, *Spectrochim. Acta, Part A*, 2018, **200**, 51–57.

79. A. E. Gerdon, D. W. Wright and D. E. Cliffel, Quartz crystal microbalance detection of glutathione-protected nanoclusters using antibody recognition, *Anal. Chem.*, 2005, **77**(1), 304–310.
80. A. Wang, K. Vangala, T. Vo, D. Zhang and N. C. Fitzkee, A three-step model for protein–gold nanoparticle adsorption, *J. Phys. Chem. C*, 2014, **118**(15), 8134–8142.
81. T. Cedervall, I. Lynch, S. Lindman, T. Berggård, E. Thulin and H. Nilsson, *et al.*, Understanding the nanoparticle–protein corona using methods to quantify exchange rates and affinities of proteins for nanoparticles, *Proc. Natl. Acad. Sci. U. S. A.*, 2007, **104**(7), 2050–2055.
82. A. Patra, T. Ding, G. Engudar, Y. Wang, M. M. Dykas and B. Liedberg, *et al.*, Component-Specific Analysis of Plasma Protein Corona Formation on Gold Nanoparticles Using Multiplexed Surface Plasmon Resonance, *Small*, 2016, **12**(9), 1174–1182.
83. J. Hühn, C. Fedeli, Q. Zhang, A. Masood, P. del Pino and N. M. Khashab, *et al.*, Dissociation coefficients of protein adsorption to nanoparticles as quantitative metrics for description of the protein corona: A comparison of experimental techniques and methodological relevance, *Int. J. Biochem. Cell Biol.*, 2016, **75**, 148–161.
84. R. Capomaccio, I. O. Jimenez, P. Colpo, D. Gilliland, G. Ceccone and F. Rossi, *et al.*, Determination of the structure and morphology of gold nanoparticle–HSA protein complexes, *Nanoscale*, 2015, **7**(42), 17653–17657.
85. K. Fischer and M. Schmidt, Pitfalls and novel applications of particle sizing by dynamic light scattering, *Biomaterials*, 2016, **98**, 79–91.
86. S. Dominguez-Medina, S. Chen, J. Blankenburg, P. Swanglap, C. F. Landes and S. Link, Measuring the hydrodynamic size of nanoparticles using fluctuation correlation spectroscopy, *Annu. Rev. Phys. Chem.*, 2016, **67**, 489–514.
87. G. U. Nienhaus, P. Maffre and K. Nienhaus, Studying the protein corona on nanoparticles by FCS, *Methods in Enzymology*, Elsevier, 2013, vol. 519, pp. 115–137.
88. C. Carrillo-Carrion, M. Carril and W. J. Parak, Techniques for the experimental investigation of the protein corona, *Curr. Opin. Biotechnol.*, 2017, **46**, 106–113.
89. A. Jedlovszky-Hajdú, F. B. Bombelli, M. P. Monopoli, E. Tombácz and K. A. Dawson, Surface coatings shape the protein corona of SPIONs with relevance to their application in vivo, *Langmuir*, 2012, **28**(42), 14983–14991.
90. C. Y. Tay, M. I. Setyawati, J. Xie, W. J. Parak and D. T. Leong, Back to basics: exploiting the innate physico-chemical characteristics of nanomaterials for biomedical applications, *Adv. Funct. Mater.*, 2014, **24**(38), 5936–5955.
91. T. L. Doane, C.-H. Chuang, R. J. Hill and C. Burda, Nanoparticle ζ-potentials, *Acc. Chem. Res.*, 2011, **45**(3), 317–326.
92. E. L. Blundell, M. J. Healey, E. Holton, M. Sivakumaran, S. Manstana and M. Platt, Characterisation of the protein corona using tunable resistive

pulse sensing: determining the change and distribution of a particle's surface charge, *Anal. Bioanal. Chem.*, 2016, **408**(21), 5757–5768.
93. E. D. Kaufman, J. Belyea, M. C. Johnson, Z. M. Nicholson, J. L. Ricks and P. K. Shah, *et al.*, Probing protein adsorption onto mercaptoundecanoic acid stabilized gold nanoparticles and surfaces by quartz crystal microbalance and ζ-potential measurements, *Langmuir*, 2007, **23**(11), 6053–6062.
94. M. Gianneli, E. Polo, H. Lopez, V. Castagnola, T. Aastrup and K. Dawson, Label-free in-flow detection of receptor recognition motifs on the biomolecular corona of nanoparticles, *Nanoscale*, 2018, **10**(12), 5474–5481.
95. D. Walczyk, F. B. Bombelli, M. P. Monopoli, I. Lynch and K. A. Dawson, What the cell "sees" in bionanoscience, *J. Am. Chem. Soc.*, 2010, **132**(16), 5761–5768.
96. C. Gräfe, A. Weidner, M. vd Lühe, C. Bergemann, F. H. Schacher and J. H Clement, *et al.*, Intentional formation of a protein corona on nanoparticles: serum concentration affects protein corona mass, surface charge and nanoparticle–cell interaction, *Int. J. Biochem. Cell Biol.*, 2016, **75** 196–202.
97. U. Pyell, A. H. Jalil, D. A. Urban, C. Pfeiffer, B. Pelaz and W. J. Parak Characterization of hydrophilic coated gold nanoparticles via capillary electrophoresis and Taylor dispersion analysis. Part II: Determination of the hydrodynamic radius distribution–comparison with asymmetric flow field-flow fractionation, *J. Colloid Interface Sci.*, 2015, **457**, 131–140.
98. N. Li, S. Zeng, L. He and W. Zhong, Probing nanoparticle–protein interaction by capillary electrophoresis, *Anal. Chem.*, 2010, **82**(17), 7460–7466.
99. J. Ashby, S. Schachermeyer, S. Pan and W. Zhong, Dissociation-based screening of nanoparticle–protein interaction via flow field-flow fractionation, *Anal. Chem.*, 2013, **85**(15), 7494–7501.

CHAPTER 7

Computer Simulations for Understanding Nanoparticle-biomolecule Corona Formation†

LOKESH BAWEJA*

Department of Biochemistry, Molecular Biology and Biophysics, Iowa State University, Ames, Iowa 50011, USA
*E-mail: baweja.lokesh@gmail.com, lbaweja@iastate.edu

7.1 Introduction

Nanoparticles (NPs) are promising candidates for biomedical applications due to their superior/enhanced physico–chemical properties compared to their bulk counterparts. Nanoscale dimensions provide high surface–volume ratios, and enhanced optical and reactive properties, which make NPs suitable candidates for drug delivery, and *in vivo* and *in vitro* imaging.[1–4] However, once administered in biological fluids, proteins spontaneously form adsorption layers on NPs, termed "protein coronas"[5,6] (Figure 7.1). Physico–chemical properties such as shape, size, and surface chemistry of the NPs, and their biological environment, play important roles in deciding the composition of the protein corona.[7,8] Layers of protein that do not readily desorb from the surface of the NPs are part of the hard corona,

†Electronic supplementary information (ESI) available. See DOI: 10.1039/9781788016308

Issues in Toxicology No. 40
Nanoparticle–Protein Corona: Biophysics to Biology
Edited by Ashutosh Kumar and Alok Dhawan

Published by the Royal Society of Chemistry, www.rsc.org

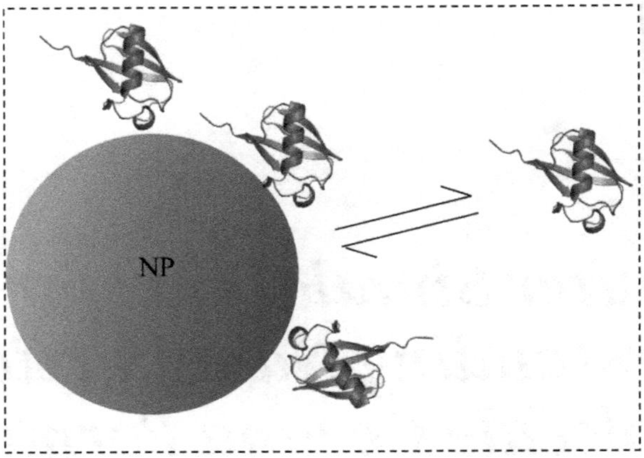

Figure 7.1 A pictorial representation of NP-protein corona formation.

whereas loosely bound proteins involved in a dynamic exchange with the proteins present in the surroundings are the soft corona.[9,10] Proteins with a high affinity for NPs tend to bind strongly and remain adsorbed during biophysical events such as endocytosis, and even after translocation to a new physiological environment.[11] Interaction of NPs with protein may lead to hazardous consequences, such as an abnormal conformational change of proteins, enough to cause protein aggregation,[12] exposure of cryptic epitopes[13] and structural modifications resulting in impaired activity of enzymes.[14] Conversely, there are studies that have demonstrated that adsorption of proteins on NPs reduces cytotoxicity,[15] facilitates receptor-mediated uptake,[16] and improves the pharmacokinetic profiles, which could be advantageous outcomes of protein adsorption on NPs. Biomolecular coronas form readily as NPs enter the physiological environment, and the coronas can be highly heterogeneous, comprising 300 different proteins at a time.[17] The affinity of coronas on NPs is governed by electrostatic, hydrophobic and van der Waals interactions.[18]

Understanding NP-induced changes in biomolecules is crucial for predicting their safety in a biological system and further advances in nanotechnological applications. Formation of protein coronas around NPs has been investigated mainly by mass spectrometry techniques.[19,20] It is inherently difficult to understand the dynamics of NP-corona formation and predict the affinity of the protein for the NPs' surface using experimental techniques. With the advances in computational techniques and the increase in computational power, molecular dynamics (MD) simulations have proved highly beneficial in predicting the initial steps of protein corona formation, which involves the adsorption and orientation of protein around NPs.[21] In this chapter, we have summarized MD simulation studies aimed at understanding NP-protein corona formation and discussed the strength and limitations of these studies.

7.2 Conformation of Protein in Hard Coronas

The primary layer of protein in a protein corona is highly susceptible to undergo conformational changes due to its direct interaction with the nanoparticle surface. The surface chemistry and curvature of NPs determine the extent of conformational changes in the adsorbed proteins.[22–24] Curved NPs were known to cause an irreversible change in the secondary structures of proteins.[25] This may have implications for the NPs' clearance and their immunoreactivity. Experimental as well as computational studies have demonstrated the effect of physico–chemical properties. There are several pieces of evidence where NP-induced conformational change modulates the interaction and function of protein or enzymes. It has been reported that superparamagnetic iron oxide nanoparticles (SPIONs) led to a conformational change in transferrin from a closed to an open conformation, causing the release of iron by the protein.[26] The SPION-introduced change is irreversible and the open state of the transferrin is maintained after the removal of the NPs. Structural changes not only affect the function of the protein but also modulate the interactions of proteins with their physiological binding partners. Fleischer and Payne demonstrated that the secondary structure of adsorbed BSA is strongly responsible for the interaction of the complexes with the receptors.[27] Polystyrene nanoparticle (NP)-protein complexes bind with different biological receptors based on the conformation of the adsorbed protein, cationic NPs perturb the secondary structure of bovine serum albumin (BSA) and bind with scavenger receptors, whereas BSA adsorbed on anionic NPs retains its native structure and binds to albumin receptors. Further, it has been reported that a NP-induced conformational change in fibrinogen promotes Mac-1 receptor activation and inflammation and the unfolding induced by the NPs.[28] These could be plausible mechanisms by which NPs can have adverse biological responses once administered in a physiological system. Smaller nanoparticles may also occupy the active site in the protein. It has been shown that C_{60} fullerene can plug several proteins and inhibits the activity of vital enzymes.[29] Solution NMR studies also predicted the formation lysozyme–fullerene conjugates.[30] Experimental techniques such as circular dichroism (CD), fluorescence spectroscopies and NMR studies have revealed the secondary structural changes of adsorbed proteins.[31] Along with experimental techniques, molecular dynamics (MD) simulations have provided invaluable information on the conformation of adsorbed protein and the forces responsible for causing the structural changes. NPs have a high tendency to aggregate in the physiological environment and adsorb chemical entities present in reaction systems.[32] CD is used to characterize the structure of the molecule through the different absorptions of circularly polarized light in the left-handed direction and the right-handed direction on the asymmetric molecule.[33,34] Structures of drug-delivery nanocarriers and the interactions of nanocarriers with biomolecules have been investigated using CD techniques. However, the actual contribution of amino acid residue and its conformation cannot be obtained by CD. Moreover, the analysis of CD spectra is a challenging task in a complex of chiral molecules adhering to a chiral receptor. Like CD, fluorescence spectroscopy also poses challenges to probe the secondary structural changes induced by NPs in the biomolecule. This can be explained by the fact that secondary structural changes within the protein are measured by the fluorescence of tyrosine and tryptophan residues, which are buried in the hydrophobic environment. The interaction of NPs with these residues can quench their fluorescence. Therefore, it is difficult to measure the change in the fluorescence of tyrosine and tryptophan when NPs interact with biomolecules.[21] MD simulation may circumvent these and provide a comprehensive understanding of NP-induced conformational changes in biomolecules. With the advances in computational power and the development of a coarse-grained technique, computer simulations have also been applied to understand biomolecule corona formation.

7.3 MD Simulations to Understand the Protein Corona Around Gold Nanoparticles (AuNPs)

All-atom simulations have been widely applied to understand the interactions of proteins with nanoparticles at a unimolecular level. However, simulation of nanoparticle-protein using all-atom resolution is computationally intensive. Studies using coarse-grained models of NPs and proteins were performed to understand NP-protein corona formation.[35] A coarse-grained model of gold nanoparticles (AuNPs) was built using 48 positively charged beads, which were randomly placed on a spherical surface 2.5 nm from the center of the NP to maintain a uniform distribution of the beads. A neutral bead was at the center of the sphere with a repulsive wall providing a spherical shape throughout the simulations and avoiding the penetration of the citrate and protein molecule.[35] The citrate molecules were randomly assigned near the surface of AuNPs to mimic citrate-coated gold nanoparticles. The competition between insulin and fibrinogen for binding to the AuNPs was studied with different concentrations of insulin and fibrinogen in a simulation box. The authors used relative concentrations of insulin–fibrinogen of 10 : 1, 20 : 2 and 34 : 4. This study reported that the 10 : 1 ratio of insulin adsorbed before the binding of fibrinogen and the rearrangement of proteins on the AuNPs' surface allowed a third insulin molecule to adsorb on the AuNPs. This simulation also captured the dissociation of the insulin molecule by the fibrinogen molecule. This could be attributed to the structural organization and high affinity of fibrinogen for AuNPs. Also, upon increasing the relative concentration of fibrinogen, the study reported a decrease in the binding of insulin with AuNPs. This study also reported an interesting finding – as the fibrinogen approaches, the insulin-bound AuNPs experience repulsion from the bound insulin molecule and move away from the surface. However, in another attempt, fibrinogen in contact with the AuNPs leads to the dissociation of the insulin molecule. This simulation captures the dynamic nature of protein corona formation. A less affine protein may bind more easily than a large high-affinity protein; the latter requires a structural arrangement to adsorb on the surface of the NPs. The study also probed the preferred binding site of fibrinogen with gold nanoparticles.

7.4 Ubiquitin Corona Around Silver Nanoparticles (AgNPs) Using Multiscale Modeling Approaches

Coarse-grained simulations have limited predictive power for studying NP interactions with other proteins and the role of amino acids in NP–protein interactions. Combining a high-resolution all-atom simulation with a coarse-grained model could be an alternative. Researchers have combined all-atom discrete molecular dynamics (DMD) simulations[36] with coarse-grained models to understand the formation of NP-ubiquitin coronas. In this study, the all-atom resolution was used to investigate the possible binding modes of ubiquitin on silver nanoparticles (AgNPs).[37] The knowledge of AgNP-ubiquitin was then incorporated into the coarse-grained models. The authors used multiple ubiquitin molecules to observe the formation of AgNP-ubiquitin coronas. These simulations predicted the specific binding between ubiquitin and AgNP, and corona formation involved non-specific binding between adsorbed and free protein molecules. The authors reported an exponential decay of the protein concentration and suggested that the association rate decreased as the time increased and as more proteins bound to the NP surface. This could be attributed to anti-cooperativity, which arises due to steric hindrance from the absorbed protein. Simulations performed by the authors at a ubiquitin–AgNP stoichiometry of 50 : 1 revealed the presence of multiple layers of ubiquitin on the AgNPs. The first layer in the complex is formed by the specific interaction of ubiquitin with the NP surface, whereas the outer layers involve protein–protein interactions. Furthermore, the authors found a close agreement with dynamic light scattering experiments performed on ubiquitin–AgNP stochiometric ratios of 100 : 1 and 500 : 1. This study also reported unfolded ubiquitin molecules, which could cause nanotoxicity (Figure 7.2). Unfolded protein molecules in coronas may expose cryptic epitopes and could initiate immune cascades.[13]

7.5 Prediction of the Protein Corona Compositions of Poly(ethylene glycol)- and Poly(phosphoester)-coated Nanoparticles by MD Simulation

Poly(ethylene glycol) (PEG) conjugated nanoparticles act as antifouling agents and retard the adsorption of proteins.[38,39] PEGylation also increases the circulation life of biologics as well as nanocarriers. Experimental studies

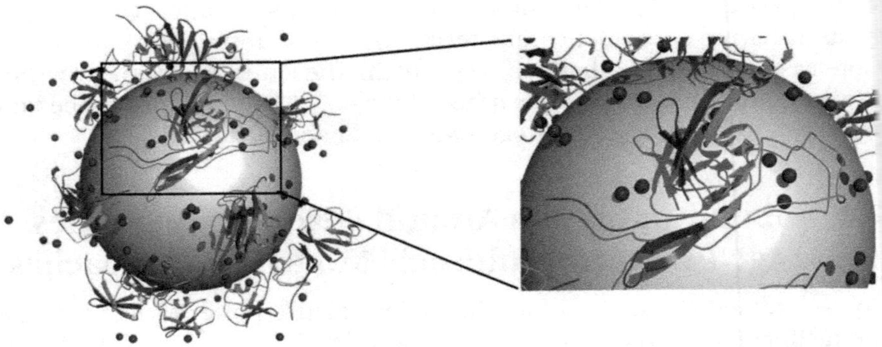

Figure 7.2 Ubiquitin-AgNP corona formation obtained from a discrete molecular dynamics (DMD) simulation. Unfolded ubiquitin can be seen in the inset. Reproduced from ref. 37 with permission from the Royal Society of Chemistry.

have reported that the protein corona of PEGylated nanoparticles has a very specific protein composition.[40] Settanni *et al.* predicted the composition of the protein corona around PEGylated nanoparticles and the results of the study were in close agreement with label-free mass proteomic approaches.[41] The authors, using a MD simulation, revealed the possible origin of the selective adsorption of protein on PEGylated nanoparticles. To understand the adsorption of selective plasma proteins on PEG-coated NPs, the authors simulated proteins in PEG–water mixtures and studied the distribution of PEG around the proteins. All the simulations were performed at an ionic strength of 0.15 M NaCl with an all-atom CHARMM27 force field.[42] The authors selected three plasma proteins – human serum albumin (HSA), transferrin (Tf), and complement Cq1 subcomponent subunit (CQ1C) – based on experimentally observed concentrations in the protein coronas of PEG-coated NPs.[40] The simulation studies showed the different affinities of proteins for PEG molecules, which is consistent with the experimental observation of the PEGylated nanoparticle corona.[40] PEG molecules tend to bind specifically to the hydrophobic region of the proteins; this information helped the authors to devise a method to predict the protein corona around the PEGylated nanoparticle based on the solvent accessible surface area (SASA) of the proteins. The model was derived and tested on the data derived from proteomic mass spectrometry experiments. The derived model was applied to provide a simple and accurate description of the adsorption process of plasma proteins on the surface of PEGylated proteins. This study highlighted a simple yet powerful approach for effectively predicting the adsorption of protein around functionalized NPs as the functional groups make initial contact with the adsorbed protein. Nevertheless, computer simulations were also proved useful in predicting the coronas of pulmonary surfactants as discussed in the next section of this chapter.

7.6 Molecular Structure of Pulmonary Surfactant Coronas on Nanoparticles

Inhalation is one of the major routes by which nanoparticles can enter physiological systems. NPs may preferentially interact with pulmonary surfactant present in the lungs and due to its hydrophobicity, it can preferentially adsorb on the surface of NPs.[43] PS is mainly composed of approximately 90% lipids, cholesterol, and 10% proteins such as SP-A, SP-B, SP-C and SP-D by weight.[44] Among the surfactant proteins, SP-B and SP-C are small hydrophobic membrane proteins that are essential for reducing alveolar surface tensions and important for maintaining normal respiratory physiology.[45] Formation of PS coronas on NPs may result in toxicity. A recent study by Hu *et al.* shows the mechanism of formation of a PS corona around NPs using coarse-grained models of hydrophilic silver and hydrophobic polystyrene NPs.[46] The authors used dipalmitoylphosphatidylcholine (DPPC) and palmitoyloleoylphosphatidylglycerol (POPG) and cholesterol at a molecular ratio of 14 : 6 : 5 to mimic the PS. The choice of model was dependent on the MD simulation study of PS monolayers using the same composition of lipids. The MD studies were in close agreement with the topographic structures of PS monolayers obtained from atomic force microscopy.[47] NPs with a diameter of 5 and 15 nm were studied for their interactions with PS and the formation of PS coronas. The process of corona formation occurs rapidly on the NPs with the smaller diameter. The assembly of PS molecules is dependent on the chemical properties of NPs. The lipid molecules adopt a bilayer-like conformation on hydrophilic AgNP and form a monolayer-like conformation on hydrophobic polystyrene PST-NP. The authors then performed coarse-grained simulations at larger timescales. After enough equilibration, the authors found that the formation of the PS corona is relatively conserved despite the different physico–chemical properties of the NPs. These simulation results were in good agreement with a recent lipidomic and proteomic characterization of the PS corona using high-resolution mass spectrometry.

The study also highlighted that the PS-associated proteins, SP-B and SP-C, appear to mediate the formation of the corona by interacting directly with the lipids in the corona (Figure 7.3). During the formation of the corona, SP-B and SP-C adopt different orientations. SP-B orients parallel to the lipid bilayer, whereas SP-C adopts a transmembrane orientation. SP-C is more hydrophobic than SP-B and is able to incorporate in the hydrophobic core of the lipid membrane. Despite their different orientations in PS, both proteins promoted the formation of the PS corona around the nanoparticles. Compared to SP-B and SP-C, SP-A is associated with PS corona formation to a lesser extent due to its hydrophilic nature.

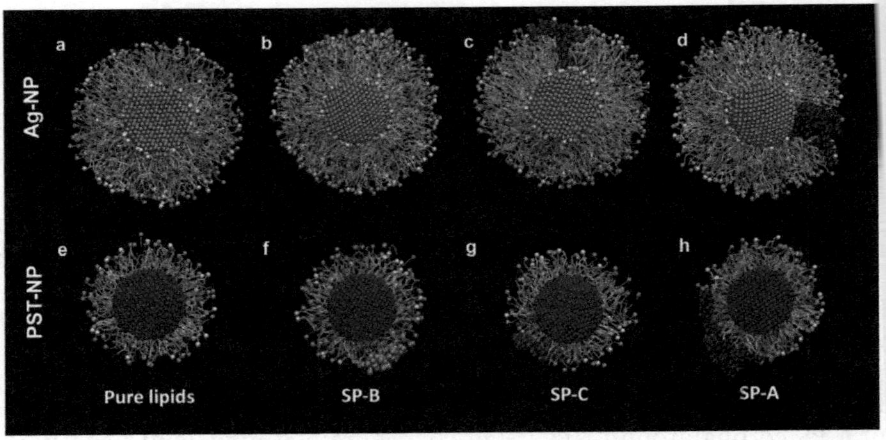

Figure 7.3 Molecular structures of the PS corona on silver (Ag-NP) (a)–(d) and polystyrene (PST-NP) (e)–(h). Reproduced from ref. 46 with permission from American Chemical Society, Copyright 2017.

7.7 Formation of an Amino Acid Corona Around TiO_2 Nanoparticles

Adsorption of amino acids on NPs has a major implication in understanding their toxicity towards the cellular system. It has been reported that NPs deplete the amino acids that are important for the support of cellular growth. This could be one of the indirect mechanisms by which NPs induce toxicity. Liu *et al.* have developed TiO_2-specific force fields to understand the interactions of 20 amino acids with TiO_2 nanoparticles.[48] The adsorption mechanism of amino acids on amorphous spherical nanoparticles was studied by molecular dynamics (MD) simulations and the affinities of the amino acids for the TiO_2 surface were further investigated by estimating free-energy profiles through calculating the potential of mean force (PMF).[49] Due to the hydrophilic nature of NPs, the author investigated the role of water around the NPs. TiO_2 showed three distinct layers of water around the NPs based on radial distribution function (RDF) analysis, which further governs the orientation of amino acids around TiO_2 NPs. The total adsorption probability for each amino acid was calculated by integrating the probability distribution over a distance range of 0–5 Å. Charged residues such as arginine, lysine, aspartate, and glutamate showed high adsorption probabilities. Uncharged amino acids showed very weak adsorption probabilities, and the amino acids tyrosine, serine and threonine interacted with the first layer of water around the TiO_2 NPs, which involved the formation of hydrogen bonds between the hydroxyl groups of the amino acids and water molecules. The findings from this MD study are in close agreement with experimental findings. Using simple models of NPs, one can predict the adsorption of amino acids accurately. These findings could be used further for understanding the types of proteins that can preferentially adsorb on the surface of NPs and predicting the NP-protein coronas and this knowledge could also be applied to understand the indirect mechanism of toxicity of NPs.

7.8 Conclusion

The safety of nanoparticles is a major concern while realizing their potential as nanomedicines. NPs are immediately covered by proteins as they enter the bloodstream. NP-protein complexes are biological entities that the cell sees. It has been reported that proteins adsorbed on the surface of NPs determine the biological outcome rather than NP alone. Therefore, studying NP-protein complex formation and its dynamics is essential for determining the toxicity, off-target effects, and efficiency of nano-based medicines. Experimental methods to study NP-protein corona formation have several limitations. Multi-disciplinary approaches have been required to obtain a comprehensive understanding of NP-protein complex formation. Computer simulations have proved beneficial in understanding NP-induced conformational changes in biomolecules. However, studying NP-protein corona formation using MD simulations is highly computationally intensive. In this chapter, we have summarized computer-simulation-based studies performed to study NP-biomolecule corona formation using multiscale computational methods. MD simulation has proved beneficial in predicting the protein corona around NPs. Combining computational and experimental approaches provides much more information about PC formation and its properties. This will prove beneficial in the safe and conscious application of nanotechnology in medicine.

Acknowledgements

The author would like to thank Iowa State University and support from the Roy J. Carver foundation.

References

1. L. Dykman and N. Khlebtsov, Gold Nanoparticles in Biomedical Applications: Recent Advances and Perspectives, *Chem. Soc. Rev.*, 2012, **41**, 2256–2282.
2. F. Danhier, E. Ansorena, J. M. Silva, R. Coco, A. Le Breton and V. Préat, PLGA-Based Nanoparticles: An Overview of Biomedical Applications, *J. Controlled Release*, 2012, **161**, 505–522.
3. J. Gao, H. Gu and B. Xu, Multifunctional Magnetic Nanoparticles: Design, Synthesis, and Biomedical Applications, *Acc. Chem. Res.*, 2009, **42**, 1097–1107.
4. G. L. Prasad, Biomedical Applications of Nanoparticles, in *Safety of Nanoparticles*, Springer, 2009, pp. 89–109.

5. T. Cedervall, I. Lynch, S. Lindman, T. Berggård, E. Thulin, H. Nilsson, K. A. Dawson and S. Linse, Understanding the Nanoparticle–protein Corona Using Methods to Quantify Exchange Rates and Affinities of Proteins for Nanoparticles, *Proc. Natl. Acad. Sci. U. S. A.*, 2007, **104**, 2050–2055.
6. E. Casals, T. Pfaller, A. Duschl, G. J. Oostingh and V. Puntes, Time Evolution of the Nanoparticle Protein Corona, *ACS Nano*, 2010, **4**, 3623–3632.
7. A. Kurtz-Chalot, C. Villiers, J. Pourchez, D. Boudard, M. Martini, P. N. Marche, M. Cottier and V. Forest, Impact of Silica Nanoparticle Surface Chemistry on Protein Corona Formation and Consequential Interactions with Biological Cells, *Mater. Sci. Eng., C*, 2017, **75**, 16–24.
8. M. Lundqvist, J. Stigler, G. Elia, I. Lynch, T. Cedervall and K. A. Dawson, Nanoparticle Size and Surface Properties Determine the Protein Corona with Possible Implications for Biological Impacts, *Proc. Natl. Acad. Sci. U. S. A.*, 2008, **38**, 14265–14270.
9. O. Vilanova, J. J. Mittag, P. M. Kelly, S. Milani, K. A. Dawson, J. O. Rädler and G. Franzese, Understanding the Kinetics of Protein–nanoparticle Corona Formation, *ACS Nano*, 2016, **10**, 10842–10850.
10. W. Liu, J. Rose, S. Plantevin, M. Auffan, J.-Y. Bottero and C. Vidaud, Protein Corona Formation for Nanomaterials and Proteins of a Similar Size: Hard or Soft Corona?, *Nanoscale*, 2013, **5**, 1658–1668.
11. S. Milani, F. Baldelli Bombelli, A. S. Pitek, K. A. Dawson and J. Rädler, Reversible *versus* Irreversible Binding of Transferrin to Polystyrene Nanoparticles: Soft and Hard Corona, *ACS Nano*, 2012, **6**, 2532–2541.
12. R. Parveen, T. N. Shamsi and S. Fatima, Nanoparticles-Protein Interaction: Role in Protein Aggregation and Clinical Implications, *Int. J. Biol. Macromol.*, 2017, **94**, 386–395.
13. I. Lynch, K. A. Dawson and S. Linse, Detecting Cryptic Epitopes Created by Nanoparticles, *Sci. Signaling*, 2006, **327**, 1–6.
14. N. S. Wigginton, A. d. Titta, F. Piccapietra, J. A. N. Dobias, V. J. Nesatyy, M. J. F. Suter and R. Bernier-Latmani, Binding of Silver Nanoparticles to Bacterial Proteins Depends on Surface Modifications and Inhibits Enzymatic Activity, *Environ. Sci. Technol.*, 2010, **44**, 2163–2168.
15. C. Ge, J. Du, L. Zhao, L. Wang, Y. Liu, D. Li, Y. Yang, R. Zhou, Y. Zhao and Z. Chai, Binding of Blood Proteins to Carbon Nanotubes Reduces Cytotoxicity, *Proc. Natl. Acad. Sci. U. S. A.*, 2011, **41**, 16968–16973.
16. S. R. Saptarshi, A. Duschl and A. L. Lopata, Interaction of Nanoparticles with Proteins: Relation to Bio-Reactivity of the Nanoparticle, *J. Nanobiotechnol.*, 2013, **11**, 1–26.
17. V. H. Nguyen and B.-J. Lee, Protein Corona: A New Approach for Nanomedicine Design, *Int. J. Nanomed.*, 2017, **12**, 3137.
18. S. Rana, Y.-C. Yeh and V. M. Rotello, Engineering the Nanoparticle–protein Interface: Applications and Possibilities, *Curr. Opin. Chem. Biol.*, 2010, **14**, 828–834.

19. M. Magro, M. Zaccarin, G. Miotto, L. Da Dalt, D. Baratella, P. Fariselli, G. Gabai and F. Vianello, Analysis of Hard Protein Corona Composition on Selective Iron Oxide Nanoparticles by MALDI-TOF Mass Spectrometry: Identification and Amplification of a Hidden Mastitis Biomarker in Milk Proteome, *Anal. Bioanal. Chem.*, 2018, **410**, 2949–2959.
20. L. Baweja and A. Dhawan, Computational Approaches for Predicting Nanotoxicity at the Molecular Level, in *Nanotoxicology*, Royal Society of Chemistry, Cambridge, 2017, pp. 304–327.
21. T. Mesarič, L. Baweja, B. Drašler, D. Drobne, D. Makovec, P. Dušak, A. Dhawan and K. Sepčić, Effects of Surface Curvature and Surface Characteristics of Carbon-Based Nanomaterials on the Adsorption and Activity of Acetylcholinesterase, *Carbon*, 2013, **62**, 222–232.
22. L. Baweja, K. Balamurugan, V. Subramanian and A. Dhawan, Hydration Patterns of Graphene-Based Nanomaterials (GBNMs) Play a Major Role in the Stability of a Helical Protein: A Molecular Dynamics Simulation Study, *Langmuir*, 2013, **29**, 14230–14238.
23. M. Kurylowicz, H. Paulin, J. Mogyoros, M. Giuliani and J. R. Dutcher, The Effect of Nanoscale Surface Curvature on the Oligomerization of Surface-Bound Proteins, *J. R. Soc., Interface*, 2014, **11**, 20130818.
24. M. Lundqvist, I. Sethson and B.-H. Jonsson, Protein Adsorption onto Silica Nanoparticles: Conformational Changes Depend on the Particles' Curvature and the Protein Stability, *Langmuir*, 2004, **20**, 10639–10647.
25. J. W. E. Worrall, A. Verma, H. Yan and V. M. Rotello, "Cleaning" of Nanoparticle Inhibitors *via* Proteolysis of Adsorbed Proteins, *Chem. Commun.*, 2006, **22**, 2338–2340.
26. M. Mahmoudi, I. Lynch, M. R. Ejtehadi, M. P. Monopoli, F. B. Bombelli and S. Laurent, Protein– Nanoparticle Interactions: Opportunities and Challenges, *Chem. Rev.*, 2011, **111**, 5610–5637.
27. C. C. Fleischer and C. K. Payne, Nanoparticle–cell Interactions: Molecular Structure of the Protein Corona and Cellular Outcomes, *Acc. Chem. Res.*, 2014, **47**, 2651–2659.
28. Z. J. Deng, M. Liang, M. Monteiro, I. Toth and R. F. Minchin, Nanoparticle-Induced Unfolding of Fibrinogen Promotes Mac-1 Receptor Activation and Inflammation, *Nat. Nanotechnol.*, 2011, **6**, 39.
29. M. Calvaresi and F. Zerbetto, Baiting Proteins with C60, *ACS Nano*, 2010, **4**, 2283–2299.
30. M. Calvaresi, F. Arnesano, S. Bonacchi, A. Bottoni, V. Calo, S. Conte, G. Falini, S. Fermani, M. Losacco and M. Montalti, C60@ Lysozyme: Direct Observation by Nuclear Magnetic Resonance of a 1: 1 Fullerene Protein Adduct, *ACS Nano*, 2014, **8**, 1871–1877.
31. F. Pederzoli, G. Tosi, M. A. Vandelli, D. Belletti, F. Forni and B. Ruozi, Protein Corona and Nanoparticles: How Can We Investigate On?, *Wiley Interdiscip. Rev.: Nanomed. Nanobiotechnol.*, 2017, **9**, e1467.

32. L. Guo, A. Von Dem Bussche, M. Buechner, A. Yan, A. B. Kane and R. H. Hurt, Adsorption of Essential Micronutrients by Carbon Nanotubes and the Implications for Nanotoxicity Testing, *Small*, 2008, **4**, 721–727.
33. P.-C. Lin, S. Lin, P. C. Wang and R. Sridhar, Techniques for Physicochemical Characterization of Nanomaterials, *Biotechnol. Adv.*, 2014, **32**, 711–726.
34. B. Ranjbar and P. Gill, Circular Dichroism Techniques: Biomolecular and Nanostructural Analyses-a Review, *Chem. Biol. Drug Des.*, 2009, **74**, 101–120.
35. F. Tavanti, A. Pedone and M. C. Menziani, Competitive Binding of Proteins to Gold Nanoparticles Disclosed by Molecular Dynamics Simulations, *J. Phys. Chem. C*, 2015, **119**, 22172–22180.
36. E. A. Proctor, F. Ding and N. V. Dokholyan, Discrete Molecular Dynamics, *Wiley Interdiscip. Rev.: Comput. Mol. Sci.*, 2011, **1**, 80–92.
37. F. Ding, S. Radic, R. Chen, P. Chen, N. K. Geitner, J. M. Brown and P. C. Ke, Direct Observation of a Single Nanoparticle–ubiquitin Corona Formation, *Nanoscale*, 2013, **5**, 9162–9169.
38. Q. Yu, Y. Zhang, H. Wang, J. Brash and H. Chen, Anti-Fouling Bioactive Surfaces, *Acta Biomater.*, 2011, **7**, 1550–1557.
39. C. Boyer, M. R. Whittaker, M. Luzon and T. P. Davis, Design and Synthesis of Dual Thermoresponsive and Antifouling Hybrid Polymer/gold Nanoparticles, *Macromolecules*, 2009, **42**, 6917–6926.
40. S. Schöttler, G. Becker, S. Winzen, T. Steinbach, K. Mohr, K. Landfester, V. Mailänder and F. R. Wurm, Protein Adsorption Is Required for Stealth Effect of Poly (Ethylene Glycol)-and Poly (Phosphoester)-Coated Nanocarriers, *Nat. Nanotechnol.*, 2016, **11**, 372.
41. G. Settanni, J. Zhou, T. Suo, S. Schöttler, K. Landfester, F. Schmid and V. Mailänder, Protein Corona Composition of Poly (Ethylene Glycol)-and Poly (Phosphoester)-Coated Nanoparticles Correlates Strongly with the Amino Acid Composition of the Protein Surface, *Nanoscale*, 2017, **9**, 2138–2144.
42. K. Vanommeslaeghe, E. Hatcher, C. Acharya, S. Kundu, S. Zhong, J. Shim, E. Darian, O. Guvench, P. Lopes and I. Vorobyov, CHARMM General Force Field: A Force Field for Drug-like Molecules Compatible with the CHARMM All-atom Additive Biological Force Fields, *J. Comput. Chem.*, 2010, **31**, 671–690.
43. M. J. Ellenbecker and C. S.-J. Tsai, *Exposure Assessment and Safety Considerations for Working with Engineered Nanoparticles*, John Wiley & Sons, 2015.
44. Y. Y. Zuo, R. A. W. Veldhuizen, A. W. Neumann, N. O. Petersen and F. Possmayer, Current Perspectives in Pulmonary Surfactant—inhibition, Enhancement and Evaluation, *Biochim. Biophys. Acta, Biomembr.*, 2008, **1778**, 1947–1977.
45. J. A. Whitsett and T. E. Weaver, Hydrophobic Surfactant Proteins in Lung Function and Disease, *N. Engl. J. Med.*, 2002, **347**, 2141–2148.

46. Q. Hu, X. Bai, G. Hu and Y. Y. Zuo, Unveiling the Molecular Structure of Pulmonary Surfactant Corona on Nanoparticles, *ACS Nano*, 2017, **11**, 6832–6842.
47. E. Lopez-Rodriguez and J. Pérez-Gil, Structure-Function Relationships in Pulmonary Surfactant Membranes: From Biophysics to Therapy, *Biochim. Biophys. Acta, Biomembr.*, 2014, **1838**, 1568–1585.
48. S. Liu, X.-Y. Meng, J. M. Perez-Aguilar and R. Zhou, An *In Silico* Study of TiO_2 Nanoparticles Interaction with Twenty Standard Amino Acids in Aqueous Solution, *Sci. Rep.*, 2016, **6**, 37761.
49. B. Roux, The Calculation of the Potential of Mean Force Using Computer Simulations, *Comput. Phys. Commun.*, 1995, **91**, 275–282.

CHAPTER 8

In Silico *Approaches to Design and Characterize Peptide-based Nanostructures*

CHRISTOPH GLOBISCH[a], MARC ISELE[a], CHRISTINE PETER[a] AND ALOK JAIN*[a,b]

[a]University of Konstanz, Department of Chemistry, Universitätsstraße 10, Konstanz, 78464, Germany; [b]National Institute of Pharmaceutical Education and Research Ahmedabad, Department of Biotechnology, Palaj, Gandhinagar-382355, Gujarat, India
*E-mail: alokjain@niperahm.ac.in

8.1 Introduction

Molecular dynamics (MD) simulations are highly valuable to identify critical interactions and mechanistic details in biomolecular processes, biomaterials, nanotechnology, tissue engineering and material science.[1–3] Conventional atomistic MD simulations have been successfully applied to many problems, however, they often do not cover the necessary timescales to sufficiently explore conformational phase and reach convergence. In such cases, one uses advanced simulation methods such as umbrella sampling, meta dynamics, coarse-grained simulations or replica exchange MD (REMD) techniques.[4–6] Nature is always a great source of inspiration for researchers as it provides elegant and complex examples of adaptive and intelligent processes and systems. Self-assembly is one such process where the fundamental

Issues in Toxicology No. 40
Nanoparticle–Protein Corona: Biophysics to Biology
Edited by Ashutosh Kumar and Alok Dhawan

Published by the Royal Society of Chemistry, www.rsc.org

building blocks of life such as proteins, peptides, lipids, saccharides and nucleic acids build complex three-dimensional systems with well-defined functionality. Self-assembly is defined as spontaneous organization of molecules or nanoparticles under near thermodynamic equilibrium conditions into structurally well-defined and stable arrangements through non-covalent interactions.

8.1.1 Experimental Studies on Peptide-based Nanoparticles

The use of self-assembling peptide molecules to create ordered templates on the nanoscale or components for functional devices is an emerging area in biomedicine and bionanotechnology.[7] Peptides adopt diverse supramolecular architectures such as vesicles, micelles, nanorods, nanoribbons or nanotubes through various non-covalent interactions and provide a new platform for a bottom-up approach to obtain distinct morphologies with the additional benefit of biocompatibility.[8] It is worth noticing that peptides crafted of a few amino acids form a new approach in bottom-up technology, which provides simple, predictable structures together with easy functionalization, chemical diversity, relative chemical/physical stability and low-cost production.[9,10] Furthermore, the physicochemical properties of peptide-based supramolecular structures can be altered relatively easily with slight modifications in the sidechain composition. Therefore, peptides were used as major building blocks to develop novel supramolecular structures more than a decade ago. Reza Ghadiri and coworkers utilized the properties of D- and L-amino acids to form an extended β-sheet-like structure that eventually formed hollow and extended cylinders by stacking on top of each other.[11] Extracting the active part of a protein to generate a functional nanostructure was also utilized extensively by other researchers. Zhang *et al.* designed a self-assembling 16mer peptide based on the yeast protein zuotin.[12] The octapeptide NSGAITIG and its variants were designed based on a fibre protein of the adenovirus to fabricate various nanostructures with different functionality.[13–15] Similarly, the most extensively studied diphenylalanine (FF) was extracted from Alzheimer's β-amyloid protein and was shown to form physically and chemically stable molecular assemblies. Gazit and Reches further studied and demonstrated the potential application of FF self-assembling peptides by casting nanowires from them.[16] This study opened a wide range of applications for the FF motif containing peptides to develop new structures and functionality. Addition of a third residue, cytosine, at the C-terminus or converting diphenylalanine to diphenylglycine changed the morphology from nanotubes to nanospheres. On the other hand, self-assembly of the tripeptide FFF resulted in formation of nanoplatelets.[17] Electrostatic interactions between charged termini, the flexibility of sidechains and H-bond interactions between the peptides were a few factors attributed to the stability of different nanostructures. It should be noted that it is extremely difficult to identify these dominating factors from existing experimental techniques. Other reports studied the influence of chirality of amino acids, the sequence length or the modification of terminal groups and other factors that alter the self-assembly ability of peptides. Resultant nanostructures are diverse in their structures and surface properties. Moreover, nanostructures were also modified by introduction of functional groups. Ulijn and coworkers utilized the fluorenylmethyloxycarbonyl (Fmoc) aromatic group and its derivatives.[18] They fabricated the Fmoc functional peptide-based structure using co-assembly of Fmoc-FF and Fmoc-S and demonstrated its application in a cell culture. Growing interest in this field is dedicated to designing dynamic systems that can assemble or disassemble triggered by internal or external factors. Temperature, pH value, light, solvent polarity, mechanical force and ionic strength are some factors that were explored. The most common approach was to generate pH-sensitive nanomaterials by introduction of protonable residues in the peptide sequence, which, by a local or global change in the pH value can affect or even reverse the self-assembly process. A particularly promising group of amphiphilic peptides that are rich in acidic residues was reported by Rapaport and coworkers;[19] these peptides form pH-sensitive hydrogels in bulk that can serve as a scaffold for cultures of bone tissue. Examples of other stimuli, such as light, were reported by the van Hest and Stupp groups.[20–22] Also, the effect of different ions and solvent types on the assembly process was studied.[23,24] Many peptides were shown to undergo structural transformation while changing the solvent, which affects the interaction of peptide residues with the surrounding molecules.

8.1.2 Computational Studies

As discussed, the rational design of biomaterials with desired properties requires a deep understanding of both microscopic and macroscopic processes. However, the intrinsic complexity involved in self-assembly processes and the intricacy in getting high resolution structures using X-ray or NMR restrict the characterization of such processes using experimental techniques. Therefore, even with recent advancements in experimental techniques, fine-tuning of physicochemical properties of a nanostructure is often tedious, expensive and connected with a trial and error approach. Herein, computational approaches along with experimental techniques helped to reveal the microscopic properties of these nanomaterials. This knowledge is not only necessary to understand the governing factors for already available nanomaterials but is also important for the design of more specialized materials. Such a preknowledge helps to control the morphology of the final structure. Computational approaches assist to decode the building blocks, *i.e.*, monomer structures and the mechanism of molecular assembly that further helps to design programmed architectures and controlled material properties with advanced functions, such as the controlled release of drugs loaded or embedded in a nanoparticle matrix.

All-atom molecular simulation methods (where all atoms are treated explicitly according to Newton's laws of motion) were extensively employed to gain insight about the early and final stages of self-assembly processes and provide a detailed picture of atomic contacts. They not only make it feasible to disentangle the dominant interactions between monomers but also make the stability of final aggregates accessible. Nguyen and coworkers studied the molecular assembly of peptide amphiphiles to elucidate the kinetic mechanism as a function of varying solvent conditions. They observed and characterized a multistep assembly process *via* transiently stable intermediates that eventually form different nanostructures.[25] Kaltofen *et al.* developed a computational *de novo* design protocol that allowed them to design self-assembled peptides with predefined structures. Using various experimental techniques, they established computational approaches that can be used to engineer peptides that form predefined three-dimensional structures and can further act as a scaffold for the development of functional biomaterials with desired properties and conformations.[26] Hansmann and coworkers thoroughly studied the effect of various temperatures and concentrations on the self-assembly of phenylalanine and discussed the potential toxicity mechanisms using molecular dynamics.[27] The Xu group extended their experimental study by employing the REMD simulation technique and revealed how microscopic interaction patterns dictate their mesoscopic self-assembled morphology, which was out of reach from their experimental study. An important role of atomistic simulation studies for the identification of driving forces in nanostructure assemblies with intrinsic detail is evident in many studies.[28–33] Recently, multiscale simulation methods (systems are modeled at more than one representation, *e.g.*, sub atomistic to field approaches) were also successfully employed to predict aggregation properties of a big dataset of peptides.[34,35] Tuttle and group have screened all possible dipeptides and tripeptides for their aggregation tendency.[34,35] With their novel scoring method, they screened for peptides that could potentially form supramolecular nanostructures. The resultant morphologies in coarse-grained (CG) simulations were in good agreement with experimental data. In this book chapter, we present two examples where different computational approaches were combined to improve the understanding of the molecular mechanism of nanostructure formation, especially the influences of various factors that are a prerequisite to intelligent design of nanostructures with predefined properties.

8.2 Hydrogel Forming Peptides as Mineralization Templates

8.2.1 Classical MD

Biomineralization is the intricate process employed by living organisms to form minerals on pre-formed biological aggregates to build skeletal structures and shells. Orthopedic conditions, such as osteoporosis, are a direct consequence of poorly orchestrated biomineralization. Deciphering the molecular mechanism of this vital yet poorly understood process is thus essential for the development of therapeutic approaches. In a first example, we will demonstrate how atomistic simulation fills this void by revealing the factors that might promote the formation of stable aggregates, akin to the extracellular matrix (ECM), and subsequent biomineralization events. The system consists of a series of amphiphilic peptides that were designed by our experimental collaborators, Hanna Rapaport and coworkers.[19] The peptides consist of alternating hydrophobic and acidic residues and form hydrogels in solution and β-sheet rich aggregates at hydrophobic interfaces. Their sequence follows the general pattern of X-Y-(Z-Y)$_n$-X where X represents the termini group and Y and Z denote hydrophilic and hydrophobic residues, respectively. The number of repetitive units of Z-Y is denoted by n. Throughout this book chapter, we have abbreviated the peptide names corresponding to the nature of the terminal residues followed by the nature of acidic residues (while phenylalanine was always employed as the hydrophobic residue) and finally the length of the peptide repeat unit, *e.g.*, ProGlu-5, which corresponds to Pro-Glu-(Phe-Glu)$_5$-Pro. We used molecular simulations to investigate the effects of sequence modification on aggregate stability, and how ions and peptides mutually affect each other during the early stages of biomineralization. Our simulation study provides insight into the underlying molecular mechanism, *i.e.*, the effect of terminal residues (proline *vs.* phenylalanine), sidechain types (aspartic *vs.* glutamic acid) and various ions (mono- and divalent ones) on aggregate stability and mineral nucleation. We carried out simulations with preassembled aggregates initially arranged in 3 × 3 arrays and followed the process of 9 × 1 aggregate formation at an air–water interface (mimicking the experimental mineralization conditions). Figure 8.1 shows a comparison of peptides [PheGlu-2, ProGlu-2], which differ only in terminal residues, *i.e.*, proline and phenylalanine, where they organized themselves into differently ordered aggregates. Peptide systems with proline termini arrange into perfect fibrils within 50 ns of simulation time and remain intact for the rest of the simulation. In the case of peptides with phenylalanine termini, they reorganize faster but form a less perfect fibril arrangement with defects (Figure 8.1). To identify the reasons for such distinct behavior, we have analyzed the main chain–main chain (MC–MC) hydrogen bonding pattern for these two systems. For the ProGlu-2 peptide system, we observed an alternate arrangement of four and six MC–MC hydrogen bonds that we labeled as weak and strong strand pairs, respectively. It should be noted that newly formed strand pairs during the 3 × 3 to 9 × 1 transition also followed the same alternating pattern of weak and strong pairs. On the contrary, the PheGlu-2 peptide system always formed six inter-peptide MC–MC hydrogen bonds. Importantly, one of the newly formed (4 ns simulation time) strand pairs did not align into a perfect fibril arrangement, causing a shift in the 9 × 1 pattern. We also observed such a shift in the longer peptide chain (PheGlu-5). As proline lacks the H-bond donor capability, the optimal number of H-bonds can only be achieved by alternate arrangements of weak and strong strand pairs. Such arrangements ensure a high order in the final structure, which serves subsequently as a template for ion growth. On the other hand, phenylalanine can act as both H-bond donor and acceptor,

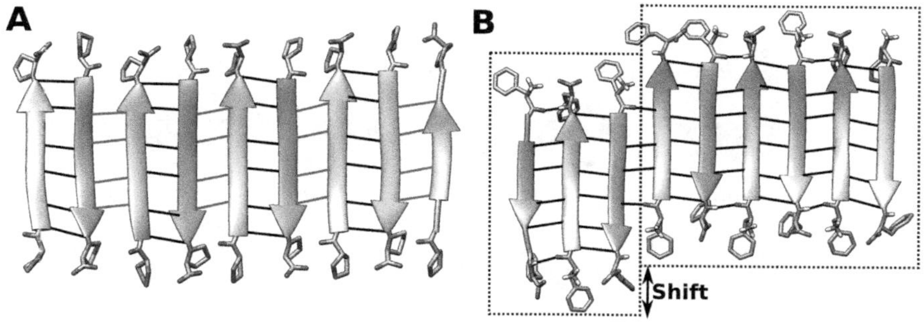

Figure 8.1 Cartoon representations of the secondary structure after 100 ns of simulation. (A) ProGlu-2 peptide forming a β-sheet aggregate with an alternating pattern of six (strong strand pairs, black lines) and four (weak strand pairs, grey lines) hydrogen bonds between the main chains of the peptides. (B) PheGlu-2 peptide forming a β-sheet aggregate with a uniform pattern of six bonds between the main chains of the peptides. The aggregate contains a shift in the orientation of the initial 3 × 3 peptide building block, resulting in a non-perfect aggregate. Reproduced from ref. 54 with permission from American Chemical Society, Copyright 2014.

making the strand stickier, consequently leading to unordered fibril formation as observed during the simulations. Our results not only reproduce the experimental observations but also explain the molecular mechanism of the ordered/unordered fibril formation.

Further simulations were executed to study the influence of sidechains (aspartic *vs.* glutamic acid) in the presence of various ions (see Figure 8.2) for the ProAsp-5 and ProGlu-5 systems. Both sidechains provide a framework for salt interactions (cation/acidic sidechain) that serve as a template in the early stage of bone-tissue formation. However, the sidechains differ regarding the stability of the peptide template. The peptide with aspartic acid sidechains offers a flexible template with relatively lower stability, *i.e.*, weak strand pairs may disentangle after some time. This effect was independent of the length of the peptide and the type of the involved cations (data not shown). A closer examination of different molecular structures shows that sidechains of the weak strand pairs of ProAsp-5 systems are oriented towards each other. Such an arrangement provides optimal interactions with Ca^{2+} ions and was not observed for the strong strand pairs. The shorter sidechains of the aspartic acid residues could not accommodate the Ca^{2+} ions without disturbing the MC–MC H-bonds. Consequently, to form the electrostatically favorable ion–protein interactions, weak strand pairs were pushed apart. One such example where several ions collectively break up the weak strand pair of the ProAsp-5 peptide system is shown in Figure 8.2. In contrast, peptides with glutamic acid sidechains do not exhibit similar behavior because of the presence of an extra methylene group in their sidechains that allows them to accommodate Ca^{2+} ions without affecting the structural integrity of the system. It provides

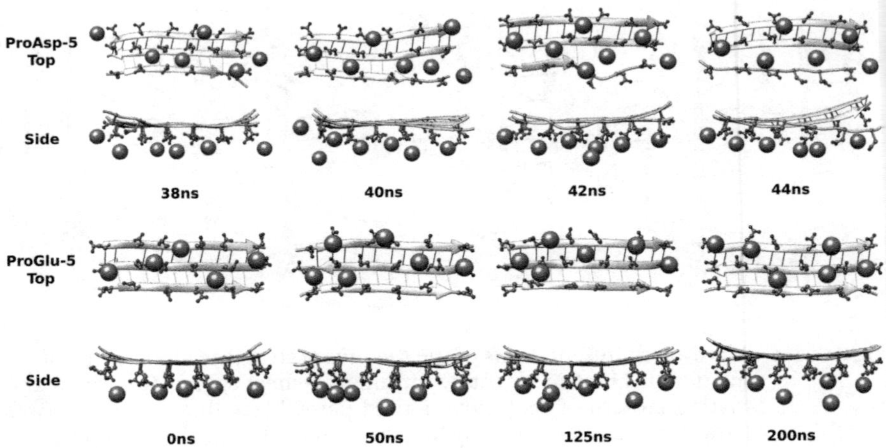

Figure 8.2 Chain separating effect of ions in the β-sheet formation of ProAsp-5 at the air–water interface and attachment of ion layers to the ProGlu-5 β-sheet. For clarity, only three strands of each system together with the interacting ions are shown at the air–water interface with ions being in the aqueous solution. Snapshots from the interface and solution perspective. Ions are shown in space filling representation whereas the acidic sidechains are displayed as balls and sticks. The backbone is shown as a cartoon and the hydrogen bonds between the main chains of the peptides are indicated as grey and black lines for the weak and strong strand pairs, respectively. Time steps of the snapshots are indicated below. Reproduced from ref. 54 with permission from American Chemical Society, Copyright 2014.

a more stable, well-organized layer of carboxylic sidechains where Ca^{2+} ions can uniformly assemble hence providing a scaffold that can initiate the mineralization process. In summary, aspartic and glutamic acid sidechains provide two different ways for peptide templated mineralization. Peptides with aspartic acid sidechains are more advantageous when the goal is to support the mineral formation while being able to adjust the template. On the other hand, if the peptide template rigidity is more important for the mineralization, peptides with glutamic acid sidechains seem to be the appropriate choice as they not only maintain the structural integrity but also support uniform and efficient mineral growth.

8.2.2 H-REMD as a Tool to Overcome Metastable Intermediate Structures

Originally, we wanted to extend the atomistic MD study to simulations of the full self-assembly process with initially separated peptides. The aim was to get a better mechanistic insight into the relationship between sequence (terminal residues, exact nature of the repeat sequence, *etc.*) and the fibrils formed. Unfortunately, conventional MD simulations, as you will see later, only lead to metastable states of the aggregation process where the system is arrested at intermediates. Therefore, we tried to utilize an advanced sampling method, *i.e.*, Hamiltonian replica exchange MD (H-REMD), to overcome these problems.

8.2.2.1 MD Reference Simulations

A conventional MD simulation of nine – initially separated and randomly oriented – peptides of ProGlu-5 at the air–water interface without ions was performed. As expected, the peptides very rapidly formed β-strand-rich aggregates. Based on the experimental results, these aggregates should form highly ordered fibrillar structures, given enough time. However, after a few nanoseconds of simulation time, the system was stuck in a metastable state where only one pair of strands was perfectly aligned. The other peptides were bent, formed hairpins or strand pairs aligned with some displacements.

The time evolution of secondary structures during the first 40 ns is shown in Figure 8.3, together with characteristic snapshots of the aligned strands. Peptide structures are presented as a top view on the water surface. For a later comparison with the results from H-REMD simulations, we have extended this MD simulation up to 1 μs, but the results were unchanged. Thus, a self-healing of the system and a formation of well-ordered fibrils cannot be observed in brute-force (without any advanced sampling techniques) MD simulations.

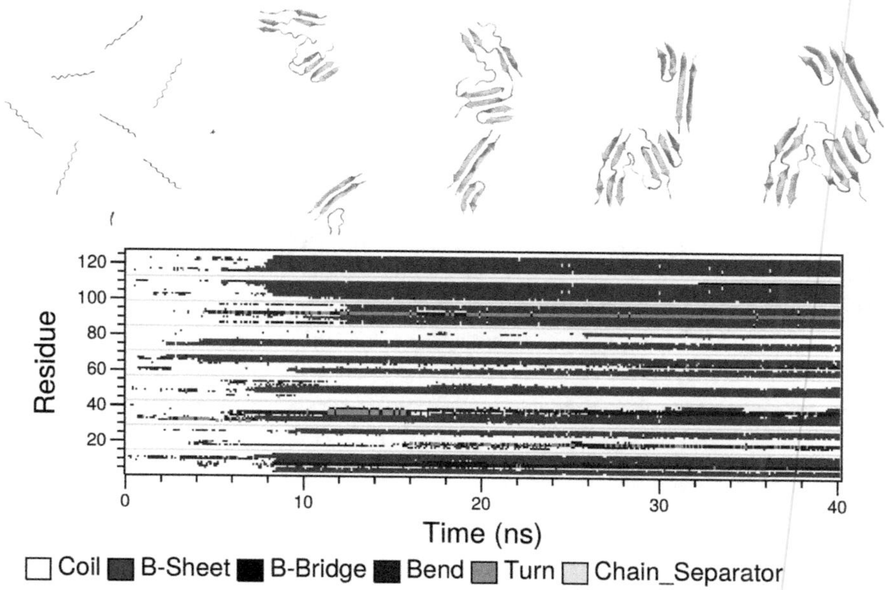

Figure 8.3 Time evolution of secondary structures (DSSP) for a conventional MD simulation. Snapshots of structures at every 10 ns are shown in the upper panel.

8.2.2.2 H-REMD

Replica exchange molecular dynamics (REMD) is an algorithm developed for more efficient sampling of conformational phase space.[36] Several simulations (replicas) are conducted in parallel, but with slightly differing values of an exchange variable.[36–38] An exchange variable can consist of, *e.g.*, any interaction potential or the temperature. It has to be chosen so that higher replicas are able to overcome transition barriers. At defined time intervals, exchange attempts are tried out between adjacent replicas. If successful, this exchange leads to the swapping of the atomic coordinates between both thermodynamic states. The acceptance or rejection of exchanges is defined by a Metropolis criterion,[36] ensuring a proper statistical mechanical ensemble. For an illustration, see Figure 8.4. The system will sample phase space according to the Boltzmann distribution, but at each exchange, the trajectories are becoming discontinuous. Therefore, the frequency of exchange attempts is a trade between the chance to overcome a barrier in a higher energy replica *versus* the continuity of the dynamic processes. The special feature of H-REMD is that the Hamiltonian (*i.e.*, the interaction potential) gets changed over the course of the replicas while the temperature may stay unchanged.[39–41] There are numerous possibilities as to how one can modify the Hamiltonian, and depending on the system, one has to find optimal settings that weaken the interactions that mostly contribute to the system getting stuck. At the same time, one has to ensure that exchanges between neighboring replicas have a sufficiently high acceptance rate. Otherwise, exchanges from the lowest replica (at the state point of interest) up to the

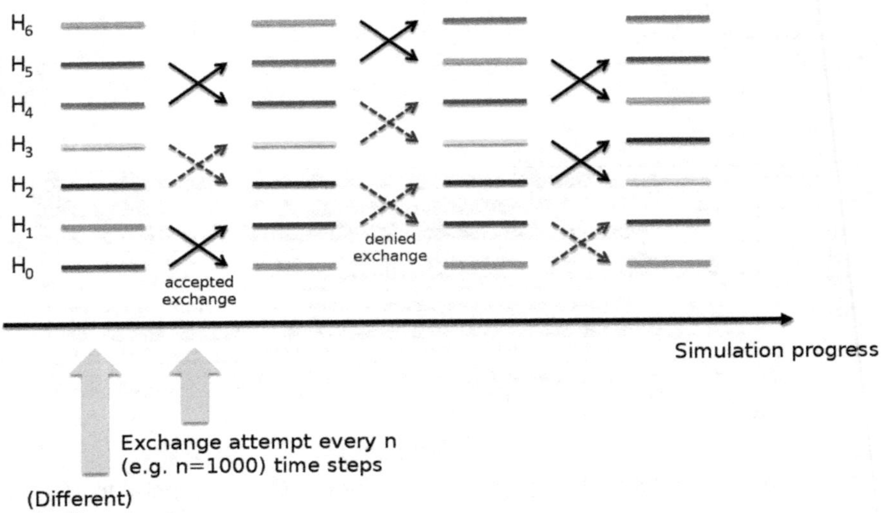

Figure 8.4 Schematic representation of the H-REMD simulations illustrating the exchange attempts throughout the replicas.

high replicas (where transitions across barriers are sped up) are hindered, thus defeating the purpose of the method. Therefore, we have extensively tested which settings and Hamiltonian modifications efficiently break up the barriers of the aggregated peptide system and speed up transitions while still maintaining good replica exchange rates.

8.2.2.3 Optimizing Suitable H-REMD Settings

We have tested several REMD parameters to find out the suitable settings that can be used to accelerate the sampling. In order to optimize the REMD settings for the system mentioned, we first tested which modifications to the simulation setup have the desired effect of weakening the interactions between the peptide strands in such a way that ill-formed aggregates that appear during the aggregation process are loosened so that the system can continue exploring phase space. This was done by running MD simulations (starting from a structure taken from the MD simulation described earlier at 300 K after 20 ns where the system had already reached an arrested state, see Figure 8.3) at elevated temperatures and with different ways of modifying the Hamiltonian. For all these test simulations, the secondary structure, the inter-strand H-bonding pattern and the terminal distances were monitored. The latter property was used to identify bent monomers (hairpins), which are very important since they are difficult to break and likely to introduce defects in fibrillar aggregates. Once promising modifications had been identified, we needed to optimize the spacing between replicas for a sufficiently high exchange rate together with a minimal number of replicas.

The following methods were examined to weaken the peptide–peptide interactions. Increasing the temperature: to test if a conventional temperature REMD (T-REMD) approach is suitable to enhance the sampling of this system, a simulation at 340 K was performed. However, not much difference was observed compared to the behavior at 300 K. This suggests that for the system under investigation, T-REMD would not be very effective at improving the sampling and at breaking up the β-strands connected by multiple H-bonds.

Changing the Hamiltonian: since simulations at elevated temperature didn't help much to accelerate transitions from one metastable state to another, we have modified the Hamiltonian of the system and weakened the interactions between the peptide strands (H-REMD). We tested four different approaches: (i) reducing the inter-strand H-bonding by weakening the electrostatic interactions between amide groups, (ii) increasing the (Lennard–Jones) repulsion between the amide groups, (iii) by a combination of both, and (iv) by weakening the electrostatic interactions between amide groups in combination with modifying the terminal charges in such a manner that terminal residues remain neutral in contrast to the zwitterionic form which corresponds to the "normal" state of the peptide. Details about the different simulation settings are given in Table 8.1.

Table 8.1 H-REMD modifications tested for this system.

Run	Type of modification	Regular parameter(s)	Modified parameter(s)
A	Elevated temperature	300 K	340 K
B	Decreased electrostatic interaction between amide groups	Amide partial charges: C (0.45), O (−0.45), N (−0.31), H (0.31)	Amide partial charges: C (0.20), O (−0.20), N (−0.15), H (0.15)
C	Increased repulsive interaction between amide group	C^{12} Lennard Jones parameter between H and O: 0 kJ nm^{12} mol^{-1}	C^{12} Lennard Jones parameter between H and O: 10^{-5} kJ nm^{12} mol^{-1}
D	Combination of B and C		
E	As in B, additionally modified charges on terminal amino acids	Total terminal charge: ±1	Total terminal charge ±0

8.2.2.4 Improved Sampling in H-REMD of Peptide Aggregates at an Air–Water Interface

We observed that decreased electrostatic interactions between the amide groups in combination with modified partial charges on the termini sufficiently reduce the inter-strand hydrogen bonding and the electrostatic attraction between oppositely charged termini so that intermediates in the aggregation process are destabilized. Based on these results, we started 200 ns H-REMD simulations (setting E in Table 8.1) with 40 replicas from the pre-aggregated structure used before. We monitored the stability of the initial contacts between the peptide strands and within the peptide strands to identify whether the chosen settings could promote healing of defects and the formation of a well-aligned fibrillar aggregate structure.

In Figure 8.5, the total number of intra-strand and inter-strand peptide H-bonds formed by peptide aggregates and the respective final structures are shown side-by-side for the conventional MD simulation and the H-REMD simulation (lowest replica). The figure clearly illustrates that many (not all) of the defects (*i.e.*, the not perfectly aligned β-strand pairs) that had formed during the conventional MD simulation have been healed during the H-REMD simulation. Some pairs that were misaligned during conventional MD simulations are realigned and formed a nearly perfect pair in H-REMD. A particularly striking difference was observed in the development of the intra-molecular H-bonds (which are indicative of hairpin structures). A significant number of intra-peptide H-bonds were broken in the H-REMD, a necessary requirement for the healing of hairpin defects and towards a formation of a perfect fibril. On the contrary, few changes were observed during 1 μs of conventional MD simulation (as was seen already in Figure 8.3). These results suggest that the H-REMD approach can contribute to the healing of defect structures, but nevertheless, the final structure was far from being a perfect fibril. Figure 8.5B suggests that the

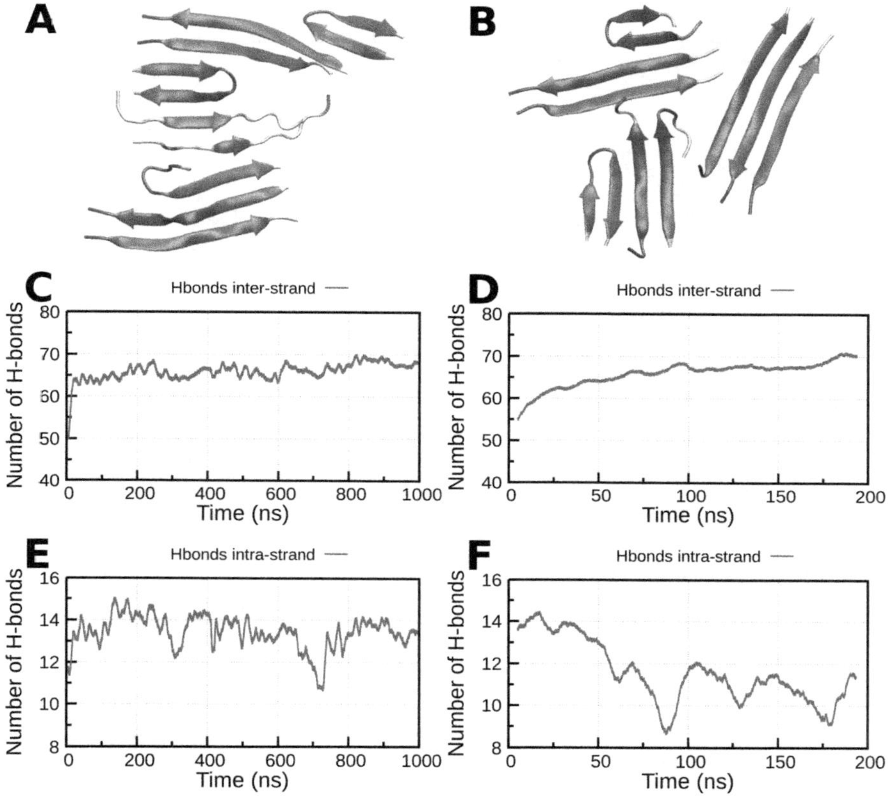

Figure 8.5 Final structures, total number of inter-strand peptide H-bonds and intra-strand peptide H-bonds for 1 μs conventional MD simulation (A, C, E, see also this series in Figure 8.3) and the lowest replica of the H-REMD simulation (B, D, F).

presence of stable hairpins is a possible factor inhibiting the formation of perfectly aligned fibrillar structures. A similar effect of hairpins inhibiting perfect fibril formation was also observed for the amyloid-β peptide by Shea and coworkers.[42]

While Figure 8.5 shows the comparison of the lowest replica (*i.e.*, the state point of interest) with the respective conventional MD simulation, Figure 8.6 shows representative examples of how individual replica threads move through the replica Hamiltonians, and how the exchanges into higher Hamiltonians indeed have the desired effect of speeding up transitions and breaking up arrested intermediate structures. Figure 8.6A shows three replica threads moving across the replica Hamiltonians. We see that the transitioning through Hamiltonian space is not perfect; thread 33, for example, has not been handed down to lower Hamiltonians during the entire simulation time. For threads 0 and 5, eventually, higher replicas are visited. Thread 0

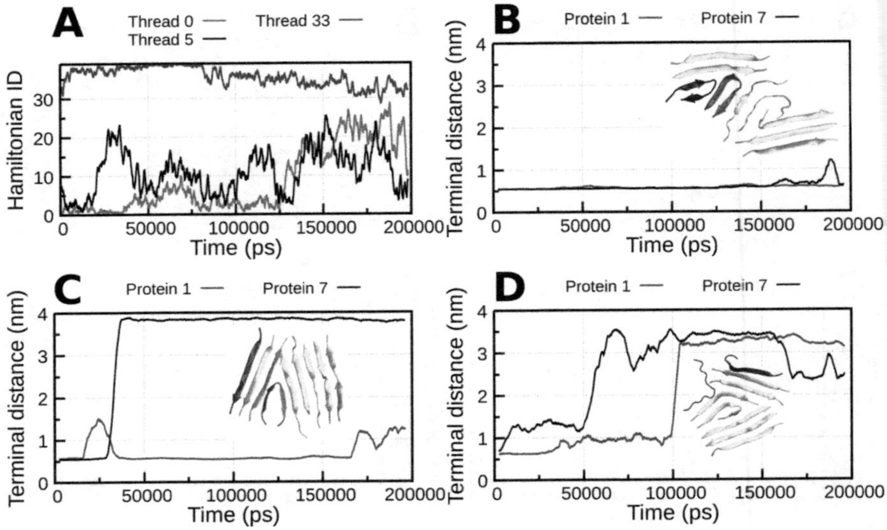

Figure 8.6 (A) Time trajectory of representative threads in different Hamiltonian spaces. Terminal distances involving the C-alpha atoms of the first and last residues of the hairpin forming peptides for (B) thread 0, (C) thread 5 and (D) thread 33. Final structures of each thread are also shown in the respective sub-figures. Hairpins are colored in dark grey and black.

remains in the lower Hamiltonian space for the first 130 ns while thread 5 made repeated transitions between lower and higher Hamiltonians. Figure 8.6B–D shows, for these three replica threads, the time evolution of the intra-strand end-to-end distances of two peptide strands that had initially been in a hairpin structure, thus impeding the formation of very ordered fibrillar aggregates. We find that transitions in the end-to-end distances, *i.e.*, breaking of the hairpins, can be correlated to the system being in higher Hamiltonians. For thread 0 (Figure 8.6B), no real breaking of the hairpins is observed. Towards the end of the simulation time, when the system has left the lower Hamiltonians, a short instance of loosening of the black hairpin is observed, but this has no lasting structural effect. Thread 5 visited the higher Hamiltonian space already after approximately 30 ns and at the same time we observed the breaking of the hairpin formed by peptide 7 (black line in Figure 8.6C) and a temporary increase of the terminal distance for the other hairpin (bottom line in Figure 8.6C). Finally, for thread 33, which remained for most of the time in the higher Hamiltonian space, we observed unfolding and stretching of both hairpins (Figure 8.6D). These results suggest that the parameters of our highest replicas are suitable to loosen intra-peptide H-bonds, enabling the system to heal defects and form ordered structures. However, we also find that the overall duration of the H-REMD simulations of 200 ns is still insufficient to observe well-ordered structures and to find perfect transitions of the replicas through the Hamiltonian space.

The latter observation is related to still imperfect exchange probabilities between the replicas. In the beginning of the H-REMD simulation, replicas exchanged substantially but after some time, they did not exchange as efficiently anymore (data not shown). Furthermore, replicas that were started in the upper/lower half of the Hamiltonian space mostly remained there and did not exchange very efficiently with the lower/upper half. Additionally, we also observed that replicas in the upper half of the replica space have a higher exchange rate compared to the replicas in the lower half of the Hamiltonian space. To overcome this problem, one would have to further optimize the spacing between the Hamiltonians. In summary, our data indicate that the H-REMD approach looks promising to overcome the limitation of conventional MD where a system gets stuck in a metastable aggregate state.

8.3 Penta-peptides

8.3.1 Atomistic Study

The second studied system consists of penta-peptides that exhibit significant changes in morphology following a proline to glycine substitution.[43] Here, we aim to provide a mechanistic interpretation of nanoparticle formation, which by identification of general principles may also help in the design of novel, highly-specialized materials with pre-defined properties. The basic penta-peptide system (FLXLF; where X can be proline or glycine) is derived from the R1 peptide that binds with the Apical membrane Antigen 1 (AMA1) and inhibits the malaria parasite invasion. An experimental study observed that penta-peptides with proline at the center position show nice spherical morphology while a substitution of proline to glycine disrupts the morphology and leads to an ill-defined structure. To understand the molecular mechanism, we have carried out several atomistic simulations starting with 125 peptides arranged in a 5 × 5 × 5 array. Visual inspection of the structure suggested a characteristic structural motif in the glycine-containing peptide. The presence of this bent structure can be attributed to the increased amount of intra-peptide H-bonds observed only in FLGLF. Similarly, the amount of inter-peptide hydrogen bonds increased and many instances with more than five bonds could be observed. This unique arrangement of H-bonds was responsible for the formation of the characteristic dimers in case of the FLGLF peptide. Further analysis demonstrated the transition of a bent (monomer) structure to stretched (dimers) structures. However, during this process some of the bent structures were trapped between the dimers and many dimers were randomly aggregated. Such an arrangement leads to a disorder in the systems and is a possible reason why it does not form a uniform and well-defined structure. Figure 8.7 gives a schematic overview of the suggested model, derived from atomistic simulations. On the other hand, peptides with a central proline residue remain stretched and rigid during the simulations. The presence of relatively fewer inter-peptide H-bonds allows this system to rearrange

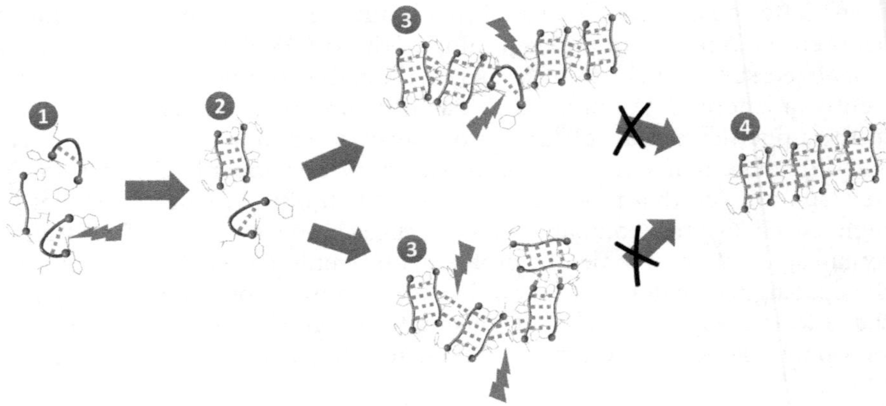

Figure 8.7 Proposed model for the FLGLF peptide aggregation. Step 1: single monomers in solution can exist in a bent or extended structure. Step 2: dimer formation with breaking of intramolecular H-bonds and conformational transitions. Step 3: random arrangement of dimers or bigger pre-aggregation blocks in the systems as observed during the simulations. The huge amount of interpeptide H-bonds and slow diffusion of dimers inhibits the formation of perfect fibrils as indicated in Step 4. The intrapeptide and interpeptide H-bonds are indicated in grey and dark grey, respectively. Lightning bolts point to positions where H-bonds would have to be broken in order to achieve reorganization. Reproduced from ref. 43 with permission from American Chemical Society, Copyright 2017.

and form a specific shape. Figure 8.8A shows a representative snapshot of an FLPLF peptide aggregate. The analysis of the solvent accessible surface area (see Figure 8.8B) of the hydrophobic moieties allows the conclusion that aggregation of this peptide is driven by hydrophobic interactions and the spherical shape of the rather mobile nanostructure is caused by surface tension.

Although this study very clearly explained why the glycine-containing peptide cannot form any uniform, well defined structure and also described the possible mechanism of the spherical morphology formation by the proline-containing peptide, we could not observe the actual formation of a spherical nanoparticle as seen in experiments. The reason probably lies in the limitations of an atomistic study.

8.3.2 Coarse-grained MD

In a recently submitted manuscript, we report an approach to simulate the penta-peptide with a coarse-grained model.[44] The preceding atomistic simulation study suggested many important characteristics of the system that determine the different morphologies. Due to the atomistic representation of the system, the size of the simulated aggregates and the timescales were

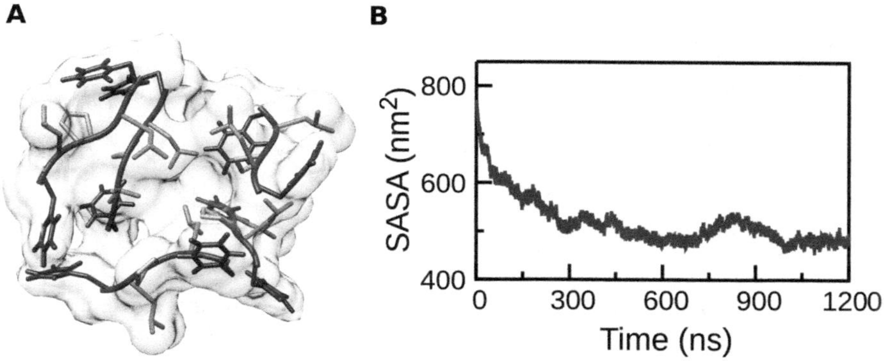

Figure 8.8 (A) A representative structure showing the characteristic patterns for hydrophobic interactions in clusters of FLPFL peptides. The peptide surface is displayed in white. The backbone together with the Phe sidechains is presented in dark grey whereas Leu and Pro sidechains are shown in light grey. Panel (B) shows the time series of the solvent accessible surface area (SASA) of the hydrophobic sidechains for the FLPFL peptide system at a 50:50 (water–methanol) solvent composition. Reproduced from ref. 43 with permission from American Chemical Society, Copyright 2017.

rather restricted, so we could not observe a real nanoparticle formation. Therefore, we developed a scheme to transfer the system to a coarse-grained representation allowing for much bigger system sizes and timescales. In a coarse-grained model, several atoms are combined in one bead, therefore minimizing the number of simulated particles and additionally smoothing the interaction potentials, which together results in a drastic speedup of the system. We selected the popular MARTINI model for the simulations; it combines up to four heavy atoms in one bead (for example, four water molecules) and it has previously been shown to be able to predict aggregation propensities for a series of di- and tri-peptides.[34,35] For this general scoring of short peptides, default parameters of MARTINI have been working very nicely. For our penta-peptide system, we already knew of the special role of local structure on the aggregation propensities. In such a case, the plain MARTINI model is not optimally suitable to properly address the critical factors that dominate the aggregation process. Due to its coarse resolution, and no explicit treatment of hydrogen bonding, it has known shortcomings in the prediction of local conformational preferences. For peptides, it is mostly able to maintain a predefined state after preassigning – and therefore fixing – the secondary structure and the corresponding non-bonded parameters. One approach to overcome this restriction was assigning different secondary structures as coming from atomistic simulations, either as a single average structure[45,46] or an ensemble of three representative structures.[47] We follow another approach presented by Tieleman and coworkers who have studied amyloid- and elastin-like peptides.[48] There they imposed a backbone dihedral potential derived from atomistic simulation on the

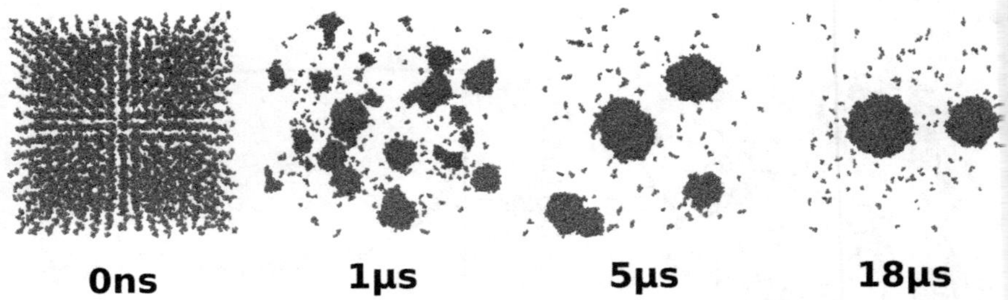

Figure 8.9 Nanoparticle formation as observed in a CG simulation of FLPFL peptides. The snapshots are showing the formation and growth of nanoparticles. The single peptides in solution are pointing towards the still high exchange rate of the peptides with the nanoparticle allowing for the spherical shape.

MARTINI CG model to accurately describe the peptide conformational dynamics. We apply this method of a supportive backbone dihedral to the bonded interactions of our model and supplement it by careful adjustment of the bead types to find the right interplay of our solvent methanol with the penta-peptides. Finally, we have to introduce additional bonds between the monomers to mimic the dimers observed in the atomistic simulations. By stepwise addition of the modifications, we were able to disentangle the influences of the different contributions, like flexibility of the peptides, the dimer formation, and the non-bonded properties of the beads, on the overall aggregation propensity and the morphology of the nanoparticles. The obtained models reproduce the experimental and atomistic behavior, provide a mechanistic interpretation of nanoparticle formation and demonstrate the importance of a carefully defined model. In Figure 8.9, you see the formation of nanoparticles for the FLPFL penta-peptide. The simulated system consists of 3000 peptides and was simulated for 18 μs, which is way beyond atomistic timescales on regular computing machines, but even now due to the slow diffusion process of the big aggregates the system did not fuse to one big nanoparticle. The resulting spherical morphology nicely matches with the experiment.

8.4 Methods

8.4.1 Computational Details: Classical MD and H-REMD

8.4.1.1 MD

MD simulations were run using the Gromos54a7[49] force field and the Gromacs 4.6.5 package.[50] Electrostatic interactions were treated with particle mesh Ewald (PME) with a coulomb and van der Walls (vdW) cut-off of 0.9 nm and 1.4 nm, respectively.[51] The neighbor list was updated every five steps. Bonds were constrained with the LINCS algorithm.[52] Periodic boundary conditions were used and the velocity-rescale thermostat was applied.[53]

The initial structure of a single peptide Pro-(Glu-Phe)$_5$-Glu-Phe was taken from a previous study.[54] The system was simulated for 40 ns at an air–water interface to mimic the experimental conditions.[19,55] It served to generate starting structures for the nine peptides, which were randomly arranged/rotated with the genconf tools of Gromacs. The starting structure for simulations at a higher temperature (340 K) and modified electrostatic/repulsion interactions was taken from the 300 K simulation at 20 ns. The Hamiltonians of the systems were modified either by changing the partial charges of C (0.45), N (−0.31), O (−0.45), H (0.31) to C (0.20), N (−0.15), O (−0.20), H (0.15) or modifying the C12 parameters of the Lennard–Jones potential between hydrogen and oxygen to 1×10^{-5}. Additionally, a combination of modified partial charges together with modified terminal residues was tested (for details, see Table 8.1). Our motivation for choosing such settings was based on the assumption that decreasing the partial charges will affect the strength of H-bonding, increased repulsion will prevent the amide groups coming close enough for H-bonding and neutral termini will reduce hairpin stability.

8.4.1.2 H-REMD

Exchange attempts were made every 1000 steps. For the H-REMD simulations, the starting structure was taken from a conventional MD simulation at 300 K after 20 ns. 40 replicas were simulated for 200 ns. Partial charges were modified from C (0.45), N (−0.31), O (−0.45), H (0.31) to C (0.20), N (−0.15), O (−0.20), H (0.15) with equal spacing. These simulations were carried out employing the PLUMED software plug-in combination with the Gromacs package.[56,57]

8.4.1.3 Analysis

Visualizations were produced by VMD[58] and Chimera.[59] The backbone of a peptide is shown in cartoon representation and most of the time colored based on the secondary structure. The secondary structures for the molecule representations were assigned by the STRIDE algorithm,[60] whereas in the secondary structure plots, a Gromacs tool was used utilizing the DSSP algorithm.[61]

8.5 Summary and Conclusions

Peptide-based nanostructures have a wide range of applications in the field of biomedical science because of their intrinsic biocompatibility and relatively easy synthesis. However, the design of nanostructures with pre-defined properties is not a trivial task. Often it involves trial and error approaches. Therefore, the first step towards achieving the goal of designing intelligent nanostructures is to characterize the intrinsic details of the assembly processes. In this study, we have demonstrated with two examples how computational approaches can complement the experimental findings with a detailed description and identification of influencing factors that are either difficult to find or out of reach for experimental techniques. However, it should be noted that computational techniques have their own limitations and we have to carefully select methods/parameters depending on the system under investigation. For the first system, we have used plain atomistic MD and H-REMD approaches. The plain atomistic study successfully provided a detailed description of the assembly process and also influences of the termini, sidechains, ions and sequence length. However, we could not simulate the system starting from a totally random conformation because of the limitations of the atomistic system. The system was stuck in a metastable state and could not overcome the local minima using brute-force MD. Therefore, we employed the advanced sampling method H-REMD. Although our result from H-REMD suggests that the employed parameters have worked satisfactorily, they still need further optimization to promote exchanges between replicas. Currently, we are optimizing these parameters. In the case of the second system – by employing atomistic simulation – we successfully identified the reasons why a single residue substitution affects the morphology of the resulting nanostructure so drastically. In the end, the major limitation of our atomistic study was the system size and simulation time as we could not simulate the larger system for a longer time and failed to detect the experimentally observed spherical morphology. To reach the required time and length scale, we extended the study with coarse-grained simulations. Here, using our new approach to deal with small peptide systems, we not only observed a morphology comparable to the experiment but were also able to disentangle the various contributing factors to control morphologies (detailed descriptions can be found in an upcoming paper). In summary, computational approaches are very useful for the design of novel nanostructures with pre-defined properties for a wide range of applications if they are employed carefully with appropriate methods and settings.

Acknowledgements

We gratefully acknowledge Tobias Lemke for constructive discussions on the book chapter. This work was performed on computational resources (bwUniCluster and ForHLR Phase I) funded within the framework program bwHPC by the State Baden-Württemberg and the German Research Foundation (DFG) (through grant INST 35/1134-1 FUGG BwForCluster MLS&WISO).

References

1. A. Hospital, J. R. Goñi, M. Orozco and J. L. Gelpí, Molecular dynamics simulations: advances and applications, *Adv. Appl. Bioinf. Chem.*, 2015, **8**, 37–47.
2. W. Shinoda, R. DeVane and M. L. Klein, Computer simulation studies of self-assembling macromolecules, *Curr. Opin. Struct. Biol.*, 2012, **22**, 175–186.

3. A. Tilocca, in *Nanotechnology in Regenerative Medicine*, ed. M. Navarro and J. A. Planell, Humana Press, 2012, pp. 285–301.
4. V. Spiwok, Z. Sucur and P. Hosek, Enhanced sampling techniques in biomolecular simulations, *Biotechnol. Adv.*, 2015, **33**, 1130–1140.
5. U. Doshi and D. Hamelberg, Towards fast, rigorous and efficient conformational sampling of biomolecules: Advances in accelerated molecular dynamics, *Biochim. Biophys. Acta, Gen. Subj.*, 2015, **1850**, 878–888.
6. M. Christen and W. F. van Gunsteren, On searching in, sampling of, and dynamically moving through conformational space of biomolecular systems: A review, *J. Comput. Chem.*, 2008, **29**, 157–166.
7. L. Sun, C. Zheng and T. J. Webster, Self-assembled peptide nanomaterials for biomedical applications: promises and pitfalls, *Int. J. Nanomed.*, 2017, **12**, 73–86.
8. J. J. Panda and V. S. Chauhan, Short peptide based self-assembled nanostructures: implications in drug delivery and tissue engineering, *Polym. Chem.*, 2014, **5**, 4418–4436.
9. Y. Zhao, L. Deng, J. Wang, H. Xu and J. R. Lu, Solvent controlled structural transition of KI4K self-assemblies: from nanotubes to nanofibrils, *Langmuir*, 2015, **31**, 12975–12983.
10. H. Cui, A. G. Cheetham, E. T. Pashuck and S. I. Stupp, Amino acid sequence in constitutionally isomeric tetrapeptide amphiphiles dictates architecture of one-dimensional nanostructures, *J. Am. Chem. Soc.*, 2014, **136**, 12461–12468.
11. T. D. Clark, J. M. Buriak, K. Kobayashi, M. P. Isler, D. E. McRee and M. R. Ghadiri, Cylindrical β-sheet peptide assemblies, *J. Am. Chem. Soc.*, 1998, **120**, 8949–8962.
12. S. Zhang, T. Holmes, C. Lockshin and A. Rich, Spontaneous assembly of a self-complementary oligopeptide to form a stable macroscopic membrane, *Proc. Natl. Acad. Sci. U. S. A.*, 1993, **90**, 3334–3338.
13. K. Papanikolopoulou, G. Schoehn, V. Forge, V. T. Forsyth, C. Riekel, J. F. Hernandez, R. W. H. Ruigrok and A. Mitraki, Amyloid fibril formation from sequences of a natural beta-structured fibrous protein, the adenovirus fiber, *J. Biol. Chem.*, 2005, **280**, 2481–2490.
14. P. Tamamis, E. Kasotakis, A. Mitraki and G. Archontis, Amyloid-like self-assembly of peptide sequences from the adenovirus fiber shaft: insights from molecular dynamics simulations, *J. Phys. Chem. B*, 2009, **113**, 15639–15647.
15. P. Tamamis, E. Kasotakis, G. Archontis and A. Mitraki, Combination of theoretical and experimental approaches for the design and study of fibril forming peptides, *Methods Mol. Biol.*, 2014, **1216**, 53–70.
16. M. Reches and E. Gazit, Casting metal nanowires within discrete self-assembled peptide nanotubes, *Science*, 2003, **300**, 625–627.
17. H. Görbitz, The structure of nanotubes formed by diphenylalanine, the core recognition motif of Alzheimer's beta-amyloid polypeptide, *Chem. Commun.*, 2006, 2332–2334.

18. V. Jayawarna, S. M. Richardson, A. R. Hirst, N. W. Hodson, A. Saiani, J. E. Gough and R. V. Ulijn, Introducing chemical functionality in Fmoc peptide gels for cell culture, *Acta Biomater.*, 2009, **5**, 934–943.
19. H. Rapaport, H. Grisaru and T. Silberstein, Hydrogel Scaffolds of Amphiphilic and Acidic β-Sheet Peptides, *Adv. Funct. Mater.*, 2008, **18**, 2889–2896.
20. D. W. P. M. Löwik, E. H. P. Leunissen, M. van den Heuvel, M. B. Hansen and J. C. M. van Hest, Stimulus responsive peptide based materials *Chem. Soc. Rev.*, 2010, **39**, 3394–3412.
21. T. Muraoka, H. Cui and S. I. Stupp, Quadruple helix formation of a photoresponsive peptide amphiphile and its light-triggered dissociation into single fibers, *J. Am. Chem. Soc.*, 2008, **130**, 2946–2947.
22. T. Muraoka, C.-Y. Koh, H. Cui and S. I. Stupp, Light-triggered bioactivity in three dimensions, *Angew. Chem., Int. Ed.*, 2009, **48**, 5946–5949.
23. J. Kim, T. H. Han, Y.-I. Kim, J. S. Park, J. Choi, D. G. Churchill, S. O. Kim and H. Ihee, Role of water in directing diphenylalanine assembly into nanotubes and nanowires, *Adv. Mater.*, 2010, **22**, 583–587.
24. J. Wang, K. Liu, R. Xing and X. Yan, Peptide self-assembly: thermodynamics and kinetics, *Chem. Soc. Rev.*, 2016, **45**, 5589–5604.
25. I. W. Fu, C. B. Markegard and H. D. Nguyen, Solvent Effects on Kinetic Mechanisms of Self-Assembly by Peptide Amphiphiles via Molecular Dynamics Simulations, *Langmuir*, 2015, **31**, 315–324.
26. S. Kaltofen, C. Li, P.-S. Huang, L. C. Serpell, A. Barth and I. André, Computational De Novo Design of a Self-Assembling Peptide with Predefined Structure, *J. Mol. Biol.*, 2015, **427**, 550–562.
27. H. W. German, S. Uyaver and U. H. E. Hansmann, Self-assembly of phenylalanine-based molecules, *J. Phys. Chem. A*, 2014, **119**, 1609–1615.
28. L. Deng, P. Zhou, Y. Zhao, Y. Wang and H. Xu, Molecular Origin of the Self-Assembled Morphological Difference Caused by Varying the Order of Charged Residues in Short Peptides, *J. Phys. Chem. B*, 2014, **118**, 12501–12510.
29. S. Emamyari and H. Fazli, All-atom molecular dynamics study of EAK16 peptide: the effect of pH on single-chain conformation, dimerization and self-assembly behavior, *Eur. Biophys. J.*, 2014, **43**, 143–155.
30. Y. Cote, I. W. Fu, E. T. Dobson, J. E. Goldberger, H. D. Nguyen and J. K. Shen, Mechanism of the pH-Controlled Self-Assembly of Nanofibers from Peptide Amphiphiles, *J. Chem. Phys. C*, 2014, **118**, 16272–16278.
31. J. Jeon, C. E. Mills and M. S. Shell, Molecular Insights into Diphenylalanine Nanotube Assembly: All-Atom Simulations of Oligomerization, *J. Phys. Chem. B*, 2013, **117**, 3935–3943.
32. Y. Zhao, J. Wang, L. Deng, P. Zhou, S. Wang, Y. Wang, H. Xu and J. R. Lu, Tuning the Self-Assembly of Short Peptides via Sequence Variations, *Langmuir*, 2013, **29**, 13457–13464.
33. O.-S. Lee, S. I. Stupp and G. C. Schatz, Atomistic Molecular Dynamics Simulations of Peptide Amphiphile Self-Assembly into Cylindrical Nanofibers, *J. Am. Chem. Soc.*, 2011, **133**, 3677–3683.

34. P. W. J. M. Frederix, R. V. Ulijn, N. T. Hunt and T. Tuttle, Virtual Screening for Dipeptide Aggregation: Toward Predictive Tools for Peptide Self-Assembly, *J. Phys. Chem. Lett.*, 2011, **2**, 2380–2384.
35. P. W. J. M. Frederix, G. G. Scott, Y. M. Abul-Haija, D. Kalafatovic, C. G. Pappas, N. Javid, N. T. Hunt, R. V. Ulijn and T. Tuttle, Exploring the sequence space for (tri-) peptide self-assembly to design and discover new hydrogels, *Nat. Chem.*, 2014, **7**, 30–37.
36. Y. Sugita and Y. Okamoto, Replica-exchange molecular dynamics method for protein folding, *Chem. Phys. Lett.*, 1999, **314**, 141–151.
37. W. Zhang and J. Chen, Efficiency of Adaptive Temperature-Based Replica Exchange for Sampling Large-Scale Protein Conformational Transitions, *J. Chem. Theory Comput.*, 2013, **9**, 2849–2856.
38. R. C. Bernardi, M. C. R. Melo and K. Schulten, Enhanced sampling techniques in molecular dynamics simulations of biological systems, *Biochim. Biophys. Acta, Gen. Subj.*, 2015, **1850**, 872–877.
39. D. R. Roe, C. Bergonzo and T. E. Cheatham, Evaluation of Enhanced Sampling Provided by Accelerated Molecular Dynamics with Hamiltonian Replica Exchange Methods, *J. Phys. Chem. B*, 2014, **118**, 3543–3552.
40. K. Ostermeir and M. Zacharias, Advanced replica-exchange sampling to study the flexibility and plasticity of peptides and proteins, *Biochim. Biophys. Acta, Proteins Proteomics*, 2013, **1834**, 847–853.
41. T. Mori, R. J. Hamers, J. A. Pedersen and Q. Cui, Integrated Hamiltonian Sampling: A Simple and Versatile Method for Free Energy Simulations and Conformational Sampling, *J. Phys. Chem. B*, 2014, **118**, 8210–8220.
42. L. Larini and J.-E. Shea, Role of *β*-hairpin formation in aggregation: the self-assembly of the amyloid-*β* (25-35) peptide, *Biophys. J.*, 2012, **103**, 576–586.
43. N. K. Mishra, A. Jain, C. Peter and S. Verma, Combining Experimental and Simulation Techniques to Understand Morphology Control in Pentapeptide Nanostructures, *J. Phys. Chem. B*, 2017, **121**, 8155–8161.
44. A. Jain, C. Globisch, S. Verma and C. Peter, Coarse grained simulations of peptide nanoparticle formation – the role of local structure and non-bonded interactions, *J. Chem. Theory Comput.*, 2019, **15**, 1453–1462.
45. Y. Sun, Z. Qian, C. Guo and G. Wei, Amphiphilic Peptides A6K and V6K Display Distinct Oligomeric Structures and Self-Assembly Dynamics: A Combined All-Atom and Coarse-Grained Simulation Study, *Biomacromolecules*, 2015, **16**, 2940–2949.
46. S. Emamyari, F. Kargar, V. Sheikh-hasani, S. Emadi and H. Fazli, Mechanisms of the self-assembly of EAK16-family peptides into fibrillar and globular structures: molecular dynamics simulations from nano- to microseconds, *Eur. Biophys. J.*, 2015, **44**, 263–276.
47. T. Yu and G. C. Schatz, Free-Energy Landscape for Peptide Amphiphile Self-Assembly: Stepwise versus Continuous Assembly Mechanisms, *J. Phys. Chem. B*, 2013, **117**, 14059–14064.
48. M. Seo, S. Rauscher, R. Pomès and D. P. Tieleman, Improving Internal Peptide Dynamics in the Coarse-Grained MARTINI Model: Toward Large

Scale Simulations of Amyloid- and Elastin-like Peptides, *J. Chem. Theory Comput.*, 2012, **8**, 1774–1785.
49. N. Schmid, A. P. Eichenberger, A. Choutko, S. Riniker, M. Winger, A. E. Mark and W. F. van Gunsteren, Definition and testing of the GROMOS force-field versions 54A7 and 54B7, *Eur. Biophys. J.*, 2011, **40**, 843–856.
50. S. Pronk, S. Páll, R. Schulz, P. Larsson, P. Bjelkmar, R. Apostolov, M. R. Shirts, J. C. Smith, P. M. Kasson, D. van der Spoel, B. Hess and E. Lindahl, GROMACS 4.5: a high-throughput and highly parallel open source molecular simulation toolkit, *Bioinformatics*, 2013, **29**, 845–854.
51. U. Essmann, L. Perera, M. L. Berkowitz, T. Darden, H. Lee and L. G. Pedersen, A smooth particle mesh Ewald method, *J. Chem. Phys.*, 1995, **103**, 8577–8593.
52. B. Hess, H. Bekker, H. J. C. Berendsen and J. G. E. M. Fraaije, LINCS: A linear constraint solver for molecular simulations, *J. Comput. Chem.*, 1997, **18**, 1463–1472.
53. G. Bussi, D. Donadio and M. Parrinello, Canonical sampling through velocity rescaling, *J. Chem. Phys.*, 2007, **126**, 014101.
54. A. Jain, M. Jochum and C. Peter, Molecular Dynamics Simulations of Peptides at the Air-Water Interface: Influencing Factors on Peptide-Templated Mineralization, *Langmuir*, 2014, **30**, 15486–15495.
55. H. Rapaport, K. Kjaer, T. R. Jensen, L. Leiserowitz and D. A. Tirrell, Two-Dimensional Order in β-Sheet Peptide Monolayers, *J. Am. Chem. Soc.*, 2000, **122**, 12523–12529.
56. M. Bonomi, D. Branduardi, G. Bussi, C. Camilloni, D. Provasi, P. Raiteri, D. Donadio, F. Marinelli, F. Pietrucci, R. A. Broglia and M. Parrinello, PLUMED: A portable plugin for free-energy calculations with molecular dynamics, *Comput. Phys. Commun.*, 2009, **180**, 1961–1972.
57. G. Bussi, Hamiltonian replica exchange in GROMACS: a flexible implementation, *Mol. Phys.*, 2014, **112**, 379–384.
58. W. Humphrey, A. Dalke and K. Schulten, VMD: Visual molecular dynamics, *J. Mol. Graphics*, 1996, **14**, 33–38.
59. E. F. Pettersen, T. D. Goddard, C. C. Huang, G. S. Couch, D. M. Greenblatt, E. C. Meng and T. E. Ferrin, UCSF Chimera – A visualization system for exploratory research and analysis, *J. Comput. Chem.*, 2004, **25**, 1605–1612.
60. D. Frishman and P. Argos, Knowledge-based protein secondary structure assignment, *Proteins: Struct., Funct., Bioinf.*, 1995, **23**, 566–579.
61. W. Kabsch and C. Sander, Dictionary of protein secondary structure: Pattern recognition of hydrogen-bonded and geometrical features, *Biopolymers*, 1983, **22**, 2577–2637.

CHAPTER 9

Nanomaterial–Blood Interactions: A Biomedical Perspective

PRITI SINGH[a] AND SUNIL KUMAR SINGH*[b]

[a]Department of Biotechnology, Motilal Nehru National Institute of Technology, Allahabad-211004, Uttar Pradesh, India; [b]Department of Animal Sciences, School of Basic and Applied Sciences, Central University of Punjab, Bathinda-151001, Punjab, India
*E-mail: singh.sunil06@gmail.com

9.1 Introduction

Over the past few years, nanotechnology has attracted tremendous global attention and gained recognition in important biomedical areas like targeted drug delivery, photodynamic therapy and bio-imaging due to the potential to create novel structures with increased solubility, stability and enhanced ability to translocate through cell membranes.[1–5] At the nanoscale, nanomaterials (NMs) possess unique optical, electrical and mechanical properties owing to their high surface–volume ratio.[6] Nanostructures can be divided into various categories based on their dimensionality like 0D (quantum dots, nanospheres, fullerenes), 1D (nanowires, nanorods, nanotubes), 2D (nanoprisms, nanoplates, nanosheets, nanowalls, and nanodisks) and 3D (nanoballs, nanocoils, nanopillars and nanoflowers) as shown in Figure 9.1.[7–11]

Issues in Toxicology No. 40
Nanoparticle–Protein Corona: Biophysics to Biology
Edited by Ashutosh Kumar and Alok Dhawan

Published by the Royal Society of Chemistry, www.rsc.org

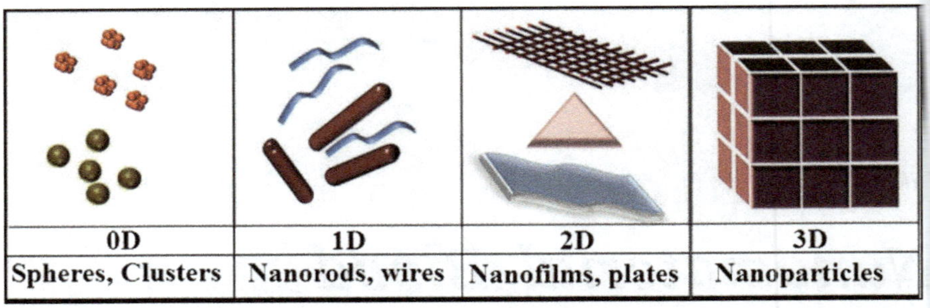

Figure 9.1 Schematic illustration of different dimensional nanomaterials. Reproduced from ref. 11 with permission from Elsevier, Copyright 2017.

Carbon-based nanomaterials (CBNs), metallic nanoparticles (NPs), silica-NPs (Si-NPs), quantum dots (QDs), zinc oxide (ZnO) and nanopolymers have been exploited extensively for bio-sensing, drug delivery and bio-imaging applications.[12–19] NMs encounter the complex physiological environment of blood when NMs are intravenously administered during targeted drug delivery and bioimaging. Different components of blood including cells, proteins, lipoproteins, and the coagulation system can interact with NMs and trigger patho-physiological events such as complement activation and thrombosis. In addition, there is the possibility that such interactions might modify the NPs' physicochemical properties, affecting their *in vivo* distribution, pharmacokinetics, and therapeutic efficacy. Different factors including the size, shape, nature and composition of NMs decide their fate in the blood vascular system.[20–22] In light of this observation, the evaluation of blood–NM interactions is extremely important for their *in vivo* applications. Blood is a connective tissue consisting of blood cells including erythrocytes or red blood cells (RBCs), white blood cells, platelets (PLTs) and plasma proteins (see Figure 9.2). Blood maintains the haemostasis of the body by transport of oxygen, nutrients, hormones *etc.*, and regulating body temperature, pH and the water content of cells.

A good coordination is established between its components, RBCs, platelets, macrophages, complement system, endothelial cells and plasma proteins for maintaining haemostasis. Any dysregulation that may disturb the balance between its components leads to thrombosis and produces an inflammatory response. Thus, a comprehensive study of the interactions between nanomaterials and blood components is important. This chapter focuses on the effects of NMs on blood constituents, and is divided into subsections according to the different components of blood, their importance in the body and the different haemocompatibility assays performed. The mechanism of the interactions between NMs and blood plasma proteins and cellular components with a focus on CBNs, metal/metal oxide NPs and Si-NPs is discussed in detail. Finally, in the concluding remarks, different forthcoming challenges involved in the design of nanomaterials for future nanomedicines will be discussed.

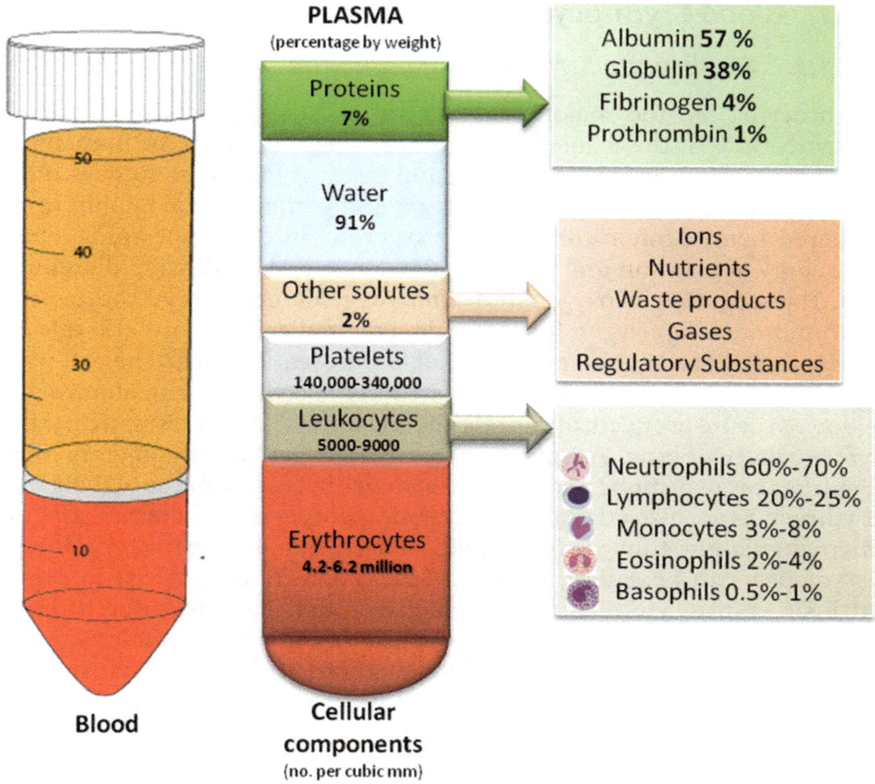

Figure 9.2 Different components of human blood.

9.2 Blood–Nanomaterial Interactions

For the development of blood-compatible NMs, the mechanism of blood–nano interactions is extremely important. Blood generally considers NMs as foreign particles and different immune cells and proteins are stimulated to bring opsonisation. Further, as NPs travel within the blood circulation, different blood proteins get adsorbed on the NM's surface forming a 'protein corona' and this interface may bring denaturation/conformational changes to tertiary/secondary structures of proteins. The specific orientation of a nanostructure and its surface chemistry generally influences the extent of protein alterations. This dysregulated functionality of proteins leads to loss of protein biological activity, influencing cell–cell interactions. Interactions between NMs and erythrocytes and the blood coagulation system are mediated by electrostatic attraction, hydrophobic interaction, H-bonds, and van der Waals forces *etc.*, which exist in linking NMs and cell lipid layers.[23] Hence, it is vital to address in detail the nature of blood–NMs interactions.

9.2.1 Blood Erythrocyte–Nanomaterial Interactions

9.2.1.1 *Erythrocyte Structure and Function*

Erythrocytes are the major constituents of human blood contributing to nearly 45% of the whole blood volume. RBCs circulate in humans for about three months (115–120 days) and in mice for about 40 days. RBCs lack organelles, have a biconcave shape and contain haemoglobin (a red coloured heme protein that carries oxygen to all the tissues of the body for energy generation and also functions as a carbon dioxide disseminator). Haemoglobin deoxygenation and oxygenation in erythrocytes may end up producing oxygen free radicals. Free radicals at normal levels can be cleared by erythrocyte enzymes but will pose a threat to the cell when abnormal levels of radicals are deposited in the cells. Such an abnormality is due to the toxic effects of various chemicals, which can result in erythrocyte breakage or lysis. Further, erythrocytes are perfect osmometers where disturbances in osmotic and physical changes in blood can cause their lysis. Erythrocytes are also critical in maintaining proper flow characteristics in a microcirculation, thus preventing vessel occlusion. In general, erythrocyte destruction can be caused by several factors such as bacterial infection, intrinsic erythrocyte membrane defects, mechanical shear due to blood flow, osmotic and pH changes in blood, drug-induced haemolysis, excessive complement activation, and platelet activation.[24] Erythrocyte destruction causes release of haemoglobin and its excretion in urine, leading to haemoglobinuria. Erythrocytes may be compromised due to breakage and clumping of erythrocytes, which are referred to as haemolysis and haemagglutination, respectively. Hence, erythrocyte lysis and agglutination are good indicators of toxicity of any foreign material to the cells in circulation. Since RBC membranes accelerate coagulation reactions, NM-induced damage is essential for its *in vivo* application.

9.2.1.2 *Erythrocyte-carbon-based Nanomaterial Interactions*

CBNs such as CNTs, carbon dots (CDs), fullerenes, NDs, graphene and its derivatives have unique electronic, optical, thermal, and mechanical properties that are exploited extensively in the biomedical field. Hence, a detailed understanding of erythrocyte interactions with CBNs is critical before employing them in drug delivery and therapeutics.

Graphene oxide (GO) and graphene sheets (GS) cause membrane disruption of RBCs in a dose-dependent manner leading to free haemoglobin release in the supernatant. It is found that the haemolytic activity of GO is increased after sonication due to the greater extent of exfoliation and small GO particle size. GS has a lower hemolytic activity than GO because it forms aqueous aggregations, yielding few cell-contactable reactive oxygen group surfaces. The RBC membrane disruption is likely attributed to the strong electrostatic interactions between negatively charged oxygen groups on the GO/GS surface and positively charged phosphatidylcholine lipids present on the RBCs' outer membrane. Hence, both particle size and surface charge/oxygen content influence GO/GS blood compatibility.[25] Besides, amine-modified graphene (G-NH_2) is highly hemocompatible and does not affect the biology of circulating RBCs.[26] Similarly, the effect of two different types of single-walled CNTs (SWCNTs), such as pristine and acid-treated, on RBCs was observed and it was found that carboxylated SWCNTs could induce haemolysis by binding to the erythrocyte membrane due to exposure of phosphatidylserine, and microscopic images clearly reflected the internalisation of CNTs in the RBCs, in a dose- and time-dependent manner, while the pristine form showed no significant effect.[27] The impact of six different types of multi-walled CNTs (MWCNTs) (long 50 µm: long-aminated and long-carboxylated; or short 0.5–2 µm: short-aminated and short-carboxylated) on human erythrocytes was observed, which resulted in cell morphology distortions, loss of membrane smoothness, and cell shrinkage leading to disruption. Among them all, aminated MWCNTs are more severe than carboxylated, and short MWCNTs are more damaging than long MWCNTs.[28] Carbon dots also showed a haemolytic effect; they are adsorbed on RBC membranes in concentration-dependent manner, resulting in loss of membrane structure and integrity.[29] The mechanical properties of RBC membranes change in the presence of C60 NPs. The structural changes on the membranes' lipid bilayer takes place by the infiltration of C60, making the cell softer and easier to break.[30–32] Membrane integrity loss as well as a change in the rate of RBC oxygenation/deoxygenation were discovered with the use of high concentrations of nanodiamonds.[33,34]

9.2.1.3 *Erythrocyte–Metal/Metal Oxide Nanoparticle Interactions*

Gold nanoparticles (GNPs) have excellent optical properties and a highly bio-compatible nature, that makes them promising materials for biomedical applications like bio-imaging and drug delivery. It is established that the haemolytic activity of GNPs is size- and functional-group-dependent. Bare GNPs of size 50–70 nm showed significant haemolysis while their counterparts coated by BSA had no considerable effect on haemolysis. The large curvature of the particle is held responsible for RBCs' cell membrane disruption.[35–37] The haemolytic activity and oxidative stress induced by silver nanoparticles (Ag-NPs) is mainly due to their surface negative charge that can interact strongly with the organic cations on RBCs, leading to haemolysis. It is also believed that Ag-NPs bind to thiol groups present within the proteins or phospholipids of a membrane with a high affinity, promoting their denaturation (Figure 9.3).[38] But, it's still unclear whether the haemolytic activity of Ag-NPs is due to their direct cellular interaction, the release of free silver ions, or a combination of both events.[39,40] Titanium

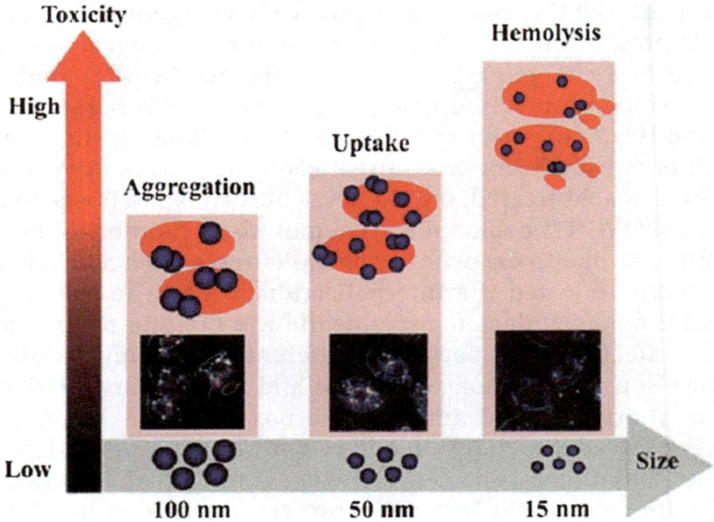

Figure 9.3 Illustration demonstrates the size effect of Ag-NPs on erythrocytes. Reproduced from ref. 38 with permission from American Chemical Society, Copyright 2015.

dioxide (TiO_2) NPs of size 20–30 nm are able to penetrate human RBCs' cell membranes by diffusion or adhesive interactions but have no haemolysis effect.[41,42]

9.2.1.4 Erythrocyte–Silica Nanoparticle Interactions

Studies have shown strong electrostatic interactions between the negatively charged silanol (–OH) groups of silica nanoparticles and positively charged trimethyl-ammonium head groups of the membrane lipids that ultimately result in RBC membrane lysis. The inverse relation between NP size and haemolysis is due to silanol (Si–OH) being less available to RBC membranes. The magnitude of binding energy of NPs with RBCs or the bending energy of the membrane to wrap around the NPs is affected by the external surface area and curvature of the particles. A large surface area and a small curvature render the haemolysis process thermodynamically favourable.[43–46] Hence, capping of mesoporous silica with neutral molecules completely prevents haemolytic activity even at high concentrations (1 mg mL^{-1}) by blocking the electrostatic interactions of silanol groups with the RBC membrane.[47] The interactions of polyethylene glycol (PEG)-attached and non-PEGylated Si-NPs with RBCs were also studied, and, at high conc. (200 μg mL^{-1}), NPs caused no appreciable release of haemoglobin.[48,49]

9.2.1.5 *Erythrocyte–Other Biomedically Important Nanomaterial Interactions*

Poly(lactic-*co*-glycolic) acid (PLGA) NPs are also extensively studied materials as colloidal drug carriers for intravenous administration of drugs. PLGA is a FDA approved polymer that does not cause haemolysis.[50] Erythrocyte membrane functional parameters are also intact in the presence of hydroxyapatite NPs.[51] Chitosan-NPs loaded with siRNA-Npr3 showed high haemocompatibility while chitosan-NPs dissolved with tri-polyphosphate to an acidic pH produce haemolysis.[52–54]

RBC interactions with different NMs have diverse effects depending on the intrinsic pathways stimulated. The interaction mechanisms described earlier give an insight into the design of NMs so that the surfaces of RBCs do not get disturbed and NMs can be successfully used for targeted drug delivery and therapeutics.

9.2.2 Blood Coagulation System–Nanomaterial Interactions

9.2.2.1 *Blood Coagulation System Components and Function*

The coagulation system serves as the first line of defence for any disruption in the vascular system. In vertebrates, this complex system consists of a cellular element (blood platelets), and endothelial cells of blood vessel and coagulation proteins (enzymes, co-factors and anticoagulant proteins). Whenever the coagulation system is triggered, platelets and coagulation proteins interact and lead to the formation of a stable clot to prevent bleeding from vascular injury that is known as haemostasis.[55] A thrombosis phenomenon, on the other hand, occurs in a pathological condition due to the formation of clots inside an intact blood vessel causing obstruction in the working of the circulatory system.[56] When vascular injury occurs, three pathways are stimulated, referred to as Virchow's Triad.[57] The first pathway is initiated due to the activation of platelets. The second pathway (extrinsic pathway) involves factors that are not found in blood and is initiated by the release of tissue factor (TF) from the sub-endothelium. The final pathway (intrinsic pathway) is triggered by factors found in whole blood. The extrinsic and intrinsic pathways of the coagulation system are presented in Figure 9.4.[58]

The extrinsic pathway is activated due to the release of TF from the sub-endothelium following rupture of the blood vessel. Binding of TF with coagulation factor VIIa forms TF-VIIa complex and activates factor IX (Christmas factor) and X (Stuart-Prower factor) to form IXa and Xa, respectively.[58] The intrinsic pathway, on the other hand, is activated by factor XII (Hageman factor) on encountering a negatively charged surface of the sub-endothelium. When factor XII is activated, it splits into equal amounts

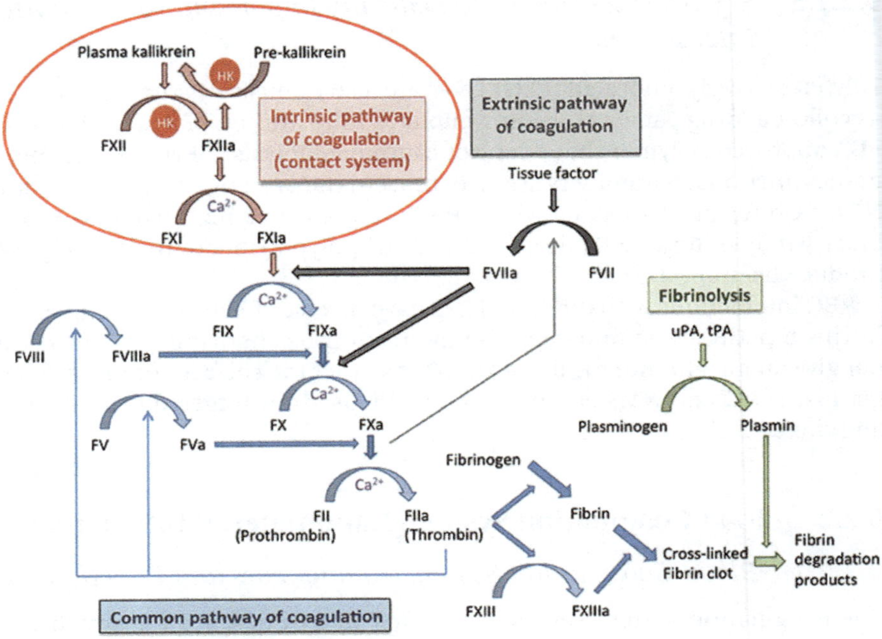

Figure 9.4 Coagulation pathways: extrinsic, intrinsic and common. Reproduced from ref. 62 [https://doi.org/10.3389/fcimb.2014.00128] under the terms of a CC BY 4.0 license [https://creativecommons.org/licenses/by/4.0/].

of α-XIIa and β-XIIa.[59] These two factors convert pre-kallikrein (preK) to kallikrein (kalli), which serves to activate more factor XII. This positive feedback loop serves as the main form of factor XII activation and greatly increases the amount of factor XIIa being produced.[60] Factor XIIa activates XI to XIa, and XIa in turn activates IX to IXa. This coagulation cascade changes a soluble plasma protein – fibrinogen – into fibrin, which forms a mesh to form the clot.

Platelets (or thrombocytes) are cellular components of the blood coagulation system; they originate from megakaryoblasts in the bone marrow. PLTs have a round to oval shape with no nucleus and measure 2–4 µm in diameter. Human blood has 150 000–450 000 platelets per mm^3 with a life span of about 10 days. Only 67% of all platelets circulate in blood, the rest is stored in the spleen. Platelets consist of a peripheral zone, structural zone and organelle zone having glycoprotein receptors, contractive microtubules, alpha granules, dense granules and lysosomes, respectively. α-Granules are 10 times more abundant than dense granules and contain large adhesive and healing proteins, haemostatic factors vWF, factor V, fibrinogen, angiogenic factors such as vascular endothelial growth factor (VEGF), angiogenin and antiangiogenic factors angiostatin, platelet factor 4 (PF-4) and β-thromboglobulin (β-TG). These granules also contain basic fibroblast growth factor, platelet-derived growth factor (PDGF), stromal-cell-derived factor 1, proteases,

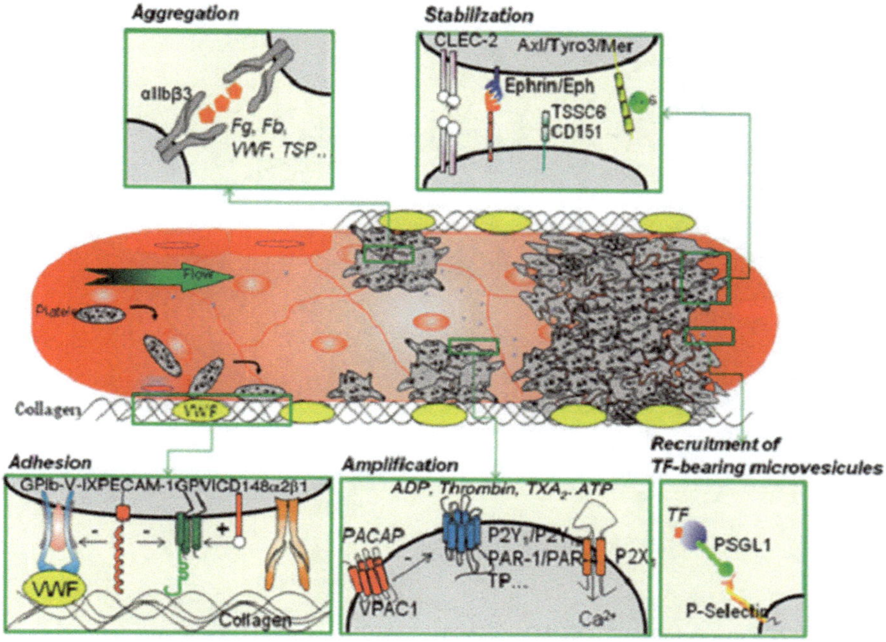

Figure 9.5 Role of platelets during vascular injury leading to platelet activation, adhesion, spreading, aggregation and stable thrombus formation. Reproduced from ref. 62 with permission from Springer Nature, Copyright 2012.

particularly metalloproteinases 2 and 9 (MMP-2, MMP-9), necrotic factors such as tumour necrosis factor alpha and beta (TNF-α, TNF-β) and other cytokines. Dense granules contain small molecules such as ADP, serotonin, Ca^{2+}, and histamine.[61] Platelets also secrete factors necessary to convert soluble fibrinogen into insoluble polymeric fibrin for forming a mesh at the site of injury that traps cells from flowing blood. Platelets are highly sensitive cells that are activated inside the body by different activators such as collagen, thrombin, adenosine diphosphate (ADP), thromboxane A2 (TXA2), adrenaline, epinephrine and shear stress. After a vessel wall is damaged, platelets attach to von Willebrand factor (vWf), collagen and fibronectin located in the sublayer of the endothelium. Four major events are initiated by platelet activation (Figure 9.5).[62]

Upon vascular injury, platelets are activated by a number of agonists like ADP, thrombin, collagen and TXA2. These agonists activate platelets through specific signalling pathways *via* their respective surface receptors (Figure 9.6).[62]

Platelet activation generally leads to the formation of platelet-derived microparticles (PMPs).[63] Microparticles (MPs) are small membrane-bound vesicles (0.1 to 1.0 μm in size range) formed due to budding of the plasma membrane during cell activation or apoptosis, and have different cellular origins including endothelial cells, platelets, leukocytes, lymphocytes

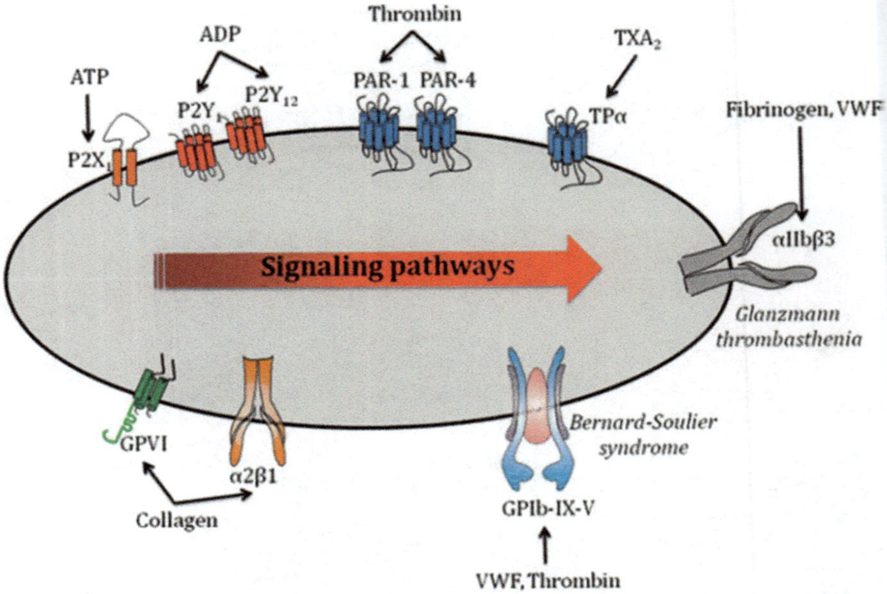

Figure 9.6 Overview of platelet surface receptors with their respective ligands. Reproduced from ref. 62 with permission from Springer Nature, Copyright 2012.

and erythrocytes.[64] MPs consist of parent-cell-derived proteins, mRNA, lipoproteins and markers. Different MPs are understood to play important physiological roles in various diseases like inflammation, coagulation and angiogenesis. PMPs represents about 70–90% of all circulating MPs.[65] Zwaal *et al.* (1997) found that a key event during platelet activation is translocation of anionic phospholipids, phosphatidylserine (PS), from the inner leaflet to the outer leaflet. PS has an important role in cell-to-cell interactions and coagulation.[66]

9.2.2.2 *Platelet/Clotting Protein–Carbon-based Nanomaterial Interactions*

The interactions of CBNs with platelets release thrombin, which triggers the intrinsic pathway of blood clotting, while the extrinsic pathway (tissue-factor-dependent pathway) is activated rapidly in response to vascular injury when the factors in adventitia are released to flowing blood. The platelet–nanocarbon interaction should be critically scrutinised before its biomedical application. CBNs generally elicit the production of pro-inflammatory cytokines, *e.g.*, TNF-α, IL-1β, IL-8, resulting in the development of inflammation. Upon entry of the CBNs into the microenvironment of the bloodstream, massive biomolecules (such as proteins and lipids) quickly adsorb onto CBNs, forming so-called corona complexes.[67] Formation of a corona complex largely reduces the surface free energy and alters the physicochemical properties of the CBNs, which results in changes in their function and surface morphology. The effects of nanodiamonds on blood clotting are well studied and it has been proved that even at high concentrations, no effect was observed on coagulation when the intrinsic pathway was followed by an activated partial thromboplastin time (APTT) test.[68,69] Platelet functional studies revealed that they induce platelet aggregation in a concentration-dependent manner with the release of α granules.[70] Fluorescent carbon dots (CDs) cause the fluorescence quenching of fibrinogen, which may be due to changed microenvironments around the tryptophan residues (*i.e.*, change from a hydrophobic to a hydrophilic environment) but weakly affect its secondary structure (α-helix). APTT and prothrombin time (PT) tests illustrated that the presence of CDs did not interfere with the intrinsic or extrinsic coagulation pathways. The thromboelastography results suggested that a 1 mg mL^{-1} CD concentration in the whole blood was a safe concentration for the blood clotting system. CD–platelet interactions lead to an expression of activation marker CD62P (P-selectin), indicating the degree of platelet activation and reflecting the risk of thrombus formation. This implies that the carbon dots have two-sided effects on platelets – platelet activator and causing cytotoxicity. The former can be attributed to the presence of hydrophobic structures on the CDs and the latter due to a disturbed platelet membrane structure.[71] The four different length MWCNTs' effects on fibrin and clot formation illustrate that all four induce earlier formation of fibrin while long-chain CNT-NH_2 increases clot hardness and long-chain CNT-COOH and short-chain CNT-NH_2 make the clots softer.[72] Inhalation studies in mice revealed that SWCNT/MWCNT exposure leads to elevated plasma levels of PAI-1.[73] Molecular modelling and different photophysical techniques were employed to characterise the binding of albumin (HSA), fibrinogen (FBG), γ-globulins (IgG) and histone-H1 (H1) on CNT-COOH. The identity of the protein forming the corona greatly affects the outcome of CNT-COOH's interaction with platelets. Bare CNT-COOH induced platelet aggregation and the release of PMPs. HSA and FBG coronas diminished the effect of CNT-COOH on platelets, while IgG and H1 coronas caused platelet fragmentation, thus potentiating the release of PMPs. Urban particular matter (SRM1648), mixed carbon nanoparticles (MCNs), fullerene derivative (C60CS), MWCNT and SWCNT were compared with their tendency to induce activation and aggregation of platelets. The results showed that MCNs activate and aggregate platelets leading to P-selectin exposure, degranulation and ATP release on platelet stimulation. Platelet exposure to carbon NPs activates the matrix-metalloproteinase-dependent signalling pathway and glycoprotein expression on the cells' surface.[74] CNT-induced platelet activation is associated with the marked release and secretion of membrane markers

CD62P and CD63.[75] Lacerda *et al.* (2011) showed that MWCNTs activate human blood platelets through store-operated calcium entry.[76] Depletion of stored intracellular Ca^{2+} induces influx of Ca^{2+} from the outside, and activates store-operated Ca^{2+} entry. In addition to the changes in intracellular Ca^{2+} storage, MWCNTs induced the release of PMPs that are capable of stimulating platelets and other cells. Carboxylated CNTs are capable of inducing coagulation *in vitro* through the intrinsic pathway dependent on coagulation factor IX.[77] Moreover, factor IXa is associated with functionalised MWCNTs. Under physiological conditions, the binding of IXa to a platelet's surface increases its enzymatic activity; MWCNTs are believed to induce coagulation by creating a platform for factor IXa binding. Amidated-MWCNTs also elicit the extrinsic pathway of coagulation through platelet activation.

Our group has critically explored the effect of atomically thin GO sheets on blood platelets and showed that GO elicited a strong aggregatory response through activation of Src kinases and release of calcium from intracellular stores. The surface chemistry of nanomaterials is a critical determinant of biocompatibility, and thus differentially functionalised graphenes exhibit varied effects in activating platelets. On comparing GO, reduced GO and amine-modified graphene (G-NH_2), G-NH_2 had no stimulatory effect on platelets (Figure 9.7).[78,79] G-NH_2 had no thrombotoxic property unlike other nanocarbons and was proven to be potentially safe for *in vivo* biomedical applications.

9.2.2.3 *Platelet/Clotting Protein–Metal/Metal Oxide Nanoparticle Interactions*

GNPs bind with fibrinogen *via* cys residues through the Au–S bond to keep the protein structure intact. Most of the cys residues form disulfide bonds to keep the protein structure intact.[80,81] Reports have shown that GNPs with sizes greater than 60 nm are found to be practically inert for platelets. The observed size dependence transport of GNPs in platelets is due to the presence of an open canalicular system.[82] Unlike the suspended blood platelets, the adhered state of platelets follows an indirect route for NP-induced activation by an additional frictional force at the adhering interface. Different phosphorylated proteins showed a characteristic increase in expression level, indicating that the GNPs can be responsible for activation. The study suggested that GNP-induced platelet activation was possibly due to a mild alteration of granule release, which under critical conditions can lead to aggregation.[83] Jun *et al.* (2011) suggested that Ag-NPs enhanced procoagulant activity through phosphatidylserine exposure in platelets.[84] Besides, our group showed that silver nanoparticles significantly delayed fibrin polymerisation. While, this effect was strongly inhibited when the experiment was performed with bovine serum albumin at a concentration below the protein concentration in the blood.[85] Martínez-Gutierrez *et al.* (2012) performed an

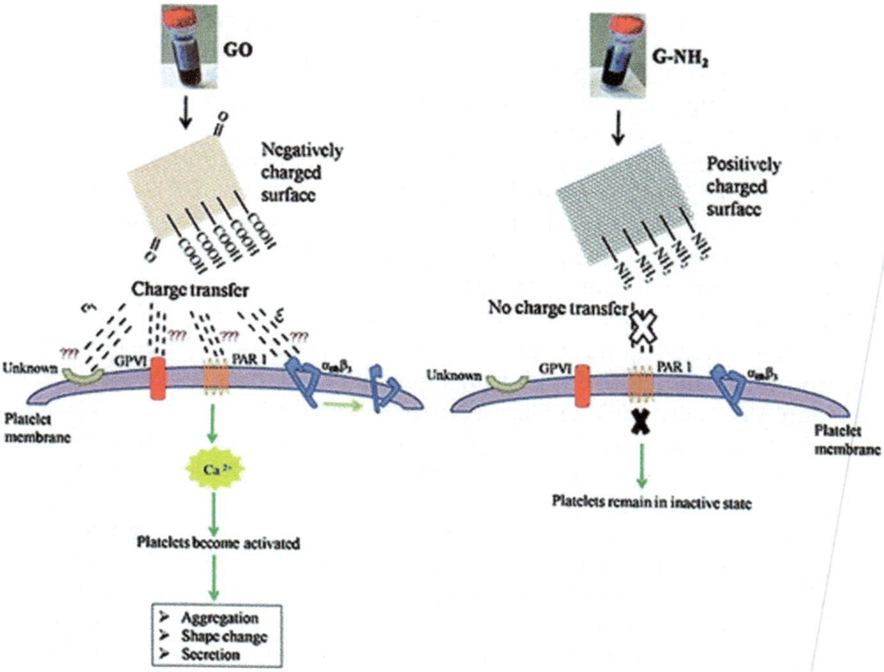

Figure 9.7 Illustrates the effect of graphene oxide and amine-modified graphene on platelets. Reproduced from ref. 79 with permission from American Chemical Society, Copyright 2012.

activated partial thromboplastin time (APTT) test and observed a moderate anticoagulant effect of Ag-NPs on the intrinsic pathway. The authors performed the same assay using a mechanical semi-automated coagulometer KC10 and observed a dose-dependent decrease in coagulation in the presence of Ag-NPs.[86] Consequently, this non-physiological interaction prevents tenase and prothrombinase complex formation, inducing an increase in the coagulation time. Moreover, procoagulant activity mediated by platelets could amplify the procoagulant effect observed on the coagulation cascade by Ag-NPs. An increase in platelet adhesion was observed at 50 $\mu g\ mL^{-1}$, with no effect on platelet activation and aggregation, while Jun *et al.* (2011) and Krajewski *et al.* (2013) demonstrated that Ag-NPs induce platelet activation and aggregation.[84,87] Our *in vitro* and *in vivo* studies showed the anti-platelet property of nanosilver. Microscopic studies revealed that Ag-NPs accumulate within platelet granules and reduce inter-platelet proximity. Studies suggested that the inhibition could be due to conformational modulation of integrins $\alpha_{II}\beta_3$, preventing interaction with fibrinogen.[88] Later, Deb *et al.* (2012) discussed that the anti-platelet effect observed by Shrivastava *et al.* (2011) was due to citric acid present at the surface of the NPs.[88–90] Further, in the case of metallic oxide NPs, it was found that the TiO_2 anatase form accelerates thrombus formation while the TiO_2 rutile form does not.[91,92]

9.2.2.4 Platelet/Clotting Protein–Silica Nanoparticle Interactions

Amorphous Si-NPs have a non-crystalline structure, high surface-area-to-volume ratio and large negative surface charge. PT and APTT tests were applied to investigate the effect of nanosilica on extrinsic and intrinsic blood coagulation pathways. The results revealed that Si-NPs at low (0.1 mg mL^{-1}) and high (1 mg mL^{-1}) concentrations showed PT and APTT values in their normal range, while some dry and porous silica materials can activate the coagulation cascade due to their high absorption capacities. On the other hand, pores of the MSN and F127-OMSN already filled with PBS significantly decrease the absorption capacity of particles and result in the observed anticoagulant behaviour of the particles.[93] The adhesion and aggregation of platelets depend on the size of the nanosilica as its effect was more pronounced with 50 nm particles compared to 500 nm particles. The microscopic studies revealed internalisation and distribution of Si-NPs within platelet cytoplasm on interacting with the platelet surface membrane.[94] Another study concludes that the penetration of Si-NPs stimulates oxidative stress that favours activation of GPIIb/IIIa and expression of markers on the platelet surface membrane.[95] Microfluidic analysis revealed that Si-NPs slightly compromise platelet adhesion to endothelial cells at a low NP dose while at a high dose, platelet adhesion increases. NP exposure leads to platelet activation and aggregation but does not result in significant loss of platelet viability.[96] Due to some disrupted events with platelets, bare silica is generally modified with different functional groups, such as using PEGylated organically-modified silica-NPs (ORMOSIL-NPs) *etc.* for *in vivo* applications.[97]

9.2.2.5 Platelet/Clotting Protein–Quantum Dot Interactions

The effect of core–shell QDs consisting of CdSe–ZnS, carboxyl- or amine-modified, was seen on platelets, and, at high doses, carboxyl-QDs were more potent in inducing pulmonary vascular thrombosis than amine-QDs.[98] CdSe, cysteine-capped QDs aggregate platelets at a concentration of 1000 μg mL^{-1}.[99] Another study demonstrated that CdTe QDs (thioglycolic acid- and cystamine-capped) had a stimulatory effect on platelets through complex mechanisms that involved activation of GPIIb/IIIa and release of a proaggregatory MMP-2.[100] However, more experimental evidence is required to figure out the exact signalling pathways involved in QD-induced platelet activation.

9.2.2.6 Platelet/Clotting Protein–Other Biomedically Important Nanomaterial Interactions

Neutral (hydroxy-terminated) and anionic (carboxy-terminated) polyamidoamine (PAMAM) dendrimers do not stimulate platelets. Similarly, anionic polystyrene and cetyl alcohol/polysorbate-based NPs do not result in platelet activation,[101] while cationic (amine-terminated) PAMAM dendrimers result in platelet aggregation in a concentration-dependent manner.[102] The dendrimer-induced platelet aggregation is accompanied by expression of activation marker P-selectin and release of intermediate signalling molecules (thrombospondin and ATP). A recent study suggested that cationic polystyrene latex particles induce platelet aggregation by disruption of membrane integrity.[103] Poly(lactic-*co*-glycolic acid) PLGA-macrogol and chitosan have no stimulatory effects on platelet aggregation.[104,105] Hence, both particle size and surface charge are important physicochemical parameters that determine NMs' interactions with platelets.

9.2.3 Plasma Protein–Nanomaterial Interactions

9.2.3.1 Plasma Protein Types and Functions

The three major plasma proteins are albumin (55.2%), globulin (38%) and fibrinogen (6.5%). Albumin is the most abundant plasma protein (2.8 to 4.5 g per 100 mL) synthesised in the liver and consists of a single polypeptide chain of 610 amino acids (69 kDa) having some essential amino acids such as lysine, leucine, valine, phenylalanine, threonine, arginine and histidine. Acidic amino acids like aspartic acid and glutamic acid are also concentrated in albumin. The presence of these residues makes the molecule highly charged with a positive and negative charge. Besides having a nutritive role, albumin acts as a transport carrier for various biomolecules such as fatty acids, trace elements, and drugs, and maintains the osmotic balance of blood. Globulins are divided into α1-, α2-, β- and γ-globulins and are synthesised in the liver, while γ-globulins are formed in the cells of the reticuloendothelial system. The average normal serum globulin concentration is 2.5 g per 100 mL. Fibrinogen is a fibrous protein with a molecular weight of 340 000, and has six polypeptide chains that are held together by disulfide linkages. Fibrinogen is an important clotting protein in the blood where it is converted to fibrin by thrombin.

NM–cell membrane interactions are complex dynamic processes in which plasma proteins play a significant role. Protein adsorption on a NM's surface results in the formation of a 'protein corona' (PC) depending on the protein–nanoparticle binding affinity and protein–protein interactions. Tightly binding proteins with a high affinity for the NM surface form a 'hard' corona, while loosely bound proteins form a 'soft' corona (Figure 9.8).[106,107] The 'hard' corona proteins generally interact directly with the NM's surface, while the 'soft' corona proteins interact with the 'hard' corona *via* weak protein–protein interactions.[108] The corona formation occurs spontaneously upon exposure to proteins and may consist of multiple layers that have different affinities for the NP surface.[109] The thickness of the corona depends on different parameters like protein concentration, particle size, and surface properties of the nanoparticles. Simberg *et al.* (2009) proposed a model for protein corona formation

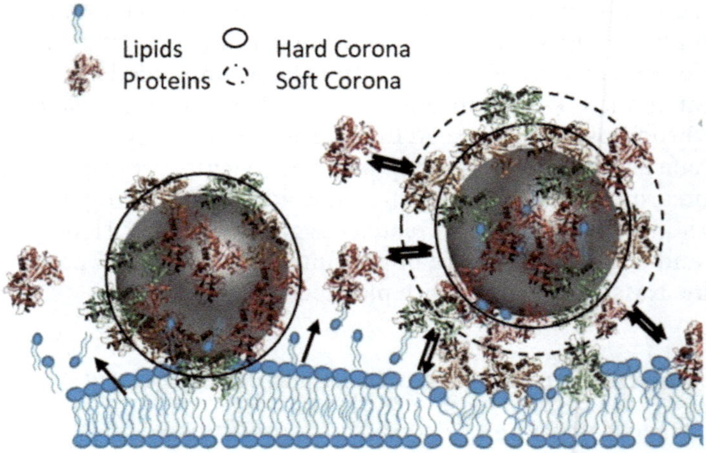

Figure 9.8 Illustration demonstrating the formation of 'hard' and 'soft' coronas on a NM's surface. Reproduced from ref. 116 with permission from Elsevier, Copyright 2017.

consisting of two kinds of binders: ones that recognise the surface of a NM, known as 'primary binders', and the 'secondary binders' that bind to the primary binders.[110] Walkey and Chan (2012) have summarised a subset of 125 plasma proteins, called adsorbomes, that were identified in the PCs of at least one NM.[111] Major plasma proteins like albumin, immunoglobulin G (IgG), fibrinogen, and apolipoproteins are mainly involved in the formation of PC.[112] The competitive adsorption of proteins on the surfaces of nanoparticles is the collective effect of incubation time, concentration of protein, and adsorption affinity between the protein–nanoparticle surface also known as the 'Vroman effect'. Protein adsorption onto nanoparticles occurs due to an increase in the collective entropy of the proteins on the NP surface and non-specific interactions between the NP surface and the proteins. The increase in the entropy is due to the rearrangement of the protein structure to a more stable position on the uncoated NP surface. Strong interactions occur between uncoated NPs and proteins due to the increase in entropy, which in turn leads to a significant decrease in the Gibbs free energy. This decrease in energy is primarily due to the relaxation of protein secondary structures.[113]

Apolipoproteins (Apo) are major proteins adsorbed on liposomes and polymeric NPs. Hydrophobic domains of nanomaterials are held responsible for protein adsorption processes. Gessner *et al.* (2000) illustrated this in an experiment where a NM's hydrophobicity affected the adsorption of ApoA-I, ApoA-IV, ApoC-III, and ApoJ. Cationic lipoplexes and polyplexes have strong affinities towards albumin binding, probably because albumin is negatively charged.[114]

Molecules that are loosely bound to a nanoparticle's surface or have weak interactions with the 'hard' corona form a 'soft' corona. In some nanoparticles, especially those with preformed functional groups such as PEGylated NPs, only a 'soft' corona is observed (Figure 9.8).[115,116]

9.2.3.2 Plasma Protein–Carbon-based Nanomaterial Interactions

Carbon nanotubes are sp^2-hybridised materials that act as universal sorbents for organic compounds in aqueous phases. Different non-covalent interactions of CNTs are favoured in biological fluids. The main forces involved in protein binding to non-functionalised CNTs are π–π stacking and hydrophobic interactions. Important proteins adsorbed on CNT surfaces are fibrin, albumin, transferrin, and IgG. Protein binding and its selectivity are highly influenced by the surface features of CNTs. The adsorption rates of proteins for different CNTs were calculated and found to be in this order: SWCNT > aligned MWCNT > carboxyl MWCNT > short MWCNT > short carboxyl MWCNT, and the calculated adsorption capacities were in the order: SWCNT > aligned MWCNT > carboxyl MWCNT > short MWCNT > short carboxyl MWCNT.[117] Bovine fetal globulin (BFG), gamma (γ) globulin, transferrin and BSA are competitively adsorbed on CNTs to form protein-coated CNT coronas. The PC fundamentally alters CNT–cellular interaction pathways and reduces their cytotoxicity.[118] Graphene interactions with blood proteins, BFG and BSA, reveal that an aromatic residue, Tyr-18 in the γ-chain, and a basic residue, Arg-391 in the γ-chain of BFG, play important roles in the adsorption process. The π–π stacking interactions attract Tyr-18 to bind to the graphene surface, the strong van der Waals interactions between the long, planar side chain and graphene provides the driving force for the adsorption of Arg-391.[119] Nanodiamonds are also well studied for their range of plasma protein adsorptions such as albumin, histone, fibrinogen and IgG. *In vitro* studies have proved that NDs show a very low affinity for albumin and γ-globulin.[120]

9.2.3.3 Plasma Protein–Metal/Metal Oxide Nanoparticle Interactions

GNPs' association with proteins like albumin, fibrinogen, γ-globulin, histone, and insulin revealed that the binding constant *K* and the degree of cooperativity of particle–protein binding (Hill constant *n*) depend on particle size and the native structure of the protein. For the same surface area, 5 nm GNPs bind with significantly more proteins due to their high curvature compared to the large GNPs with less curvature.[121,122] GNPs' interaction with BSA in aqueous solution led to a red shift of plasmon resonance by 5 to 6 nm.[123] Interactions of Ag-NPs with BSA is spontaneous and mainly driven by hydrophobic forces in the microenvironment of the tryptophan residues in the protein.[124] Cellular uptake of Ag-NPs is found to be modulated by their interactions with different proteins including albumin, transferrin, and IgG.[125] Studies showed that Ag-NPs interactions with blood proteins result in alpha–beta transition, and protein aggregation.[126,127] An irreversible conformational change in the secondary structure of BSA was found to change significantly on its association with magnetic Fe_2O_3-NPs.[128,129] The quantitative thickness measurement of the protein corona was done *in situ* using the nanomagnetism of Fe_2O_3-NPs.[130] The shape of the TiO_2 particles was a determinant in protein binding and BSA adsorption.[131,132]

9.2.3.4 Plasma Protein–Silica Nanoparticle Interactions

Si-NPs are of immense potential due to their non-toxicity and inert behaviour in biological environments. The NP-protein corona is also influenced by pH, which can have an impact on aggregate size, surface charge, adsorbed protein quantity, and the protein's secondary structure. Lehman *et al.* (2016) reported that non-porous Si-NPs led to only 10% change in α-helix content of BSA while a 12% change was reported with mesoporous Si-NPs.[133] In another study, Catalano *et al.* (2015) reported a 36% α-helix content of BSA on pyrolytic Si-NPs.[134] The protein's structural changes were also influenced by the pH of the solution. The quantity of BSA adsorbed as a function of pH change is more at the isoelectric point of the BSA–nanosilica complex.[135]

9.2.3.5 Plasma Protein–Other Nanomaterial Interactions

The absorption of three different proteins – BSA, lysozyme, and fibrinogen – on alumina and hydroxyapatite particles demonstrates destabilisation and structural loss of the bound protein retaining the residual structure.[136] Six different types of polystyrene-NPs with different surface chemistry (plain, carboxyl-modified, and amine-modified), each with two sizes (50 and 100 nm), were used in the experiment. This illustrates that the nature of the proteins in the corona is determined by the surface charge of the particle.[137,138]

9.2.4 Complement System/Immune Cell–Nanomaterial Interactions

9.2.4.1 Blood Complement System Components and Functions

The complement system is part of the innate immune system and comprises a group of 35 soluble protein fragments present in blood in addition to serum proteins and cell membrane receptors. Complement proteins boost the ability of phagocytic cells and antibodies to opsonise and neutralise invading micro-organisms, foreign materials and apoptotic or necrotic cells present in the host. The complement system triggers immune functions like phagocytosis, inflammation and membrane attack.[139] Major opsonins in the bloodstream are serum proteins like laminin, fibronectin, C-reactive protein, type-I collagen, immunoglobulins (IgG and IgM), and opsonic fragments derived from C3 complement protein cleavage. The coating of foreign particles by opsonins makes them readily available for opsonisation, influencing their fate in the bloodstream. Opsonised particles are mainly taken up by the specialised phagocytic cells of the mononuclear phagocyte system (MPS), mostly represented by the Kupffer cells of the liver and the macrophages of the bone marrow and the spleen. The complement proteins exist in serum as inactive zymogens, which, upon proteolytic cleavage, become active and carry the signals further in one of three ways: the classical pathway, the alternate pathway, or the lectin pathway. The classical pathway begins with the binding of an antigen–antibody complex to the C1 protein, while the alternative pathway begins with the activation of C3 protein by a foreign antigen alone. In the lectin pathway, the lectins bind to specific carbohydrate antigen moieties leading to the activation of C1 and continue as a mannose-binding (MB) lectin pathway. All three pathways result in C5 protein activation and cleavage, leading to formation of a membrane attacking complex with the help of other complement proteins. Both classical and alternative pathways converge right at C3 cleavage in C3a and C3b, which happens even before C5 activation. Figure 9.9 depicts different complement pathways' activation on exposure to nanomaterials.[140,141]

9.2.4.2 Complement System–Carbon-based Nanomaterial Interactions

After intravenous administration, nanoparticles come into contact with blood plasma proteins and form opsonins. It has been observed that nanoparticles with hydrophobic surfaces or are positively/negatively charged show a higher tendency to opsonise, while worm-like nanoparticles are phagocyted to a lesser extent compared to spherical ones of equal volume.[142] Complement activation may be initiated by adsorption of CNTs onto the proteins.[143] Salvadore-Morales *et al.* (2006) for the first time confirmed activation of a serum complement system *via* a classical pathway by two types of carbon nanotubes – SWCNTs and double-walled carbon nanotubes (DWCNTs). DWCNTs were also capable of stimulating the alternative pathway.[144] Zymosan, which is a strong complement activator, was used to compare the level of activation in both pathways. It was observed that activation of the complement by CNTs *via* the classical pathway led to the production of inflammatory peptides and a subsequent coating with opsonins iC3b and C3b.[145]

Salvador-Morales *et al.* (2008) confirmed that chemically modified MWCNTs can affect complement activation potential differently.[146] The authors modified pristine MWCNTs with 1,8-diaminooctane and later added L-alanine by covalent attachment. It was observed that once the MWCNTs were chemically modified with varying surface features, complement activation *via*

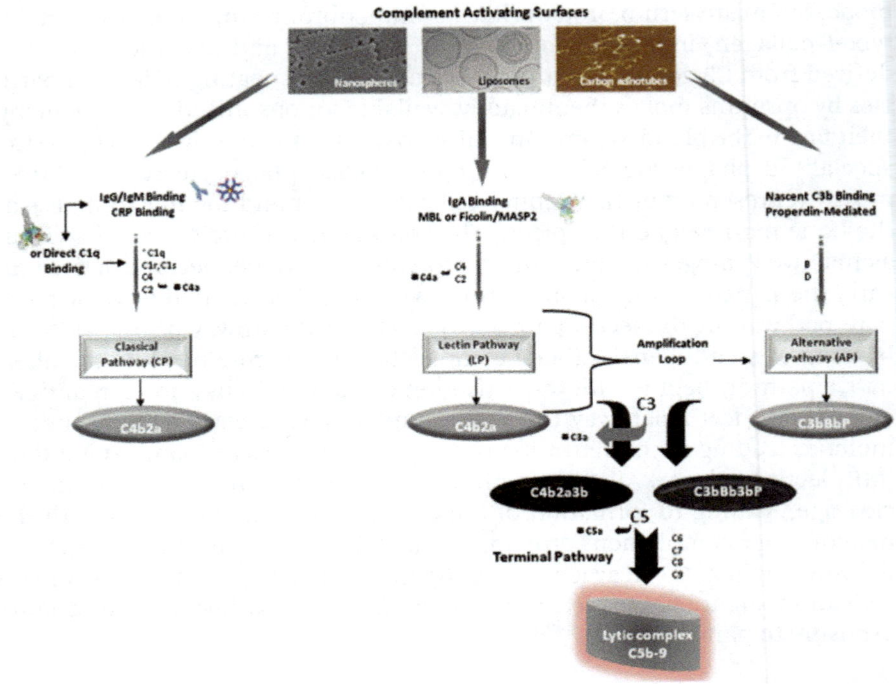

Figure 9.9 Complement pathway activation on interaction with different nanomaterials. C4b2a and C3bBbP are C3 convertase and C4b2a3b and c3bBb3bP are C5 convertase. Reproduced from ref. 140 with permission from Elsevier, Copyright 2011.

both classical and alternate pathways was changed. The chemically altered MWCNTs produced smaller complement activation in comparison to the unaltered or pristine MWCNTs and this was confirmed by the presence of fewer amounts of C1q protein binding and more factor H binding. The concentration of C3a in the blood plasma after exposure to fluorescent CDs shows elevation due to hydroxyl/carboxyl groups and the hydrophobic structures of the CDs.[147]

9.2.4.3 Complement Protein–Metal/Metal Oxide Nanoparticle Interactions

GNPs are effective complement activation evaders and can be used as valuable drug delivery systems for targeted cancer therapies.[148,149] Using a library of GNPs (10, 20, 40, and 80 nm) passivated with PEG, it was found that citrate-capped GNPs activate the whole complement system in a size-dependent manner. But PEGylated GNPs' activation level was found to be independent of both the core size of the GNPs and the molecular weight of PEG.[150] Dextran-coated superparamagnetic iron oxide (SPIO) nanoparticles are widely used as magnetic resonance imaging (MRI) contrast agents due to their magnetic susceptibility. Dextran in these particles can bind to lectin proteins and activate the complement *via* the MB-lectin pathway. The SPIO can be effectively cleared from blood due to complement activation and taken up by macrophages and carried to the liver and spleen, making it easy to image these organs.[151,152]

9.2.4.4 Complement Protein–Other Nanomaterial Interactions

Surface modification of nanoparticles with hydrophilic polymers such as PEG can be achieved by physical adsorption or covalent grafting of PEG chains at the surface of preformed NPs or using a block co-polymer where the PEG moiety is linked to the hydrophobic block like poly(lactic acid), poly(lactic-*co*-glycolic)acid, or poly(alkylcyanoacrylates) prior to the self-assembly.[153] The widely accepted theory to explain the stealth protective effect conferred by PEG coating is based on the formation of a dense, hydrophilic shell and steric stabilisation of the NPs. Hydrophilic and flexible PEG chains assume an extended conformation in aqueous media and start to be compressed in the presence of serum proteins. This compression generates repulsive forces able to counteract the attractive forces between the opsonins and the particle surface.[154] Repeated administration of PEG-coated NPs trigger an 'accelerated blood clearance' (ABC) phenomenon leading to rapid liver uptake and blood clearance. The ABC phenomenon is composed of two phases: the induction and the effectuation. The former implies the interaction between the PEGylated NPs administered as a first dose and the immune system, triggering the production of a large amount of anti-PEG immunoglobulin M antibodies (IgM) by splenic B-cells. Consequently, anti-PEG IgM present in the bloodstream would recognise and bind at the surface of PEGylated NPs administered consecutively within a short time interval, thus causing complement activation, NP opsonisation, and uptake by Kupffer cells.[155] ABC has been reported to be triggered by both PEG-modified PLA and PLGA NPs, reaching the highest extent one week after the first administration.[156] In addition to the ABC process, it has been recently observed that PEG chains could induce complement activation *via* the lectin pathway, due to the structural analogy between PEG segments and a region of the D-mannose moiety. Once again, the nanocarrier size plays a pivotal role, and it demonstrates that, due to the limited surface available for convertase deposition, NPs less than 30 nm promote weaker complement activation compared with larger ones.[157]

9.3 Toxicological Assessment

For a toxicological assessment of NMs, a number of assays is performed like lactose dehydrogenase (LDH), MTT, alamar blue, trypan blue exclusion and haemolysis *etc.* The cell viability assays measure the functional integrity of one or more organelles or the cell membrane. The NMs' surface charge, composition, shape, and size influence cell viability. Another important assay for toxicity measurements is analysis of reactive oxygen species (ROS) production in the presence of NMs. The free-radical production marks the stress response and excess free radicals/superoxides/peroxides can damage cells at genetic/cellular machinery as well as membranes by peroxidation. Hence, toxicity assays that measure the cell/organism viability or integrity form the right platform for quantifying NMs' biocompatibility.

9.3.1 Haemolysis Test

The exposure of erythrocytes to NMs may result in haemolysis and/or haemagglutination. Haemolysis results in the release of red-coloured haemoglobin; colorimetric assays can be used to estimate the extent of the damage.[158] Nanoparticles also act like antigens in bridging erythrocytes, causing haemagglutination, which results in clumping or agglutination to form a diffuse lattice. This effect has serious consequences in humans leading to obstruction of blood vessels, an imbalance in the osmolarity of blood, and the unavailability of erythrocytes to carry out normal functions. In these assays, blood from healthy human volunteers are collected and treated with nanoparticles and analysed microscopically for haemolysis or haemagglutination. Assuming erythrocyte breakdown due to treatment of cells with TritonX-100 as 100% (positive control), and untreated erythrocytes (negative control) as 0%, haemolysis can be calculated as:

$$\%\ \text{haemolysis} = (\text{abs of test sample} - \text{abs of negative control}) \div (\text{abs of control}) \times 100$$

Haemagglutination of human blood *in vitro* can be visually analysed on a scale of 0–1. A high haemagglutination reaction can be described as a granular or irregular ring occupying more than half of the well bottom by erythrocytes.

9.3.2 Thrombogenicity/Blood Clotting Assays

Platelet activation is estimated from the upregulation of the platelet receptors glycoprotein (GP) IIb/IIIa (CD41) and P-selectin (CD62P) using flow cytometry. Platelet aggregation is measured using an aggregometer. An automated platelet analyser is used to estimate changes in coagulation parameters due to nanoparticle exposure. Coagulation parameters like thrombin time (TT), prothrombin time (PT), activated partial thromboplastin time (APTT) and fibrinogen (FIB) can be analysed by the automated analyser. APTT analysis is one of the measurements that are carried out most frequently in coagulation analysis. It detects the factor deficiencies in the plasma such as a lack of decrease in factor VIII, which corresponds to the intrinsic pathway. PT represents the time taken by plasma to clot after addition of tissue factor III and this relates mainly to the extrinsic pathway. TT and FIB represent the times of formation of thrombin and fibrinogen concentrations, respectively, and FIB essentially shows soluble fibrinogen being converted into insoluble fibrin.[159]

9.3.3 Cell Viability Test

9.3.3.1 MTT/XTT Assay

This is a metabolic assay and the reagent used in this assay is (3-(4,5-dimethylthiazol-2-yl)-2,5-diphenyltetrazolium bromide) tetrazolium (MTT), which is a positively charged small molecule that undergoes NADPH-mediated conversion to formazan. Because of its positive charge, MTT can easily enter viable and non-viable cells. Upon conversion, the formazan product precipitates inside cells near the cell surface and can be detected using a spectrophotometer.[160] The percentage of viable cells is calculated using the following formula:

$$\%\text{ viable cells} = (\text{abs of test sample} - \text{abs of blank}) \div (\text{abs of control} - \text{abs of blank}) \times 100$$

9.3.3.2 ATP Assay

The amount of ATP in cells correlates with cell viability. With the loss of membrane integrity, cells lose the ability to synthesise ATP, and endogenous ATPases break down any ATP in the cell, lowering its concentration. Detection is based on using the luciferase reaction to measure the amount of ATP in viable cells. The reagent is used to lyse the cell membrane and release ATP, it inhibits endogenous ATPases, and provides ATP to luciferin and luciferase. The signal can be recorded with a luminometer.[161]

9.3.3.3 LDH Assay

Cells that have lost membrane integrity release lactate dehydrogenase (LDH) into the surrounding medium. This is a fluorescent-based method that uses coupled enzymatic reactions to measure the release of LDH from damaged cells as an indicator of cytotoxicity. The assay is designed to estimate the number of non-viable cells present in a mixed population of living and dead cells. LDH catalyses the conversion of lactate to pyruvate with the concomitant production of NADH. The reagent used contains excess substrates (lactate and NAD^+) to drive the LDH reaction and produce NADH.[160]

9.3.3.4 Calcein Assay

Calcein AM is a non-fluorescent, hydrophobic compound that can easily permeate into viable cells. The hydrolysis of calcein AM occurs by intracellular esterases, producing calcein, a hydrophilic, strongly fluorescent compound that is well-retained in the cell cytoplasm. The viable cells can be easily quantified with this assay.[162]

9.3.3.5 Trypan Blue Exclusion Assay

Trypan blue is a charged dye; it cannot permeate through living cells. So, simply incubating cells with trypan blue and looking under a microscope allows an estimation of the number of viable cells.[163]

9.3.3.6 Almar Blue Assay

This assay quantitatively measures the proliferation of cells. Resazurin is used as an oxidation–reduction (redox) indicator that undergoes a colorimetric change in response to cellular metabolic reduction. The reduced form of resorufin is pink and highly fluorescent, and the intensity of fluorescence produced is proportional to the number of respiring cells.[164]

9.3.3.7 ROS Measurement

Reactive oxygen species (ROS) such as superoxide and hydrogen peroxide are continually produced during metabolic processes. ROS generation is normally counterbalanced by the action of antioxidant enzymes and other redox molecules in the blood. However, excess ROS can lead to cellular injury in the form of damaged DNA, lipids and proteins. Different markers of the cell damage are extremely stable and, therefore, provide a more reliable way to measure oxidative stress. Assays are available to measure the activity of specific antioxidant enzymes, such as catalase and superoxide dismutase.[165]

9.4 Summary, Challenges and Future Directions

In summary, although NMs are of immense potential for biomedical applications like drug delivery, bio-imaging and sensing, their interaction with blood is poorly understood. Whenever the NMs are intravenously injected inside the biological system, they are bound to interact with blood constituents. NMs' presence within a dynamic biological environment can interact with blood constituents resulting in protein corona formation and trigger pathophysiological events such as complement and platelet activation and easily trigger the immune system, causing immunotoxicity. Thus, establishing the haemocompatibility of a nanomaterial is critical in determining the safe and successful clinical applications of NMs. In this chapter, the role of blood constituents in the biological system and interactions between NMs and blood components are critically reviewed. The surface characteristics of NMs like shape, coating, functionalisation and surface charge may be crucial factors for their adverse toxic effects. The haemocompatibility assays described in this chapter may serve as guidelines for initially establishing the suitability of a chosen nanomaterial for biomedical applications. It may also serve to improve our mechanistic understanding of the interaction between nanoparticles and blood components.

Hence, in-depth studies are needed to understand, predict, and counteract the behaviour of NMs within cellular and molecular microenvironments. While diverse mechanisms and pathways have been implicated in responding to nanoparticle properties and composition, relatively little has been known about the specific structural and biophysical designs that elicit various immune reactions. Future studies are required to design parameters that can lead to the development of more haemocompatible NMs. Since the nanomaterials are physically and chemically diverse and are structurally and functionally different from 'macro' materials, conventional assays may require modification when applied to nanomaterials. Thus, blood compatibility against nanomaterials is one of the major concerns and should be taken care of before exploiting nanomaterials for biomedical applications.

Acknowledgements

The authors gratefully acknowledge the Department of Science and Technology for providing an INSPIRE Faculty Research grant (No. IFA-LSBM-31).

References

1. E. Boisselier and D. Astruc, Gold nanoparticles in nanomedicine: Preparations, imaging, diagnostics, therapies and toxicity, *Chem. Soc. Rev.*, 2009, **38**, 1759–1782.
2. T. Jamieson, R. Bakhshi and D. Petrova, Biological applications of quantum dots, *Biomaterials*, 2007, **28**(31), 4717–4732.
3. C. Chung, Y. K. Kim, D. Shin, S. R. Ryoo, B. H. Hong and D. H. Min, Biomedical applications of graphene and graphene oxide, *Acc. Chem. Res.*, 2013, **46**, 2211–2224.
4. Z. Liu, S. Tabakman, K. Welsher and H. Dai, Carbon nanotubes in biology and medicine: *in vitro* and *in vivo* detection, imaging and drug delivery, *Nano Res.*, 2009, **2**, 85–120.
5. N. T. K. Thanh and L. A. W. Green, Functionalisation of nanoparticles for biomedical applications, *Nano Today*, 2010, **5**(3), 213–230.

6. L. C. Cheng, X. Jiang, J. Wang, C. Chen and R. S. Liu, Nano–bio effects: interaction of nanomaterials with cells, *Nanoscale*, 2013, **5**(9), 3547–3569.
7. J. N. Tiwari, R. N. Tiwari and K. S. Kim, Zero-dimensional, one-dimensional, two-dimensional and three-dimensional nanostructured materials for advanced electrochemical energy devices, *Prog. Mater. Sci.*, 2012, **57**(4), 724–803.
8. J. M. Park, K. S. Nalwa, W. Leung, K. Constant, S. Chaudhary and K. M. Ho, Fabrication of metallic nanowires and nanoribbons using laser interference lithography and shadow lithography, *Nanotechnology*, 2010, **21**(21), 215301.
9. J. Liu, J. Jiang, C. Cheng, H. Li, J. Zhang, H. Gong and H. J. Fan, Co_3O_4 Nanowire@ MnO_2 ultrathin nanosheet core/shell arrays: A new class of high-performance pseudocapacitive materials, *Adv. Mater.*, 2011, **23**(18), 2076–2081.
10. Z. Wang, X. Pan, Y. He, Y. Hu, H. Gu and Y. Wang, Piezoelectric nanowires in energy harvesting applications, *Adv. Mater. Sci. Eng.*, 2015, **2015**, 1–21.
11. T. Sadhasivam, H. T. Kim, S. Jung, S. H. Roh, J. H. Park and H. Y. Jung, Dimensional effects of nanostructured Mg/MgH_2 for hydrogen storage applications: A review, *Renewable Sustainable Energy Rev.*, 2017, **72**, 523–534.
12. J. Zuo, T. Jiang, X. Zhao, X. Xiong, S. Xiao and Z. Zhu, Preparation and application of fluorescent carbon dots, *J. Nanomater.*, 2015, **2015**, 10.
13. S. Vardharajula, S. Z. Ali, P. M. Tiwari, E. Eroğlu, K. Vig, V. A. Dennis and S. R. Singh, Functionalized carbon nanotubes: biomedical applications, *Int. J. Nanomed.*, 2012, **7**, 5361.
14. E. Perevedentseva, M. Jani and C. L. Cheng, Nanodiamond: A Promising Nanomaterial for Biomedical Applications, in *Applications of Nanomaterials*, ed. R. S. Chaughule and S. C. Watawe, American Scientific Publishers, Valencia, CA, USA, 2012, pp. 1–19.
15. D. Cabuzu, A. Cirja, R. Puiu and A. Mihai Grumezescu, Biomedical applications of gold nanoparticles, *Curr. Top. Med. Chem.*, 2015, **15**(16), 1605–1613.
16. S. Behera and A. Debata, Biomedical applications of silver nanoparticles, *J. Asian Sci. Res.*, 2011, **1**(1), 27.
17. A. K. Barui, R. Kotcherlakota and C. R. Patra, in *Inorganic Frameworks as Smart Nanomedicines*, ed. A. Grumezescu, Elsevier, Amsterdam, The Netherlands, 1st edn, 2018, pp. 239–278.
18. Z. Li, J. C. Barnes, A. Bosoy, J. F. Stoddart and J. I. Zink, Mesoporous silica nanoparticles in biomedical applications, *Chem. Soc. Rev.*, 2012, **41**(7), 2590–2605.
19. M. Elsabahy and K. L. Wooley, Design of polymeric nanoparticles for biomedical delivery applications, *Chem. Soc. Rev.*, 2012, **41**(7), 2545–2561.

20. A. Verma and F. Stellacci, Effect of surface properties on nanoparticle–cell interactions, *Small*, 2010, **6**(1), 12–21.
21. A. E. Nel, L. Mädler, D. Velegol, T. Xia, E. M. Hoek, P. Somasundaran, F. Klaessig, V. Castranova and M. Thompson, Understanding biophysicochemical interactions at the nano-bio interface, *Nat. Mater.*, 2009, **8**(7), 543–557.
22. H. Meng, W. Leong, K. W. Leong, C. Chen and Y. Zhao, Walking the line: The fate of nanomaterials at biological barriers, *Biomaterials*, 2018, **174**, 41.
23. S. O. Sowemimo-Coker, Red blood cell hemolysis during processing, *Transfus. Med. Rev.*, 2002, **16**(1), 46–60.
24. Z. H. Liu, Y. P. Jiao, T. Wang, Y. M. Zhang and W. Xue, Interactions between solubilized polymer molecules and blood components, *J. Controlled Release*, 2012, **160**, 14–24.
25. K. H. Liao, Y. S. Lin, C. W. Macosko and C. L. Haynes, Cytotoxicity of graphene oxide and graphene in human erythrocytes and skin fibroblasts, *ACS Appl. Mater. Interfaces*, 2011, **3**(7), 2607–2615.
26. S. K. Singh, M. K. Singh, P. P. Kulkarni, V. K. Sonkar, J. J. Grácio and D. Dash, Amine-modified graphene: thrombo-protective safer alternative to graphene oxide for biomedical applications, *ACS Nano*, 2012, **6**(3), 2731–2740.
27. S. Sachar and R. K. Saxena, Cytotoxic effect of poly-dispersed single walled carbon nanotubes on erythrocytes *in vitro* and *in vivo*, *PLoS One*, 2011, **6**, e22032.
28. J. Meng, X. L. Cheng, J. Liu, W. Q. Zhang, X. J. Li, H. Kong and H. Y. Xu, Effects of long and short carboxylated or aminated multiwalled carbon nanotubes on blood coagulation, *PLoS One*, 2012, **7**, e38995.
29. S. Li, Z. Guo, Y. Zhang, W. Xue and Z. Liu, Blood compatibility evaluations of fluorescent carbon dots, *ACS Appl. Mater. Interfaces*, 2015, **7**(34), 19153–19162.
30. X. Zhang, Y. Zhang, Y. Zheng and B. Wang, Mechanical characteristics of human red blood cell membrane change due to C60 nanoparticle infiltration, *Phys. Chem. Chem. Phys.*, 2013, **15**(7), 2473–2481.
31. A. Trpkovic, B. Todorovic-Markovic, D. Kleut, M. Misirkic, K. Janjetovic, L. Vucicevic, A. Pantovic, S. Jovanovic, M. Dramicanin, Z. Markovic and V. Trajkovic, Oxidative stress-mediated hemolytic activity of solvent exchange-prepared fullerene (C60) nanoparticles, *Nanotechnology*, 2010, **21**, 375102.
32. S. Bosi, L. Feruglio, T. Da Ros, G. Spalluto, B. Gregoretti, M. Terdoslavich, G. Decorti, S. Passamonti, S. Moro and M. Prato, Haemolytic effects of water-soluble fullerene derivatives, *J. Med. Chem.*, 2004, **47**, 6711–6715.
33. L. W. Tsai, Y. C. Lin, E. Perevedentseva, A. Lugovtsov, A. Priezzhev and C. L. Cheng, Nanodiamonds for medical applications: interaction with blood *in vitro* and *in vivo*, *Int. J. Mol. Sci.*, 2016, **17**(7), 1111.

34. Y. C. Lin, L. W. Tsai, E. Perevedentseva, H. H. Chang, C. H. Lin, D. S. Sun, A. Lugovtsov, A. Priezzhev, M. Jani and C. L. Cheng, The influence of nanodiamond on the oxygenation states and micro rheological properties of human red blood cells *in vitro*, *J. Biomed. Opt.*, 2012, **17**, 101512.
35. R. Purohit, N. S. Vallabani, R. K. Shukla, A. Kumar and S. Singh, Effect of gold nanoparticle size and surface coating on human red blood cells *Bioinspired, Biomimetic Nanobiomater.*, 2016, **5**(3), 121–131.
36. M. A. Dobrovolskaia, A. K. Patri, J. Zheng, J. D. Clogston, N. Ayub, P. Aggarwal, B. W. Neun, J. B. Hall and S. E. McNeil, Interaction of colloidal gold nanoparticles with human blood: effects on particle size and analysis of plasma protein binding profiles, *Nanomedicine: Nanotechnol., Bio. Med.*, 2009, **5**(2), 106–117.
37. W. Gao, C. M. Hu, R. H. Fang, B. T. Luk, J. Su and L. Zhang, Surface functionalization of gold nanoparticles with red blood cell membranes, *Adv. Mater.*, 2013, **25**(26), 3549–3553.
38. L. Q. Chen, L. Fang, J. Ling, C. Z. Ding, B. Kang and C. Z. Huang, Nanotoxicity of silver nanoparticles to red blood cells: size dependent adsorption, uptake, and hemolytic activity, *Chem. Res. Toxicol.*, 2015, **28**(3), 501–509.
39. N. Lubick, Nanosilver toxicity: ions, nanoparticles or both? *Environ. Sci. Technol.*, 2008, **42**(23), 8617.
40. F. M. Christensen, H. J. Johnston, V. Stone, R. J. Aitken, S. Hankin, S. Peters and K. Aschberger, Nano-silver – feasibility and challenges for human health risk assessment based on open literature, *Nanotoxicology*, 2010, **4**, 284–295.
41. H. Shi, R. Magaye, V. Castranova and J. Zhao, Titanium dioxide nanoparticles: a review of current toxicological data, *Part. Fibre Toxicol.*, 2013, **10**(1), 15.
42. S. Chakrabartty, A. Mondal, P. Chakrabarti, S. K. Singh, A. K. Saha and P. Singh, Synthesis of Biocompatible TiO_2 Nanodots: Glancing Angle Deposition Technique, *J. Nanosci. Nanotechnol.*, 2016, **16**(8), 8705–8710.
43. M. A. Dobrovolskaia, J. D. Clogston, B. W. Neun, J. B. Hall, A. K. Patri and S. E. McNeil, Method for analysis of nanoparticle hemolytic properties *in vitro*, *Nano Lett.*, 2008, **8**(8), 2180–2187.
44. T. Yu, A. Malugin and H. Ghandehari, Impact of silica nanoparticle design on cellular toxicity and hemolytic activity, *ACS Nano*, 2011, **5**, 5717–5728.
45. Y. Zhao, X. Sun, G. Zhang, B. G. Trewyn, I. I. Slowing and V. S. Y. Lin, Interaction of mesoporous silica nanoparticles with human red blood cell membranes: size and surface effects, *ACS Nano*, 2011, **5**, 1366–1375.
46. I. I. Slowing, C. W. Wu, J. L. Vivero-Escoto and V. S. Y. Lin, Mesoporous silica nanoparticles for reducing hemolytic activity towards mammalian red blood cells, *Small*, 2009, **5**, 57–62.

47. A. Yildirim, G. B. Demirel, R. Erdem, B. Senturk, T. Tekinay and M. Bayindir, Pluronic polymer capped biocompatible mesoporous silica nanocarriers, *Chem. Commun.*, 2013, **49**(84), 9782–9784.
48. R. Tavano, D. Segat, E. Reddi, J. Kos, M. Rojnik, P. Kocbek, S. Iratni, D. Scheglmann, M. Colucci, I. M. Echevarria, F. Selvestrel, F. Mancin and E. Papini, Procoagulant properties of bare and highly PEGylated vinylmodified silica nanoparticles, *Nanomedicine*, 2010, **5**(6), 881–896.
49. Y. S. Lin and C. L. Haynes, Impacts of mesoporous silica nanoparticle size, pore ordering, and pore integrity on hemolytic activity, *J. Am. Chem. Soc.*, 2010, **132**, 4834–4842.
50. C. Fornaguera, G. Calderó, M. Mitjans, M. P. Vinardell, C. Solans and C. Vauthier, Interactions of PLGA nanoparticles with blood components: protein adsorption, coagulation, activation of the complement system and haemolysis studies, *Nanoscale*, 2015, **7**(14), 6045.
51. Y. Han, X. Wang, H. Dai and S. Li, Nanosize and surface charge effects of hydroxyapatite nanoparticles on red blood cell suspensions, *ACS Appl. Mater. Interfaces*, 2012, **4**, 16–22.
52. B. Venkatesan, A. Tumala, V. Subramanian and E. Vellaichamy, Data on synthesis and characterization of chitosan nanoparticles for *in vivo* delivery of siRNA-Npr3: Targeting NPR-C expression in the heart, *Data in Brief*, 2016, **8**, 441–447.
53. M. Choimet, K. Hyoung-Mi, O. Jae-Min, A. Tourrette and C. Drouet, Nanomedicine: Interaction of biomimetic apatite colloidal nanoparticles with human blood components, *Colloids Surf., B*, 2016, **145**, 87–94.
54. J. M. de Lima, R. R. Sarmento, J. R. de Souza, F. A. Brayner, A. P. Feitosa, R. Padilha, L. C. Alves, I. J. Porto, R. F. Batista, J. E. de Oliveira, E. S. de Medeiros, P. R. Bonan and L. R. Castellano, Evaluation of hemagglutination activity of chitosan nanoparticles using human erythrocytes, *BioMed Res. Int.*, 2015, **2015**, 247965.
55. H. M. Spronk, J. W. Govers-Riemslag and H. Ten Cate, The blood coagulation system as a molecular machine, *Bioessays*, 2003, **25**(12), 1220–1228.
56. *Hemostasis and Thrombosis: Basic Principles and Clinical Practice*, ed. R. W. Colman, Lippincott Williams & Wilkins, 2006.
57. M. Anand, K. Rajagopal and K. R. Rajagopal, A model incorporating some of the mechanical and biochemical factors underlying clot formation and dissolution in flowing blood: review article, *J. Theor. Med.*, 2003, **5**(3–4), 183–218.
58. T. G. Loof, C. Deicke and E. Medina, The role of coagulation/fibrinolysis during Streptococcus pyogenes infection, *Front. Cell. Infect. Microbiol.*, 2014, **4**, 128.
59. S. D. Revak, C. G. Cochrane and J. H. Griffin, The binding and cleavage characteristics of human Hageman factor during contact activation: A comparison of normal plasma with plasmas deficient in factor XI, prekallikrein, or high molecular weight kininogen, *J. Clin. Invest.*, 1977, **59**(6), 1167.

60. E. Fröhlich, Action of nanoparticles on platelet activation and plasmatic coagulation, *Curr. Med. Chem.*, 2016, **23**(5), 408–430.
61. A. McNicol and S. J. Israels, Platelet dense granules: structure, function and implications for haemostasis, *Thromb. Res.*, 1999, **95**(1), 1–18.
62. A. Kauskot and M. F. Hoylaerts, Platelet receptors, in *Antiplatelet Agents*, Springer, Berlin, Heidelberg, 2012, pp. 23–57.
63. V. Leytin, Apoptosis in the anucleate platelet, *Blood Rev.*, 2012, **26**(2) 51–63.
64. M. J. VanWijk, E. VanBavel, A. Sturk and R. Nieuwland, Microparticles in cardiovascular diseases, *Cardiovasc. Res.*, 2003, **59**(2), 277–287.
65. L. L. Horstman and Y. S. Ahn, Platelet microparticles: a wide-angle perspective, *Crit. Rev. Oncol. Hematol.*, 1999, **30**(2), 111–142.
66. R. F. Zwaal and A. J. Schroit, Pathophysiologic implications of membrane phospholipid asymmetry in blood cells, *Blood*, 1997, **89**(4), 1121–1132.
67. Z. Gu, Z. Yang, L. Wang, H. Zhou, C. A. Jimenez-Cruz and R. Zhou, The role of basic residues in the adsorption of blood proteins onto the graphene surface, *Sci. Rep.*, 2015, **5**, 10873.
68. J. Mona, C. J. Kuo, E. Perevedentseva, A. V. Priezzhev and C. L. Cheng, Adsorption of human blood plasma on nanodiamond and its influence on activated partial thromboplastin time, *Diamond Relat. Mater.*, 2013, **39**, 73–77.
69. H. D. Wang, C. H. Niu, Q. Yang and I. Badea, Study on protein conformation and adsorption behaviors in nanodiamond particle-protein complexes, *Nanotechnology*, 2011, **22**, 145703.
70. S. Kumari, M. K. Singh, S. K. Singh, J. J. Grácio and D. Dash, Nanodiamonds activate blood platelets and induce thromboembolism, *Nanomedicine*, 2014, **9**(3), 427–440.
71. S. Li, Z. Guo, Y. Zhang, W. Xue and Z. Liu, Blood compatibility evaluations of fluorescent carbon dots, *ACS Appl. Mater. Interfaces*, 2015, **7**(34), 19153–19162.
72. J. Meng, X. Cheng, J. Liu, W. Zhang, X. Li, H. Kong and H. Xu, Effects of long and short carboxylated or aminated multiwalled carbon nanotubes on blood coagulation, *PLoS One*, 2012, **7**(7), e38995.
73. P. Bihari, M. Holzer and M. Praetner, Single-walled carbon nanotubes activate platelets and accelerate thrombus formation in the microcirculation, *Toxicology*, 2010, **269**(2–3), 148–154.
74. A. Radomski, P. Jurasz, D. Alonso-Escolano, M. Drews, M. Morandi, T. Malinski and M. W. Radomski, Nanoparticle-induced platelet aggregation and vascular thrombosis, *Br. J. Pharmacol.*, 2005, **146**(6), 882–893.
75. J. Semberova, S. H. De Paoli Lacerda, O. Simakova, K. Holada, M. P. Gelderman and J. Simak, Carbon nanotubes activate blood platelets by inducing extracellular Ca^{2+} influx sensitive to calcium entry inhibitors, *Nano Lett.*, 2009, **9**(9), 3312–3317.

76. S. H. Lacerda, J. Semberova, K. Holada, O. Simakova, S. D. Hudson and J. Simak, Carbon nanotubes activate store-operated calcium entry in human blood platelets, *ACS Nano*, 2011, **5**(7), 5808–5813.
77. A. R. Burke, R. N. Singh, D. L. Carroll, J. D. Owen, N. D. Kock, R. D'Agostino, F. M. Torti and S. V. Torti, Determinants of the thrombogenic potential of multiwalled carbon nanotubes, *Biomaterials*, 2011, **32**(26), 5970–5978.
78. S. K. Singh, M. K. Singh, M. K. Nayak, S. Kumari, S. Shrivastava, J. J. Grácio and D. Dash, Thrombus inducing property of atomically thin graphene oxide sheets, *ACS Nano*, 2011, **5**(6), 4987–4996.
79. S. K. Singh, M. K. Singh, P. P. Kulkarni, V. K. Sonkar, J. J. Grácio and D. Dash, Amine-modified graphene: thrombo-protective safer alternative to graphene oxide for biomedical applications, *ACS Nano*, 2012, **6**(3), 2731–2740.
80. G. Chen, N. Ni, J. Zhou, Y. J. Chuang, B. Wang and Z. Pan, Fibrinogen clot induced by gold-nanoparticle *in vitro*, *J. Nanosci. Nanotechnol.*, 2011, **11**, 74–81.
81. J. G. White, Platelets are covercytes, not phagocytes: Uptake of bacteria involves channels of the open canalicular system, *Platelets*, 2005, **16**, 121.
82. S. Deb, H. K. Patra, P. Lahiri, A. K. Dasgupta, K. Chakrabarti and U. Chaudhuri, Multistability in platelets and their response to gold nanoparticles, *Nanomedicine: Nanotechnol., Bio. Med.*, 2011, **7**(4), 376–384.
83. S. H. Lacerda, J. J. Park, C. Meuse, D. Pristinski, M. L. Becker, A. Karim and J. F. Douglas, Interaction of gold nanoparticles with common human blood proteins, *ACS Nano*, 2009, **4**(1), 365–379.
84. E. A. Jun, L. M. Lim, K. Kim, O. N. Bae, J. Y. Noh and K. H. Chung, Silver nanoparticles enhance thrombus formation through increased platelet aggregation and procoagulant activity, *Nanotoxicology*, 2011, **5**, 157–167.
85. S. Shrivastava, S. K. Singh, A. Mukhopadhyay, A. S. Sinha, R. K. Mandal and D. Dash, Negative regulation of fibrin polymerization and clot formation by nanoparticles of silver, *Colloids Surf., B*, 2011, **82**, 241–246.
86. F. Martínez-Gutierrez, E. P. Thi, J. M. Silverman, C. C. de Oliveira, S. L. Svensson, A. V. Hoek, E. M. Sánchez, N. E. Reiner, E. C. Gaynor, E. L. Pryzdial and E. M. Conway, Antibacterial activity, inflammatory response, coagulation and cytotoxicity effects of silver nanoparticles, *Nanomedicine: Nanotechnol., Bio. Med.*, 2012, **8**(3), 328–336.
87. S. Krajewski, R. Prucek, A. Panacek, M. Avci-Adali, A. Nolte, A. Straub, R. Zboril, H. P. Wendel and L. Kvitek, Hemocompatibility evaluation of different silver nanoparticle concentrations employing a modified Chandler-loop *in vitro* assay on human blood, *Acta Biomater.*, 2013, **9**(7), 7460–7468.

88. S. Shrivastava, T. Bera, S. K. Singh, G. Singh, P. Ramachandrarao and D. Dash, Characterization of antiplatelet properties of silver nanoparticles, *ACS Nano*, 2009, **3**(6), 1357–1364.
89. S. Deb, S. O. Raja, A. K. Dasgupta, R. Sarkar, A. P. Chattopadhyay and U. Chaudhuri, Surface tunability of nanoparticles in modulating platelet functions, *Blood Cells, Mol., Dis.*, 2012, **48**, 36–44.
90. J. Laloy, V. Minet, L. Alpan, F. Mullier, S. Beken, O. Toussaint, S. Lucas and J. Dogné, Impact of silver nanoparticles on haemolysis, platelet function and coagulation, *Nanobiomedicine*, 2014, **1**, 4.
91. N. Haberl, S. Hirn, M. Holzer, G. Zuchtriegel, M. Rehberg and F. Krombach, Effects of acute systemic administration of TiO_2, ZnO, SiO, and Ag nanoparticles on hemodynamics, hemostasis and leukocyte recruitment, *Nanotoxicology*, 2015, **1**(9), 963–971.
92. A. Nemmar, K. Melghit and B. H. Ali, The acute proinflammatory and prothrombotic effects of pulmonary exposure to rutile TiO_2 nanorods in rats, *Exp. Biol. Med.*, 2008, **233**(5), 610–619.
93. A. Yildirim, G. B. Demirel, R. Erdem, B. Senturk, T. Tekinay and M. Bayindir, Pluronic polymer capped biocompatible mesoporous silica nanocarriers, *Chem. Commun.*, 2013, **49**(84), 9782–9784.
94. A. Nemmar, S. Albarwani, S. Beegam, P. Yuvaraju, J. Yasin, S. Attoub and B. H. Ali, Amorphous silica nanoparticles impair vascular homeostasis and induce systemic inflammation, *Int. J. Nanomed.*, 2014, **9**, 2779–2789.
95. J. J. Corbalan, C. Medina, A. Jacoby, T. Malinski and M. W. Radomski, Amorphous silica nanoparticles aggregate human platelets: potential implications for vascular homeostasis, *Int. J. Nanomed.*, 2012, **7**, 631–639.
96. D. Kim, S. Finkenstaedt-Quinn, K. R. Hurley, J. T. Buchman and C. L. Haynes, On-chip evaluation of platelet adhesion and aggregation upon exposure to mesoporous silica nanoparticles, *Analyst*, 2014, **139**(5), 906–913.
97. R. Tavano, D. Segat, E. Reddi, J. Kos, M. Rojnik, P. Kocbek, S. Iratni, D. Scheglmann, M. Colucci, I. M. Echevarria, F. Selvestrel, F. Mancin and E. Papini, Procoagulant properties of bare and highly PEGylated vinylmodified silica nanoparticles, *Nanomedicine*, 2010, **5**(6), 881–896.
98. J. Geys, A. Nemmar, E. Verbeken, E. Smolders, M. Ratoi, M. F. Hoylaerts, B. Nemery and P. H. Hoet, Acute toxicity and prothrombotic effects of quantum dots: impact of surface charge, *Environ. Health Perspect.*, 2008, **116**(12), 1607.
99. R. Dunpall, A. A. Nejo, V. S. Pullabhotla, A. R. Opoku, N. Revaprasadu and A. Shonhai, An *in vitro* assessment of the interaction of cadmium selenide quantum dots with DNA, iron, and blood platelets, *IUBMB Life*, 2012, **64**(12), 995–1002.
100. S. P. Samuel, M. J. Santos-Martinez, C. Medina, N. Jain, M. W. Radomski, A. Prina-Mello and Y. Volkov, CdTe quantum dots induce activation of human platelets: implications for nanoparticle hemocompatibility, *Int. J. Nanomed.*, 2015, **10**, 2723–2734.

101. J. M. Koziara, J. J. Oh, W. S. Akers, S. P. Ferraris and R. J. Mumper, Blood compatibility of cetyl alcohol/polysorbate-based nanoparticles, *Pharm. Res.*, 2005, **22**, 1821–1828.
102. M. A. Dobrovolskaia, A. K. Patri, J. Simak, J. B. Hall, J. Semberova, S. H. De Paoli Lacerda and S. E. McNeil, Nanoparticle size and surface charge determine effects of PAMAM dendrimers on human platelets *in vitro*, *Mol. Pharm.*, 2012, **9**, 382–393.
103. C. McGuinnes, R. Duffin, S. Brown, N. L Mills, I. L. Megson, W. Macnee, S. Johnston, S. L. Lu, L. Tran, R. Li, X. Wang, D. E. Newby and K. Donaldson, Surface derivatization state of polystyrene latex nanoparticles determines both their potency and their mechanism of causing human platelet aggregation *in vitro*, *Toxicol. Sci.*, 2011, **119**, 359–368.
104. Z. Ramtoola, P. Lyons, K. Keohane, S. W. Kerrigan, B. P. Kirby and J. G. Kelly, Investigation of the interaction of biodegradable micro- and nanoparticulate drug delivery systems with platelets, *J. Pharm. Pharmacol.*, 2011, **63**(1), 26–32.
105. X. Li, A. Radomski, O. I. Corrigan, L. Tajber, F. De Sousa Menezes, S. Endter, C. Medina and M. W. Radomski, Platelet compatibility of PLGA, chitosan and PLGAchitosan nanoparticles, *Nanomedicine*, 2009, **4**(7), 735–746.
106. M. P. Monopoli, D. Walczyk, A. Campbell, G. Elia, I. Lynch, F. B. Bombelli and K. A. Dawson, Physical-chemical aspects of protein corona: relevance to *in vitro* and *in vivo* biological impacts of nanoparticles, *J. Am. Chem. Soc.*, 2011, **133**, 525–2534.
107. M. Rahman, S. Laurent, N. Tawil, L. Yahia and M. Mahmoudi, *Protein-nanoparticle Interactions*, Springer, Berlin, Germany, 2013.
108. C. D. Walkey and W. C. Chan, Understanding and controlling the interaction of nanomaterials with proteins in a physiological environment, *Chem. Soc. Rev.*, 2012, **41**, 2780–27992.
109. D. Walczyk, F. B. Bombelli, M. P. Monopoli, I. Lynch and K. A. Dawson, What the Cell "Sees" in Bionanoscience, *J. Am. Chem. Soc.*, 2010, **132**(16), 5761–5768.
110. D. Simberg, J. H. Park, P. P. Karmali, W. M. Zhang, S. Merkulov, K. McCrae, S. N. Bhatia, M. Sailor and E. Ruoslahti, Differential proteomics analysis of the surface heterogeneity of dextran iron oxide nanoparticles and the implications for their *in vivo* clearance, *Biomaterials*, 2009, **30**, 3926–3933.
111. C. D. Walkey and W. C. Chan, Understanding and controlling the interaction of nanomaterials with proteins in a physiological environment, *Chem. Soc. Rev.*, 2012, **41**(7), 2780–2799.
112. P. Aggarwal, J. B. Hall, C. B. McLeland, M. A. Dobrovolskaia and S. E. McNeil, Nanoparticle interaction with plasma proteins as it relates to particle biodistribution, biocompatibility and therapeutic efficacy, *Adv. Drug Delivery Rev.*, 2009, **61**, 428–437.
113. Y. Yu and G. Jin, Influence of electrostatic interaction on fibrinogen adsorption on gold studied by imaging ellipsometry combined with electrochemical methods, *J. Colloid Interface Sci.*, 2005, **283**, 477–481.

114. A. Gessner, R. Waicz, A. Lieske, B. R. Paulke, K. Mäder and R. H. Müller, Nanoparticles with decreasing surface hydrophobicities: influence on plasma protein adsorption, *Int. J. Pharm.*, 2000, **196**, 245–249.
115. M. Lundqvist, J. Stigler, T. Cedervall, T. Berggard, M. B. Flanagan, I. Lynch, G. Elia and K. Dawson, The evolution of the protein corona around nanoparticles: a test study, *ACS Nano*, 2011, **5**, 7503–7509.
116. D. Di Silvio, M. Maccarini, R. Parker, A. Mackie, G. Fragneto and F. B Bombelli, The effect of the protein corona on the interaction between nanoparticles and lipid bilayers, *J. Colloid Interface Sci.*, 2017, **504**, 741–750.
117. J. Du, C. Ge, Y. Liu, R. Bai, D. Li, Y. Yang, L. Liao and C. Chen, The interaction of serum proteins with carbon nanotubes depend on the physicochemical properties of nanotubes, *J. Nanosci. Nanotechnol.*, 2011, **11**(11), 10102–10110.
118. C. Ge, J. Du, L. Zhao, L. Wang, Y. Liu, D. Li, Y. Yang, R. Zhou, Y. Zhao, Z. Chai and C. Chen, Binding of blood proteins to carbon nanotubes reduces cytotoxicity, *Proc. Natl. Acad. Sci. U. S. A.*, 2011, **108**, 16968–16973.
119. K. P. Loh and C. T. Lim, Molecular interactions of graphene oxide with human blood plasma proteins, *Nanoscale*, 2016, **8**(17), 9425–9441.
120. S. C. Wasdo, D. S. Barber, N. D. Denslow, K. W. Powers, M. Palazuelos, S. M. Stevens Jr, B. M. Moudgil and S. M. Roberts, Differential binding of serum proteins to nanoparticles, *Int. J. Nanotechnol.*, 2008, **5**(1), 92–115.
121. S. H. Lacerda, J. J. Park, C. Meuse, D. Pristinski, M. L. Becker, A. Karim and J. F. Douglas, Interaction of gold nanoparticles with common human blood proteins, *ACS Nano*, 2009, **4**(1), 365–379.
122. M. Schäffler, M. Semmler-Behnke, H. Sarioglu, S. Takenaka, A. Wenk, C. Schleh, S. M. Hauck, B. D. Johnston and W. G. Kreyling, Serum protein identification and quantification of the corona of 5, 15 and 80 nm gold nanoparticles, *Nanotechnology*, 2013, **24**(26), 265103.
123. V. Krasovskii, I. Nagovitsyn, G. Chudinova, V. Savranskii and V. Karavanskii, Interaction of gold nanoparticles with bovine serum albumin, *Bull. Lebedev Phys. Inst.*, 2007, **34**, 321–324.
124. J. Mariam, P. Dongre and D. Kothari, Study of interaction of silver nanoparticles with bovine serum albumin using fluorescence spectroscopy, *J. Fluoresc.*, 2011, **21**, 2193–2199.
125. N. A. Monteiro-Riviere, M. E. Samberg, S. J. Oldenburg and J. E. Riviere, Protein binding modulates the cellular uptake of silver nanoparticles into human cells: implications for *in vitro* to *in vivo* extrapolations? *Toxicol. Lett.*, 2013, **220**, 286–293.
126. S. Zolghadri, A. A. Saboury, A. Golestani, A. Divsalar, S. Rezaei-Zarchi and A. A. Moosavi-Movahedi, Interaction between silver nanoparticle and bovine hemoglobin at different temperatures, *J. Nanopart. Res.*, 2009, **11**, 1751–1758.

127. D. P. Gnanadhas, M. B. Thomas, R. Thomas, A. M. Raichur and D. Chakravortty, Interaction of silver nanoparticles with serum proteins affects their antimicrobial activity *in vivo*, *Antimicrob. Agents Chemother.*, 2013, **57**(10), 4945–4955.
128. Q. Yang, J. Liang and H. Han, Probing the interaction of magnetic iron oxide nanoparticles with bovine serum albumin by spectroscopic techniques, *J. Phys. Chem. B*, 2009, **113**, 10454–10458.
129. M. Mahmoudi, M. A. Shokrgozar, S. Sardari, M. K. Moghadam, H. Vali, S. Laurent and P. Stroeve, Irreversible changes in protein conformation due to interaction with superparamagnetic iron oxide nanoparticles, *Nanoscale*, 2011, **3**(3), 1127–1138.
130. L. Lartigue, C. Wilhelm, J. Servais, C. Factor, A. Dencausse, J. C. Bacri, N. Luciani and F. Gazeau, Nanomagnetic sensing of blood plasma protein interactions with iron oxide nanoparticles: impact on macrophage uptake, *ACS Nano*, 2012, **6**(3), 2665–2678.
131. Z. Deng, G. Mortimer, T. Schiller, A. Musumeci, D. Martin and R. Minchin, Differential plasma protein binding to metal oxide nanoparticles, *Nanotechnology*, 2009, **20**, 455101.
132. A. Bouhekka and T. Bürgi, In situ ATR-IR spectroscopy study of adsorbed protein: Visible light denaturation of bovine serum albumin on TiO_2, *Appl. Surf. Sci.*, 2012, **261**, 369.
133. S. E. Lehman, I. A. Mudunkotuwa, V. H. Grassian and S. C. Larsen, Nano–Bio Interactions of Porous and Nonporous Silica Nanoparticles of Varied Surface Chemistry: A Structural, Kinetic, and Thermodynamic Study of Protein Adsorption from RPMI Culture Medium, *Langmuir*, 2016, **32**, 731.
134. F. Catalano, G. Alberto, P. Ivanchenko, G. Dovbeshko and G. Marta, Effect of silica surface properties on the formation of multilayer or submonolayer protein hard corona: Albumin adsorption on pyrolytic and colloidal SiO_2 nanoparticles, *J. Phys. Chem. C*, 2015, **119**, 26493.
135. B. E. Givens, N. D. Diklich, J. Fiegel and V. H. Grassian, Adsorption of bovine serum albumin on silicon dioxide nanoparticles: Impact of pH on nanoparticle–protein interactions, *Biointerphases*, 2017, **12**(2), 02D404.
136. N. Brandes, P. B. Welzel, C. Werner and L. W. Kroh, Adsorption-induced conformational changes of proteins onto ceramic particles: differential scanning calorimetry and FTIR analysis, *J. Colloid Interface Sci.*, 2006, **299**(1), 56–69.
137. M. Lundqvist, J. Stigler, G. Elia, I. Lynch, T. Cedervall and K. A. Dawson, Nanoparticle size and surface properties determine the protein corona with possible implications for biological impacts, *Proc. Natl. Acad. Sci. U. S. A.*, 2008, **105**, 14265–14270.
138. M. S. Ehrenberg, A. E. Friedman, J. N. Finkelstein, G. Oberdörster and J. L. McGrath, The influence of protein adsorption on nanoparticle association with cultured endothelial cells, *Biomaterials*, 2009, **30**(4), 603–610.

139. R. J. Johnson, Immunology and the complement system, in *Biomaterials Science: An Introduction to Materials in Medicine*, ed. B. Ratner, A. Hoffman, F. Schoen and J. Lemons, Elsevier Academic Press, San Diego, 1996.
140. S. M. Moghimi, A. J. Andersen, D. Ahmadvand, P. P. Wibroe, T. L. Andresen and A. C. Hunter, Material properties in complement activation, *Adv. Drug Delivery Rev.*, 2011, **63**(12), 1000–1007.
141. S. Kobayashi and K. Müllen, *Encyclopedia of Polymeric Nanomaterials – With 2021 Figures and 146 Tables*, Springer, 2015.
142. I. Lynch, T. Cedervall, M. Lundqvist, C. Cabaleiro-Lago, S. Linse and K. A. Dawson, The nanoparticle–protein complex as a biological entity; a complex fluids and surface science challenge for the 21st century, *Adv. Colloid Interface Sci.*, 2007, **134–135**, 167–174.
143. M. J. Rybak-Smith and R. B. Sim, Complement activation by carbon nanotubes, *Adv. Drug Delivery Rev.*, 2011, **63**, 1031–1041.
144. C. Salvador-Morales, E. Flahaut, E. Sim, J. Sloan, M. L. H. Green and R. B. Sim, Complement activation and protein adsorption by carbon nanotubes, *Mol. Immunol.*, 2006, **43**, 193–201.
145. W. L. Ling, A. Biro, I. Bally, P. Tacnet, A. Deniaud, E. Doris, P. Frachet, G. Schoehn, E. Pebay-Peyroula and G. J. Arlaud, Proteins of the innate immune system crystallize on carbon nanotubes but are not activated, *ACS Nano*, 2011, **5**, 730–737.
146. C. Salvador-Morales, E. V. Basiuk, V. A. Basiuk, M. L. Green and R. B. Sim, Effects of covalent functionalization on the biocompatibility characteristics of multi-walled carbon nanotubes, *J. Nanosci. Nanotechnol.*, 2008, **8**, 2347–2356.
147. S. Li, Z. Guo, Y. Zhang, W. Xue and Z. Liu, Blood compatibility evaluations of fluorescent carbon dots, *ACS Appl. Mater. Interfaces*, 2015, **7**(34), 19153–19162.
148. M. A. Dobrovolskaia, A. K. Patri, J. Zheng, J. D. Clogston, N. Ayub, P. Aggarwal, B. W. Neun, J. B. Hall and S. E. McNeil, Interaction of colloidal gold nanoparticles with human blood: Effects on particle size and analysis of plasma protein binding profiles, *Nanomedicine*, 2009, **5**(2), 106–117.
149. D. Gao, Y. Tian, F. Liang, L. Ding, S. Bi, Y. Chen, H. Zhang and A. Yu, Investigation on the interaction between colloidal gold and human complement factor 4 at different pH by spectral methods, *Colloids Surf., B*, 2006, **47**(1), 71–77.
150. Q. H. Quach, R. L. X. Kong and J. C. Y. Kah, Complement Activation by PEGylated Gold Nanoparticles, *Bioconjugate Chem.*, 2018, **29**(4), 976–981.
151. P. Aggarwal, J. B. Hall, C. B. McLeland, M. A. Dobrovolskaia and S. E. McNeil, Nanoparticle interaction with plasma proteins as it relates to

particle biodistribution, biocompatibility and therapeutic efficacy, *Adv. Drug Delivery Rev.*, 2009, **61**(6), 428–437.
152. D. Simberg, J. H. Park, P. P. Karmali, W. M. Zhang, S. Merkulov, K. McCrae, S. N. Bhatia, M. Sailor and E. Ruoslahti, Differential proteomics analysis of the surface heterogeneity of dextran iron oxide nanoparticles and the implications for their *in vivo* clearance, *Biomaterials*, 2009, **30**(23–24), 3926–3933.
153. S. M. Moghimi, A. J. Andersen, D. Ahmadvand, P. P. Wibroe, T. L. Andresen and A. C. Hunter, Material properties in complement activation, *Adv. Drug Delivery Rev.*, 2011, **63**(12), 1000–1007.
154. A. Vonarbourg, C. Passirani, P. Saulnier and J. P. Benoit, Parameters influencing the stealthiness of colloidal drug delivery systems, *Biomaterials*, 2006, **27**(24), 4356–4373.
155. A. S. Abu Lila, H. Kiwada and T. Ishida, The accelerated blood clearance (ABC) phenomenon: clinical challenge and approaches to manage, *J. Controlled Release*, 2013, **172**(1), 38–47.
156. T. Ishihara, M. Takeda, H. Sakamoto, A. Kimoto, C. Kobayashi, N. Takasaki, K. Yuki, K. I. Tanaka, M. Takenaga, R. Igarashi and T. Maeda, Accelerated blood clearance phenomenon upon repeated injection of PEG-modified PLA-nanoparticles, *Pharm. Res.*, 2009, **26**(10), 2270–2279.
157. R. Saadati, S. Dadashzadeh, Z. Abbasin and H. Soleimanjahi, Accelerated blood clearance of PEGylated PLGA nanoparticles following repeated injections: effects of polymer dose, PEG coating, and encapsulated anticancer drug, *Pharm. Res.*, 2013, **30**(4), 985–995.
158. M. A. Dobrovolskaia, J. D. Clogston, B. W. Neun, J. B. Hall, A. K. Patri and S. E. McNeil, Method for analysis of nanoparticle hemolytic properties *in vitro*, *Nano Lett.*, 2008, **8**(8), 2180–2187.
159. M. Golder, J. Mewburn and D. Lillicrap, In vitro and in vivo evaluation of the effect of elevated factor VIII on the thrombogenic process, *Thromb. Haemostasis*, 2013, **109**(1), 53.
160. G. Fotakis and J. A. Timbrell, In vitro cytotoxicity assays: comparison of LDH, neutral red, MTT and protein assay in hepatoma cell lines following exposure to cadmium chloride, *Toxicol. Lett.*, 2006, **160**(2), 171–177.
161. J. Wang, L. Wang, X. Liu, Z. Liang, S. Song, W. Li, G. Li and C. Fan, A Gold Nanoparticle-Based Aptamer Target Binding Readout for ATP Assay, *Adv. Mater.*, 2007, **19**(22), 3943–3946.
162. H. Glavinas, O. Von Richter, K. Vojnits, D. Mehn, I. Wilhelm, T. Nagy, J. Janossy, I. Krizbai, P. Couraud and P. Krajcsi, Calcein assay: a high-throughput method to assess P-gp inhibition, *Xenobiotica*, 2011, **41**(8), 712–719.
163. W. Strober, Trypan blue exclusion test of cell viability, *Curr. Protoc. Immunol.*, 2015, **111**(1), A3-B.

164. S. Al-Nasiry, N. Geusens, M. Hanssens, C. Luyten and R. Pijnenborg, The use of Alamar Blue assay for quantitative analysis of viability, migration and invasion of choriocarcinoma cells, *Hum. Reprod.*, 2007, **22**(5), 1304–1309.
165. J. Bardon, A. C. Lukaszewicz, V. Faivre, B. Huot and D. Payen, Reactive oxygen species measure for rapid detection of infection in fluids, *Ann Intensive Care*, 2016, **6**(1), 41.

CHAPTER 10

The Protein Corona: Applications and Challenges

SRUTHI ANN ALEX[a], DEBOLINA CHAKRABORTY[b], N. CHANDRASEKARAN[a] AND AMITAVA MUKHERJEE*[a]

[a]Centre for Nanobiotechnology, Vellore Institute of Technology, Vellore, India; [b]School of Advanced Sciences, Vellore Institute of Technology, Vellore, India
*E-mail: amit.mookerjea@gmail.com, amitav@vit.ac.in

10.1 Introduction

The introduction of nanomaterials into the biological environment promotes the adsorption of proteins and metabolomes in the serum, known as the 'protein corona'. The low-affinity dynamic proteins, namely the 'soft corona', have shorter residence times compared to high-affinity proteins, *i.e.*, the 'hard corona'. This protein-rich mask presented by the nanomaterials determines their biological fate based on the amount, type, orientation, and conformation of adsorbed proteins, which in turn are dependent on the surface chemistry, size, charge, solubility, and aggregation state of the nanomaterial.[1,2] The interaction of nanomaterials with the cells, the biodistribution, macrophage capture, and cellular uptake of nanoparticles (NPs) are dependent on the diverse types of proteins present in the corona layer, which strongly necessitates their extensive study in order to attain controlled interactions. From another perspective, the toxicity risk posed by bare nanomaterials can be overcome by the protective corona layer masking

Issues in Toxicology No. 40
Nanoparticle–Protein Corona: Biophysics to Biology
Edited by Ashutosh Kumar and Alok Dhawan

Published by the Royal Society of Chemistry, www.rsc.org

the nanomaterial. Though uncontrolled protein coronation is a big challenge in nanotherapeutics, nanomaterials can be designed to achieve a protein corona layer of interest that can be exploited for various applications as the corona can control their uptake by target cells and can activate/deactivate signaling cascades that impact their biodistribution/removal from the body. The safety of nanomaterials lies in tailoring their physiological response and understanding their mechanism of action, which is a complex and arduous task considering the cumulative effect of the diverse proteins in the corona and the varied phenotype of the interacting cells. This chapter discusses the impact of a protein corona on the circulation time, biodistribution, cellular uptake, response/toxicity, specific drug targeting applications, the challenges faced, and several realistic approaches that attempt to consider several physiological parameters for protein coronation, all of which are important factors to be considered for the preclinical testing of NP-based therapeutics.

10.2 The Impact of a Nanoparticle–Biocorona on Cells

10.2.1 Improved Circulatory Time

As a NP enters the biological system, our immune mechanism recognizes it as 'foreign' and facilitates its sequestration by the mononuclear phagocytic system (MPS) within a few minutes. In order to target distant organs, the nanoparticle life has to be extended, and failure to elude their clearance from the bloodstream can culminate in their failure in *in vivo* performances.[3] The fate of the nanoparticle in a complex biological milieu is gravely dependent on the adsorption of diverse protein layers on it. The binding pattern of the adsorbed proteins can also modulate the interactive interface between nanomaterials and the cell membranes of targeted cells by influencing phagocytosis and NP transport (Figure 10.1). The composition of the protein corona so formed determines the circulation time and clearance of the NPs inside the body.[2] The presence of opsonins like fibrinogen, gamma globulin, and complements enhances recognition by the macrophage receptors and their subsequent engulfment, whereas coronation with dysopsonins such as serum albumins and apolipoproteins (apo) not only increases their circulation time in the bloodstream, but also helps in specific receptor targeting. Improved NP circulation is the first key factor for improved preclinical performance of NPs, which in turn enhances the nanomaterial bioavailability and therapeutic response. Therefore, reduction of the total opsonin adsorption and enhancement of the dysopsonin binding can be achieved by tuning the NP design.[1,4]

Preliminary studies have used peptides designed from human CD47[5] and hydrophobin (a secretory substance from fungi)[6] to achieve stealth characteristics and minimize protein adsorption. Strategically circumventing the

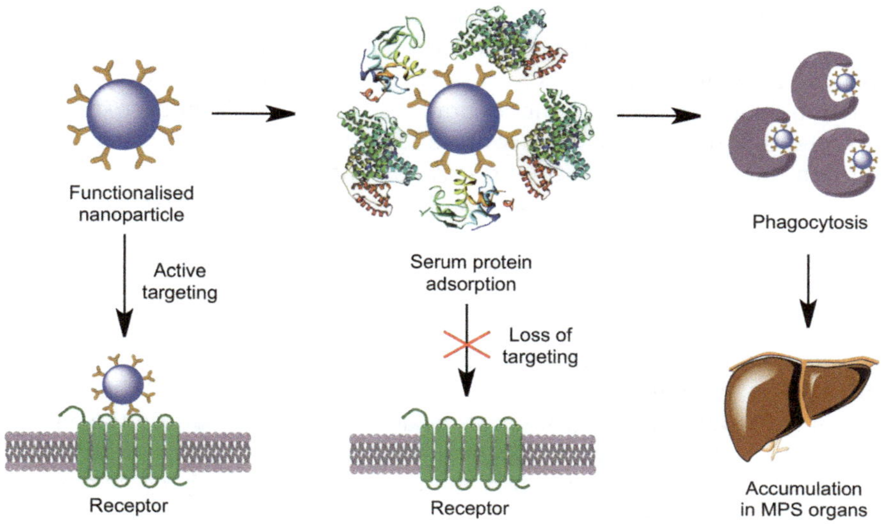

Figure 10.1 Schematic representation of biomolecular corona formation and its influence on active targeting. Under serum-free *in vitro* conditions, active targeting of functionalized nanoparticles is achieved by interaction and recognition of cell membrane-located receptors with targeting moieties grafted onto the nanoparticle surface. In a biological milieu, a biomolecular corona is formed by non-specific adsorption of biological (macro) molecules, such as proteins and lipids, to the surface of nanoparticles. This process impairs active targeting and leads to accumulation of nanoparticles in the tissues of the MPS. Reproduced from ref. 18 with permission from John Wiley and Sons, Copyright 2014 © 2014 WILEY-VCH Verlag GmbH & Co. KGaA, Weinheim.

macrophage recognition/clearance of NPs from the system and improved blood half-life circulation time can be achieved by the formation of a stealth layer of hydrophilic polymers like PEG (*i.e.*, PEGylation – pretreatment of a NP surface with PEG).[7,8] Torchilin and Trubetskoi suggested that the random cloud (hydration layer) formed by the long chain of polymers due to PEG around the NPs could prevent protein adsorption and NP recognition by the reticulo-endothelial system (RES), and thereby, impart long-circulating characteristics to the nanomaterial. Some of the hydrophilic stealth coatings that are studied include polyethylene glycol (PEG), poloxamer (block copolymer of polypropylene oxide and polyethylene oxide), pluronic F68, polyvinylpyrrolidone (PVP), and cross-linked hydrogels like dextran, of which PEG is the most commonly studied polymer.[9] However, the resistance in blood protein adsorption upon PEGylation varied for different NPs and was also influenced by several important factors such as the chain length, surface density, molecular weight charge, hydrophilic properties, and the degree of surface coverage of the polymer, which are to be considered in the preparation and design of nanomaterials.[10]

In a study by Kim *et al.* (2007), PEGylation of hexadecylcyanoacrylate NPs resulted in a decrease of the protein corona by half when compared to bare NPs and also higher apoE adsorption, which facilitated the passage of these PEGylated NPs through the blood–brain barrier (BBB).[11] Another study also reported that the surface density or molecular weight of PEG was inversely, but non-linearly, related to the protein adsorption when three NPs, namely poly(lactic acid), poly(ε-caprolactone) NPs, and poly(lactic-*co*-glycolic acid), were studied. Longer PEG chains could significantly reduce cellular uptake and prolong the NP circulation. Protein adsorption decrease was observed to be at the maximum for a PEG content of 2–5% and MW of 5000. However, PEG capping does not completely hinder protein adsorption, and lower concentrations of a few proteins like albumin, IgG, apoE, and apoA-I were observed to attach inevitably.[10] Price *et al.* (2001) also suggested that PEGylation of negatively charged phosphatidic acid liposomes reduced the adsorption of purified fibrinogen; however, PEGylation had no influence when tested with total plasma protein.[12] On the contrary, a study by Arima *et al.* (2009) suggested that dextran-immobilized surfaces could activate the complement system (*i.e.*, show decreased stealth properties) by lectin-dependent and C3-dependent signaling pathways in comparison to soluble polymers, which could be attributed to the hydroxyl groups in dextran.[13] These studies suggest the complexity of stealth coating from merely preventing protein absorption, and their interactions with macrophage receptors are also worth considering. Even though improved circulation lifetimes and reduced NP clearance could be achieved by hydrogel coatings on iron oxide nanoparticles (IONPs), their accumulation in the macrophages of the liver and spleen was inexorable, with more than 50% of the dose introduced ending up in these undesirable organs 48 h post-circulation.[14]

Superparamagnetic iron oxide nanoparticles, also called SPIONs, have diverse clinical applications for drug delivery and as contrast enhancement agents for magnetic resonance imaging (MRI); however, SPIONs with diameters >50 nm are swiftly engulfed by the RES upon intravenous injection and end up in the liver.[15] In order to circumvent this issue, PEGylated SPIONs could be employed to improve the circulatory time by reduction of protein adsorption. This stealth characteristic was also influenced by SPION surface charge, *i.e.*, positively charged SPIONs displayed enhanced breast cancer cell uptake compared to negatively charged counterparts.[16] Therefore, differential protein 'fingerprints' can not only evade their phagocytosis by macrophages, but also render the NPs available for targeting the site of interest and improve their clinical application.

Additionally, the protein corona can conceal the surface chemical functionality of NPs and hinder their targeting ability *in vivo*. Instead of using surface functionalizations that evade protein coronation, NPs themselves can be designed such as to evade protein adsorption. With this view, 'corona-free' zwitterionic NPs were proposed by Moyano *et al.* (2014), and these NPs could permit the adsorption of soft coronas alone at biological serum levels.[17–19]

Bridging the knowledge gap by comprehensive protein–NP studies and the ability to control nanomaterials for circumventing protein corona adsorption will help to develop biologically functional nanomedicine.

10.2.2 Modified Cellular Vision of Nanoparticles

'Cellular vision' refers to the various detoxification approaches taken by cells in response to a NP's entry in the biological system. The biocompatible design of any NPs must take into consideration the formation of the protein corona, which is ultimately the crucial factor that determines what the cell sees. The defense strategy (cellular response) adopted when cells are challenged with NPs is also governed by the cell type.[20] For instance, when the same concentration of SPIONs was treated against various cell types, the cellular uptake, viability, and reactive oxygen species (ROS) production observed were completely dissimilar.[16]

The fundamental basis for all sorts of NP interactions with biofluids and cells is their high surface-area-to-mass ratio. The adsorbed biomacromolecules (proteins, phospholipids, or carbohydrates) can provide a characteristic 'protein corona', which alters the biological identity of the nanomaterial, which in turn dictates the triggered cellular internalization, trafficking, and fate of the NPs.[1,2] Adsorption of opsonins serves as a molecular signature that augments NP recognition by the macrophage cells, cellular uptake, biodistribution, clearance rate from the bloodstream, and organ disposition.[21]

The intravenous administration of NPs can initiate their interaction with proteins in the bloodstream and commence the following responses:[22]

- Instantaneous NP interaction with physiological plasma/blood proteins, platelets, cells
- Time-based variations in protein–cell interaction processes
- Protein cascade activation
- Recognition, internalization, and elimination of protein-coronated nanomaterial by immunological cells
- Accumulation of nanomaterial in non-macrophages (cytotoxicity)
- Nanomaterial processing in cells and triggering of signaling cascades (cell death)

A study by Safi *et al.* (2011) suggested that polymer (polyacrylic acid)-conjugated ultra-small superparamagnetic iron oxide (USPIO) displayed higher dispersibility and stability without protein coronation; however, NP conjugation with small molecules like citrate ions resulted in protein coronation, NP aggregation, and their cell uptake. This was demonstrated by flow cytometry and TEM studies, which indicated rapid internalization of citrate-capped USPIO by human lymphoblastoids.[23] Other studies suggested that serum protein-coated dextran-functionalized iron oxide NPs showed two-fold internalization by tumor cells possibly due to the attachment of complement C3, fibronectin, and vitronectin,[24] whereas beta-2-glycoprotein adsorption on negative liposomes was observed to cause their elimination from the body.[25] However, the reduced adsorption of IgG and complement C3 on monosialoganglioside (GM1) liposomes was shown to be in good agreement with the reduced elimination time period.[19] Similarly, a study by Yan *et al.* (2005) suggested that the adsorption of apoE and beta-2-glycoprotein on phosphatidylglycerol and phosphatidylserine anionic liposomes did not impact cellular trafficking to liver macrophages but rather reduced their uptake by liver cells.[26] The intravenous injection of larger liposomes (500 nm) could be correlated with higher numbers of plasma fibronectin,[27] whereas amphotericin B liposomes were observed to be coated with high-density lipoproteins, suggesting that these NPs could attract proteins involved in their clearance.[28] On the contrary, pretreatment of poly(methyl methacrylate) (PMMA) NPs with serum proteins led to the association of dysopsonins and thereby their minimized clearance by liver macrophages.[29] Similarly, adsorption of human serum albumin (HSA) on polystyrene microparticles could hinder phagocytosis by dendritic cells.[30] Though specific protein adsorption can influence NP recognition, the direct recognition of a NP surface (in the absence of opsonins) and internalization by macrophages must also be considered, *i.e.*, even the specific type of NP could also play a role in the internalization; however, the constant presence of serum proteins *in vivo* makes it difficult to examine this hypothesis devoid of opsonins *in vivo*. Preincubation of NPs like lecithin-coated polystyrene NPs, PMMA NPs, and liposomes during *ex vivo* perfusion have been shown to reduce internalization by macrophages.[29,31,32] Other reports indicate that polyanionic ligands of phagocytic scavenger receptors (that enable elimination of foreign pathogens), such as polyinosinic acid, fucoidan, and dextran sulfate, could hinder the internalization of carbon nanotubes, quantum dots, SPIONs, and polystyrene NPs.[22]

Thus, the repetitive pattern of chemical groups and ligands on the NPs together with the unique decoration of the protein corona determine the cellular vision and their response.

All these studies suggest that the route preferred by individual cells against different NPs must by mapped, similar to the 'fingerprinting' of humans.[16]

10.2.3 Specific Cell Targeting

Successful targeting of desired NPs by a receptor-mediated mechanism is the key to attaining an effective therapeutic effect. Formation of protein coronas on such targeting nanoparticles not only masks the targeting ligand, but also interferes with the orientation, native conformation, and display of the ligand.[33] This makes protein coronas a formidable challenge for the active targeting of nanoparticles to particular diseased cells. Salvati *et al.* (2013) reported that transferrin (Tf)-coated fluorescent silica nanoparticles (SiNPs) lost their targeting ability to both soluble Tf receptors and bound Tf receptors

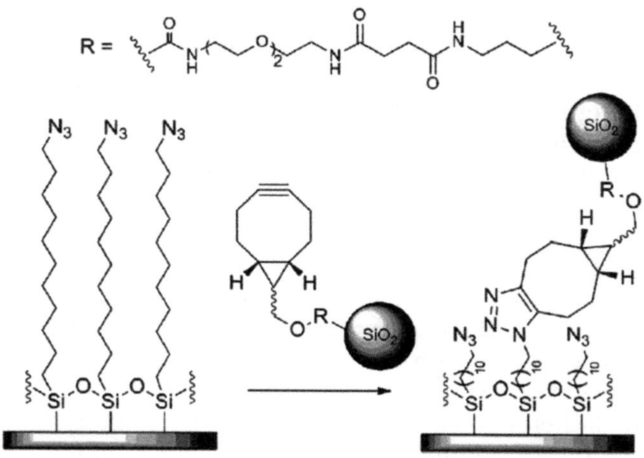

Figure 10.2 A copper-free click reaction between the BCN moieties on the NPs and the azides on the modified silicon substrate was selected as the model targeting reaction. Reproduced from ref. 35 with permission from the Royal Society of Chemistry.

when they were incubated with fetal bovine serum (FBS).[34] In another case, Mirshafiee *et al.* (2013) demonstrated this concept using a model copper-free click reaction established between bicyclononyne (BCN)-conjugated SiNPs and its azide-functionalized substrate (Figure 10.2), which was seen to be obscured in the presence of the protein corona. By fluorescence microscopic imaging, it was suggested that the attachment of BCN-SiNPs coronated with medium + 10% serum (*in vitro* mimic) or 100% serum protein (*in vivo* mimic) to the Si substrate was significantly less than the attachment of pristine BCN-SiNPs (94 and 99% reduction, respectively).[35]

Different strategies developed to overcome such inherent challenges include the use of zwitterionic coatings or anti-fouling polymers, such as polyethylene glycol. Zwitterionic polyelectrolytes have the property of forming strong ionic structures with water molecules, thus enhancing the formation of a hydrophilic surface, and also as their zeta potential is nearly zero, they do not form ion pairs with charged adsorbents.[18] Murthy *et al.* (2013) demonstrated that the attachment of cysteine or lysine to charged gold NPs could effectively eliminate non-specific protein adsorption. It was postulated that the exposure of an α-amino group and carboxy group prevented corona formation in undiluted FBS.[36] Moyano *et al.* (2014) demonstrated the use of a series of a non-fouling NP-containing outer layer of zwitterionic head groups that prevented hard corona formation on their surface.[17]

Dai *et al.* (2014) had demonstrated that the negative impact of a protein corona can be bypassed by PEGylating the targeting NPs. A higher response to the targeting cells was achieved by proper selection of the PEG molecule in terms of its chain length.[37] The hydration layer formed owing to its hydrophilic nature along with the steric hindrance due to the random coil formation of PEG chains on the NP surface inhibit non-specific protein adsorption.[10,38] More recent work by D'Hollander *et al.* (2017) used anti-HER2-fucntionalized Au nanostars, wherein the protein corona adsorption was reduced by blocking the active groups of the PEG using 2-mercaptoethanol, thereby demonstrating the active targeting *in vitro* and *in vivo* for HER2-positive tumors.[39] In another strategy, corona-directed targeting was achieved by pre-coating γ-globulin molecules on silica NPs, and this in turn enhanced the opsonin-mediated uptake by macrophages.[40]

In summary, the shielding nature of a protein corona reduces the targeting efficiency of the nanoparticles, along with their rapid clearance from the biological system. On the contrary, employing the strategy of pre-coating the nanoparticle surface with relevant proteins is seen to be promising in certain selective cases. Thus, parameters such as the type and nature of the NPs along with their surface functionalization would determine the extent of protein corona formation and their subsequent targeting efficiency. However, the protein corona can be utilized for the targeting strategy, for instance, hexadecylcyanoacrylate NPs could facilitate apoE adsorption, their passage through the BBB, and target the brain efficiently (compared to PEGylated counterparts), which can be used for the treatment of neurological problems.[11]

10.2.4 Drug Delivery

A wide range of NPs has been used for diverse drug delivery applications, under which circumstances, important considerations regarding protein corona formation on drug-carrying NPs needs to be understood thoroughly. Despite employing numerous strategies to reduce the non-specific adsorption of proteins on NP surfaces,[41,42] it has been proved that protein corona formation is an inevitable phenomenon. In recent reports, Kah *et al.* (2012) have exploited such a protein corona as a carrier for charged payloads. The protein corona formed around a CTAB-capped gold nanorod surface was seen to be efficient in carrying both negatively charged DNA and positively charged doxorubicin. Payload loading is solely dependent on several parameters such as the mode of assembly, ionic strength, and corona concentration. The capacity of the corona in terms of holding payloads was seen to be ~5–10 times higher than covalent strategies.[43] The release profile of the payload retained in protein coronas formed around gold nanorods, nanobones, and carbon nanotubes can also be modulated based on the composition of the corona. It was reported that the passive release of DNA payloads can be reduced in the presence of HSA, implying that soft corona proteins can effectively impede the 'burst release' effect. When considering nanocarrier-based drug release in a biologically complex medium, the type/size of nanocarriers becomes a major determining parameter.[44] Behzadi *et al.* (2014) proposed that the protein corona significantly reduced payload release from SPIONs and Abraxane®, whereas the thinner hard corona layer had a negligible effect on the payload release from the nanocapsules.[45]

In summary, a blend of various intricate parameters, such as composition and concentration of the corona, surfactant concentration on the nanomaterials, and ionic strength of the buffer, determines the loading efficiency of the therapeutic payloads as well as their release into the surroundings. Exploring further in this direction can give us a better understanding of *in vivo* drug delivery mechanisms as well as help us develop more efficient nanocarriers for targeted therapy by incorporating the characteristic features of the protein corona.

10.2.5 Uptake of Nanoparticles and Cellular Response

One of the prime factors that influences the cell to accept 'pristine' NPs or corona-coated NPs in a complex biological system is the nano-biointerface that the cell encounters. Under such circumstances, it is necessary to study the different parameters, as to why and how the cellular uptake and its subsequent response vary in different scenarios. Lesniak *et al.* (2012) reported that the efficiency of internalization of a silica nanoparticle by A549 cells varied depending on the presence of serum proteins, *i.e.*, bare silica nanoparticles were seen to be internalized better and appeared in the free cytosol, whereas inefficient internalization with less damage to the cells was seen in the presence of a protein corona.[46] In another study, the impact of protein glycosylation on cell adhesion was extensively studied. It was proposed that a glycosylated protein corona imparted higher steric hindrance, which attenuated cell adhesion and its subsequent uptake by the cells as compared to its deglycosylated counterpart.[47] Jiang *et al.* (2009) also demonstrated similar results wherein the presence of proteins like HSA and Tf substantially reduced the uptake of FePt NPs in live HeLa cells.[48] An important observation was proposed by Saha *et al.* (2014), wherein nanoparticles with a protein corona were seen to reduce the hemolysis of red blood cells (RBCs).[49] Even for mesoporous silica NPs, the protein coronas had a protective effect in terms of hemolysis.[50] Thus, it can be summarized that as a consequence of corona formation on the nanoparticles' surface, direct interaction with off-target cells can be minimized, which in turn reduces their cellular uptake, cytotoxicity, and improves their biocompatibility.

10.2.6 Modified Cellular Toxicity

The size-dependent property of nanoparticles makes them indispensable for various biological applications. The extensive use of diverse NPs in the biological field as DNA-probing agents, imaging agents, drug-carriers, and as bio-sensors has been routinely reported.[51] Thus, it becomes important to systematically evaluate the potential risk that these nanoparticles pose to human health. It has been proposed that protein corona formation on the nanoparticle surface mitigates its inherent toxic effect. Hu *et al.* (2011) reported that protein adsorption on graphene oxide alleviated its toxicity towards A549 cells by weakening its direct interaction with cells.[52] Alex *et al.* (2017)

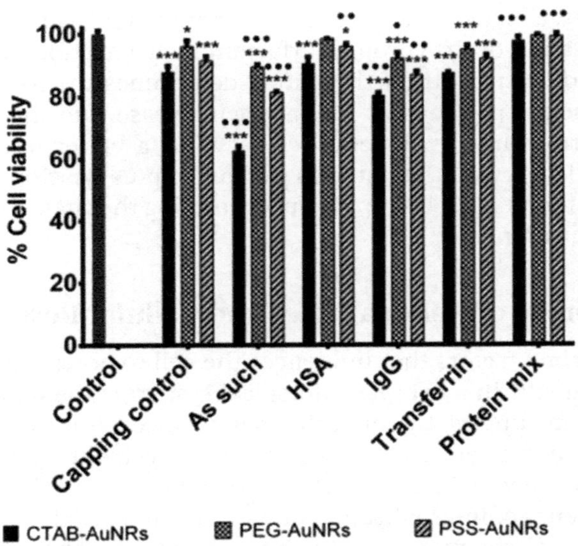

Figure 10.3 Viability studies of hcT cells treated with CTAB-AuNRs, PEG-AuNRs, and PSS-AuNRs before and after adsorption of HSA, IgG, transferrin, and a protein mixture, compared with their respective vehicle control and PBS buffer (control). *, **, and *** represent the statistically significant difference between samples and the phosphate buffer control, whereas •, ••, and ••• represent the statistically significant difference between samples and the capping control at $p < 0.05$, $p < 0.01$, and $p < 0.001$, respectively. Reproduced from ref. 53 with permission from Elsevier, Copyright 2017.

showed similar results, wherein the presence of individual serum proteins such as HSA, IgG and transferrin was seen to reduce the cytotoxicity of functionalized GNRs, and in the presence of a protein mixture, the cell viability was improved from ~62% (for CTAB-capped GNRs) to 97%, as seen in Figure 10.3, implying that the presence of serum-abundant proteins reduces the direct interaction of the cells with the bare surface of the nanoparticle.[53] Briefly, it can be stated that the cytotoxicity of protein-coronated NPs is primarily determined by the exposure of the protein corona interface, as well as by the cellular uptake of such coronated NPs, *i.e.*, increased internalization by the cells is seen to decrease the cell viability.[53] Table 10.1 presents some of this work, wherein corona formation on various nanoparticles targeting different cells was seen to reduce their toxicity depending on the nature and extent of protein adsorption.[54–56]

10.2.7 Protein Fibrillation

Protein fibrillation is a pathological hallmark of various neurodegenerative ailments, which involves the partial or complete unfolding of amyloidogenic proteins under certain conditions to form large linear

Table 10.1 A few protein-coronated NPs showing reduced toxicity *in vitro*.

Nanoparticles	Protein corona	Cell line	Reference
Carbon nanotubes (CNTs)	BFG, BSA, γ-Ig, transferrin	THP-1 and HUVEC	54
Gold nanoparticles (GNPs)	FBS-supplemented cellular media (DMEM and RPMI)	HeLa and U937	55
Amorphous silica nanoparticles	Human plasma	Caco-2	56

insoluble β-strand stacked structures called amyloid fibrils. These aggregated fibrils can precipitate as plaques, which are the primary source of protein-misfolded catastrophic diseases such as Alzheimer's, Parkinson's, and dialysis-related amyloidosis. Nanoparticles are known to significantly influence protein fibrillation.[3,57] The anti-amyloidogenic properties demonstrated by hydrated fullerene (C60) in animal studies suggested that these materials can be used as a promising remedy for neurodegenerative diseases.[58] Some hydrophobic nanoparticles (*e.g.*, polystyrene) can also indicate 'dual' fibrillation kinetics based on the physiological temperature.[59] Gold NPs were also considered as promising materials as they could noticeably hinder protein fibrillation;[60] however, recent studies suggest that AuNPs can also act as nucleation centers for amyloid fibrillation.[61] Nevertheless, since the majority of the work was performed with bare nanoparticles, the studies could not be translated to their *in vivo* behavior as these NPs get covered by protein coronas. Thus, in order to truly understand the NPs' effect on protein fibrillation in the physiological environment, the process has to be simulated using coronated NPs rather than pristine NPs.[3,57]

When the effect of protein coronation on fibrillation was tested using diverse nanoparticles (*e.g.*, polystyrene, silica, and carbon nanotubes) with various plasma concentrations and corona compositions, it was noted that protein-coronated NPs inhibited fibrillation, whereas pristine NPs accelerated the process.[57] However, AuNPs (which inhibited protein fibrillation in their bare state) exhibited a minor acceleratory tendency when coronated with serum and plasma proteins.[62] A recent study by Lotfabadi *et al.* (2018) showed that gold nanospheres and nanorods coronated with plasma proteins caused less inhibition of $A\beta_{1-42}$ fibrillation when compared with pristine and CSF-coronated NPs, whereas contrary results were observed with $A\beta_{25-35}$ fibrillation, *i.e.*, pristine NPs enhanced the fibrillation, and the fibrillation inhibitory effect of CSF-coronated NPs was less than that of plasma-coronated NPs.[63] To summarize, the NP type and protein corona composition (*i.e.*, concentration, type, and conformation of the corona protein) could regulate the impact of protein coronation on the fibrillation condition; however, further characterizations are necessary to completely address the impact of protein coronas on fibrillation.

10.3 Challenges Faced Due to Protein Coronation

Protein coronation can be considered as a hindrance for several nanotherapeutics and may sabotage the actual purpose of the system (affect targeting, efficiency of MRI contrast agent). A study involving transferrin-coated SiNPs suggested that the adsorption of the serum protein corona could impair the targeting ability of these NPs by covering the targeting ligand (transferrin) on the NP surface and reducing NP recognition by cell surface receptors.[34] This was also demonstrated in another study using a copper-free click reaction between an azide on a Si substrate (model receptor) and a bicyclononyne (BCN) moiety present on SiNPs, suggesting that protein coronation significantly reduced NP attachment to the Si substrate.[35] Another group suggested that the surface charge of a dextran coating on SPIONs can alter the adsorption of a model protein (FBS), which in turn can influence the transverse relaxivity (a parameter affecting efficiency) of negative MRI contrasts. While the relaxivity of plain SPIONs remained unaffected by protein coronation, the relaxivity of negative SPIONs was moderately enhanced and that of positive SPIONs was reduced drastically and led to NP agglomeration.[64]

Thus, a complete knowledge and control of protein corona adsorption is needed to combat their shortcomings for nanomedicine applications and to exploit their unique features for our advantage. In many cases, protein coronation can mask the NP toxicity and render protective and beneficial effects in a physiological environment; however, discrepancies in reported works can lead to unclear interpretations of their impact. Additionally, since the decoration of *in vitro* protein coronation can be remarkably different from the protein corona formed *in vivo*, the molecular complexity and biological fates of clinically developed liposomes were seen to show dissimilarities.[65] Maintaining proper controls and following standardized protocols are needed to make proper comparisons and reproducibility between different laboratories. Even the slightest detail like centrifugation speed can significantly impact protein corona decoration and exacerbate its related effects. The working of fully efficient targeted NPs requires that these unmet challenges need to be addressed, which in turn can dramatically transform and improve several therapeutic regimes.[66]

10.4 The Protein Corona: Realistic Behavior

10.4.1 *In Vivo* Protein Corona

Despite copious studies on comprehending the protein coronation of NPs *in vitro*, there are very few reports that discuss protein corona behavior for *in vivo* milieu. In order to understand the fate of nanomedicine, it is highly imperative to thoroughly study protein coronation *in vivo*. This in turn complicates *in vivo* studies as the immunological cells are not present in the

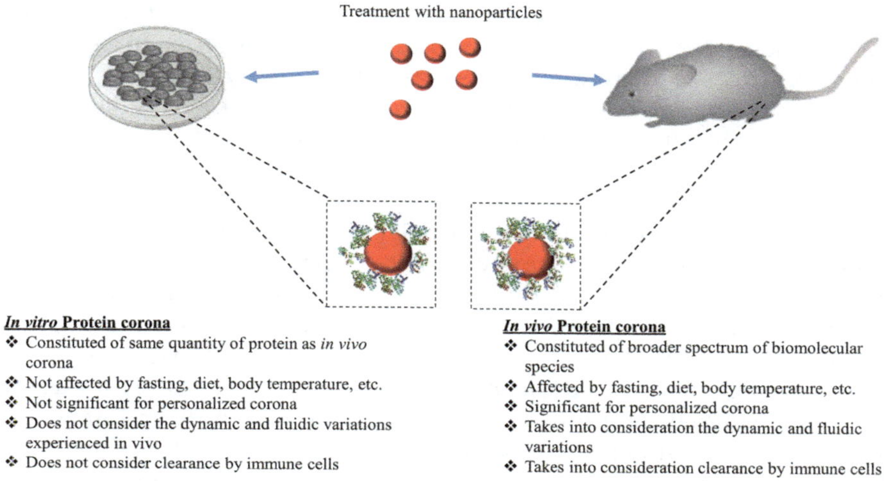

Figure 10.4 *In vitro* and *in vivo* protein coronas determine the behavior and fate of nanoparticles (NPs). Adapted from ref. 67 with permission from Elsevier, Copyright 2017.

in vitro context. Figure 10.4 indicates the differences observed between *in vitro* and *in vivo* protein coronas. The major hindrance faced in the advancement of this field is the inadequacy of sophisticated instrumentation that enable the separation of NPs from the physiological environment following their administration.[33,67] In this regard, creating 3D-printed tissues/organs and virtual model systems may help to unravel the *in vivo* protein corona.[68] Sakulkhu *et al.* (2014) used polyvinyl-alcohol-conjugated SPIONs to overcome this hindrance; hence, the SPIONs were administered to rats, and the protein-coronated NPs were separated using a strong external magnetic field by employing the inherent magnetic properties of SPIONs. The study was noteworthy as it could differentiate the protein corona profiles *in vitro* and *in vivo*, and indicated a 50% similarity for positive and neutral NPs, but only an 8% similarity for negative NPs (*i.e.*, similarity between *in vivo* and *in vitro* protein coronation).[69] Another study by Chen *et al.* (2017) suggested that dextran-coated SPION core–shell nanoworms adsorbed with human serum proteins could bind complement C3 to the adsorbed corona, which can undergo dynamic exchange; however, the association of C3 and other proteins was drastically reduced *in vivo*.[70] Stepien *et al.* (2018) also used 12 nm iron oxide NPs coated with glucose or PEG to study NP coronation, *in vitro* degradation, *in vivo* tracking, and biotransformation over four months. The study suggested that protein adsorption was governed by *in vivo* biodistribution and that PEG-coated NPs faced faster macrophage degradation.[71]

Most of the *ex vivo* perfusion studies (other than iron oxide NPs) require radio-labeled NPs that can be analyzed by scintillation counting of the organ samples. For example, lecithin-coated polystyrene NPs,[32] PMMA NPs,[29] and liposomes[31] preincubated with serum showed reduced liver uptake possibly due to the involvement of dysopsonins; however, detailed protein coronations are difficult to analyze in this case, but their biodistribution can be studied. Bertrand *et al.* (2017) studied the protein coronation of a library of radio-labeled polymer nanoparticles under *in vivo* conditions using size-exclusion chromatography and membrane ultrafiltration of plasma collected from NP-administered mice. The study indicated that NPs with higher PEG densities showed lower apoE and clusterin adsorption and that these dysopsonins could shield NPs with inadequate PEG coverage against opsonization.[72] More recently, García-Álvarez *et al.* (2018) studied the protein coronation of Au nanorods and nanostars using similar techniques. The study suggested that larger NPs showed the adsorption of more diverse proteins at higher protein abundance, highlighting the importance of NP size and shape.[73] A recent study by Müller *et al.* (2018) suggested that the translation of protein corona formation and NP fate observed will vary across different animal models to humans based on variations observed in different sera sources. The study implied that though protein corona studies have now been assessed in animal models, it must be considered that their profile and implications will vary in humans.[74] Thus, the precise interpretation of the NPs' fate requires pharmacological and imaging agents or advanced separation techniques that allow non-invasive characterization of protein coronation under realistic *in vivo* conditions.

10.4.2 Patient-specific Protein Corona

Recent studies have revealed that the protein corona is also affected by the specific disease of a patient, *i.e.*, different pathologies lead to the association of varied protein corona compositions. This gave rise to the conception of a personalized protein corona, which suggests that even with humans, the corona formation differs based on age, gender, habits, diseases, pregnancy, temporary drug consumption, and geographical origin (Figure 10.5).[75] When the effect of pathological conditions on human serum used as the protein source was studied, the type of disease played a crucial role on the tailored protein corona, which elicited specific cellular responses.[76] The relationship between the pathological state and NP bio-identity based on differential protein corona association could be studied. The specific findings will assist the advancement of personalized patient-specific nanomedicine and nanotherapeutic treatments that are disease-type-specific for efficient preclinical applications.[77] To sum up, these realistic *in vivo* and patient-specific applications can be realized only by overcoming the hurdles currently faced in protein corona studies.

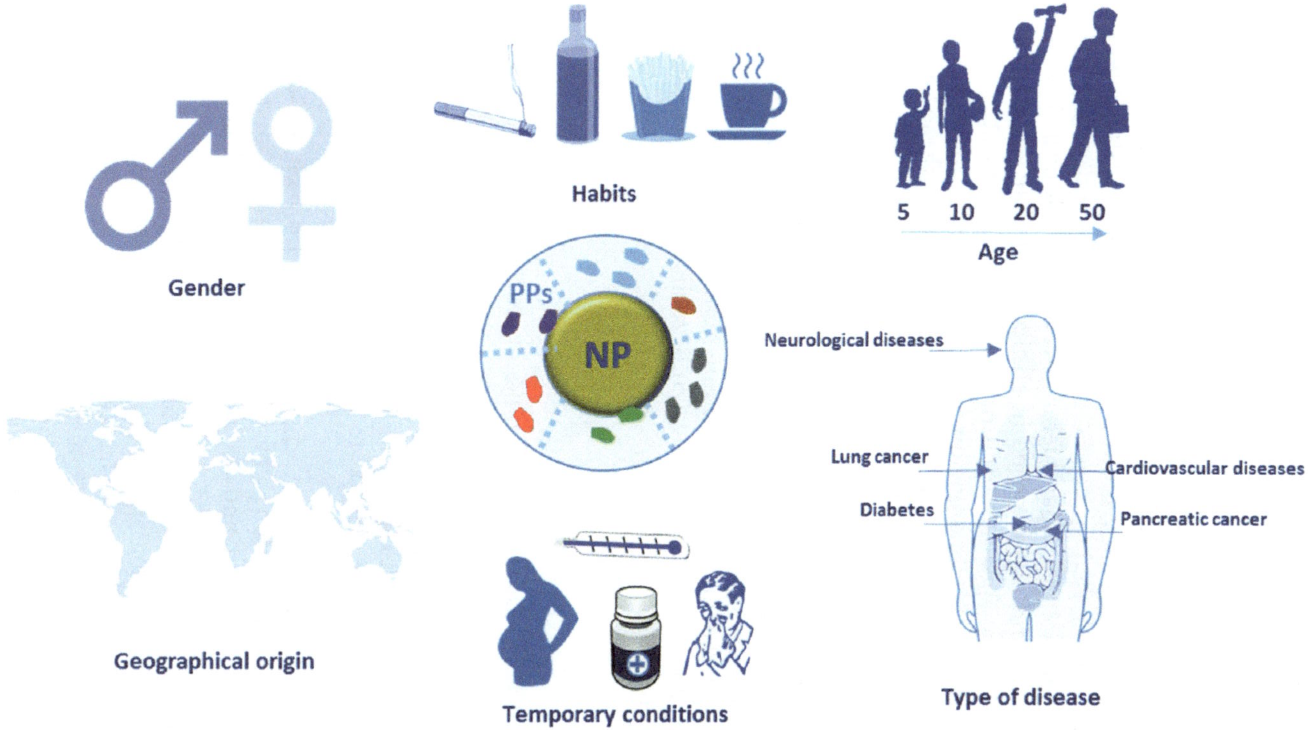

Figure 10.5 Personalized protein corona, general concept. Gender, habits, age, diseases, temporary drug consumption, pregnancy, and geographical origin all potentially affect the protein plasma composition. This, in turn, affects the protein corona composition. As a consequence, this information has to be provided when working with human plasma, not only to better interpret the results obtained but also to give useful information for the design of safe and efficient nanoparticles in a disease-specific manner. NP: nanoparticle, PPs: plasma proteins. Reproduced from ref. 75 with permission from the Royal Society of Chemistry.

10.5 Conclusion

Despite the numerous and widely commercialized non-biomedical applications of nanomaterials, their use in the biomedical field has always been looked at with skepticism, resulting in less than 30 FDA-approved nanotherapeutics. The majority of the *in vitro* and *in vivo* trials do not reach the clinics due to a conflict in the understanding that designed nanoparticles could be altered by the complex biological system presented and the formation of a protein corona. Based on the studies discussed, protein-corona-based applications are complex and very poorly understood. Several biological parameters need to be addressed to get a realistic view of the problems posed in clinical studies (*e.g.*, temperature variations, disease-specific variations in proteins presented, protein corona behavior and cellular vision of NPs *in vivo*, *etc.*), which would require detailed investigations in future studies. Addressing these knowledge gaps can revolutionize the field of modern medicine and the drug industry. Rather than using stealth techniques that repel protein corona formation, their unique properties might as well be harnessed by incorporating selective protein coronation in the design of the nanotherapeutics based on comprehensive studies to understand the protein corona composition and nanomaterial fate. A combination of chemistry, engineering, and computational modeling of the protein–nanomaterial interactions could become a game-changer in the design of realistic nanotherapeutic systems in the future.

References

1. I. Lynch and K. A. Dawson, Protein-nanoparticle interactions, *Nano Today*, 2008, **3**, 40–47.
2. M. Rahman, S. Laurent, N. Tawil, L. Yahia and M. Mahmoudi, *Protein-nanoparticle Interactions*, Springer, 2013.
3. H. H. Gustafson, D. Holt-Casper, D. W. Grainger and H. Ghandehari, Nanoparticle uptake: the phagocyte problem, *Nano Today*, 2015, **10**, 487–510.
4. M. P. Monopoli, D. Walczyk, A. Campbell, G. Elia, I. Lynch, F. Baldelli Bombelli and K. A. Dawson, Physical–chemical aspects of protein corona: relevance to *in vitro* and *in vivo* biological impacts of nanoparticles, *J. Am. Chem. Soc.*, 2011, **133**, 2525–2534.
5. P. L. Rodriguez, T. Harada, D. A. Christian, D. A. Pantano, R. K. Tsai and D. E. Discher, Minimal "self" peptides that inhibit phagocytic clearance and enhance delivery of nanoparticles, *Science*, 2013, **339**, 971–975.
6. M. Grunér, U. Kauscher, M. Linder and M. Monopoli, An environmental route of exposure affects the formation of nanoparticle coronas in blood plasma, *J. Proteomics*, 2016, **137**, 52–58.
7. S. M. Moghimi, A. C. Hunter and J. C. Murray, Long-circulating and target-specific nanoparticles: theory to practice, *Pharmacol. Rev.*, 2001, **53**, 283–318.

8. S. M. Moghimi and J. Szebeni, Stealth liposomes and long circulating nanoparticles: critical issues in pharmacokinetics, opsonization and protein-binding properties, *Prog. Lipid Res.*, 2003, **42**, 463–478.
9. V. P. Torchilin and V. S. Trubetskoy, Which polymers can make nanoparticulate drug carriers long-circulating?, *Adv. Drug Delivery Rev.*, 1995, **16**, 141–155.
10. R. Gref, M. Lück, P. Quellec, M. Marchand, E. Dellacherie, S. Harnisch, T. Blunk and R. Müller, 'Stealth'corona-core nanoparticles surface modified by polyethylene glycol (PEG): influences of the corona (PEG chain length and surface density) and of the core composition on phagocytic uptake and plasma protein adsorption, *Colloids Surf., B*, 2000, **18**, 301–313.
11. H. R. Kim, K. Andrieux, C. Delomenie, H. Chacun, M. Appel, D. Desmaële, F. Taran, D. Georgin, P. Couvreur and M. Taverna, Analysis of plasma protein adsorption onto PEGylated nanoparticles by complementary methods: 2-DE, CE and Protein Lab-on-chip® system, *Electrophoresis*, 2007, **28**, 2252–2261.
12. M. Price, R. Cornelius and J. Brash, Protein adsorption to polyethylene glycol modified liposomes from fibrinogen solution and from plasma, *Biochim. Biophys. Acta, Biomembr.*, 2001, **1512**, 191–205.
13. Y. Arima, M. Kawagoe, M. Toda and H. Iwata, Complement activation by polymers carrying hydroxyl groups, *ACS Appl. Mater. Interfaces*, 2009, **1**, 2400–2407.
14. H. Arami, A. Khandhar, D. Liggitt and K. M. Krishnan, *In vivo* delivery, pharmacokinetics, biodistribution and toxicity of iron oxide nanoparticles, *Chem. Soc. Rev.*, 2015, **44**, 8576–8607.
15. L. Lei, J. Ling-Ling, Z. Yun and L. Gang, Toxicity of superparamagnetic iron oxide nanoparticles: research strategies and implications for nanomedicine, *Chin. Phys. B*, 2013, **22**, 127503.
16. M. Mahmoudi, S. Laurent, M. A. Shokrgozar and M. Hosseinkhani, Toxicity evaluations of superparamagnetic iron oxide nanoparticles: cell "vision" versus physicochemical properties of nanoparticles, *ACS Nano*, 2011, **5**, 7263–7276.
17. D. F. Moyano, K. Saha, G. Prakash, B. Yan, H. Kong, M. Yazdani and V. M. Rotello, Fabrication of corona-free nanoparticles with tunable hydrophobicity, *ACS Nano*, 2014, **8**, 6748–6755.
18. K. P. Garcia, K. Zarschler, L. Barbaro, J. A. Barreto, W. O'Malley, L. Spiccia, H. Stephan and B. Graham, Zwitterionic-coated "stealth" nanoparticles for biomedical applications: recent advances in countering biomolecular corona formation and uptake by the mononuclear phagocyte system, *Small*, 2014, **10**, 2516–2529.
19. A. Chonn, S. Semple and P. Cullis, Association of blood proteins with large unilamellar liposomes *in vivo*. Relation to circulation lifetimes, *J. Biol. Chem.*, 1992, **267**, 18759–18765.
20. M. Mahmoudi, S. N. Saeedi-Eslami, M. A. Shokrgozar, K. Azadmanesh, M. Hassanlou, H. R. Kalhor, C. Burtea, B. Rothen-Rutishauser, S. Laurent and S. Sheibani, Cell "vision": complementary factor of protein corona in nanotoxicology, *Nanoscale*, 2012, **4**, 5461–5468.

21. P. Aggarwal, J. B. Hall, C. B. McLeland, M. A. Dobrovolskaia and S. E. McNeil, Nanoparticle interaction with plasma proteins as it relates to particle biodistribution, biocompatibility and therapeutic efficacy, *Adv Drug Delivery Rev.*, 2009, **61**, 428–437.
22. P. P. Karmali and D. Simberg, Interactions of nanoparticles with plasma proteins: implication on clearance and toxicity of drug delivery systems, *Expert Opin. Drug Delivery*, 2011, **8**, 343–357.
23. M. Safi, J. Courtois, M. Seigneuret, H. Conjeaud and J.-F. Berret, The effects of aggregation and protein corona on the cellular internalization of iron oxide nanoparticles, *Biomaterials*, 2011, **32**, 9353–9363.
24. A. Moore, R. Weissleder and A. Bogdanov, Uptake of dextran-coated monocrystalline iron oxides in tumor cells and macrophages, *J. Magn. Reson. Imaging*, 1997, **7**, 1140–1145.
25. A. Chonn, S. C. Semple and P. R. Cullis, β2-Glycoprotein I is a major protein associated with very rapidly cleared liposomes *in vivo*, suggesting a significant role in the immune clearance of "non-self" particles, *J. Biol. Chem.*, 1995, **270**, 25845–25849.
26. X. Yan, F. Kuipers, L. M. Havekes, R. Havinga, B. Dontje, K. Poelstra, G. L. Scherphof and J. A. Kamps, The role of apolipoprotein E in the elimination of liposomes from blood by hepatocytes in the mouse, *Biochem. Biophys. Res. Commun.*, 2005, **328**, 57–62.
27. H. Schreier, R. M. Abra, J. E. Kaplan and C. A. Hunt, Murine plasma fibronectin depletion after intravenous injection of liposomes, *Int. J. Pharm.*, 1987, **37**, 233–238.
28. Y. Hong, P. J. Shaw, B. N. Tattam, C. E. Nath, J. W. Earl, K. R. Stephen and A. J. McLachlan, Plasma protein distribution and its impact on pharmacokinetics of liposomal amphotericin B in paediatric patients with malignant diseases, *Eur. J. Clin. Pharmacol.*, 2007, **63**, 165–172.
29. G. Borchard and J. Kreuter, The role of serum complement on the organ distribution of intravenously administered poly (methyl methacrylate) nanoparticles: effects of pre-coating with plasma and with serum complement, *Pharm. Res.*, 1996, **13**, 1055–1058.
30. L. Thiele, J. E. Diederichs, R. Reszka, H. P. Merkle and E. Walter, Competitive adsorption of serum proteins at microparticles affects phagocytosis by dendritic cells, *Biomaterials*, 2003, **24**, 1409–1418.
31. H. Harashima, H. Matsuo and H. Kiwada, Identification of proteins mediating clearance of liposomes using a liver perfusion system, *Adv. Drug Delivery Rev.*, 1998, **32**, 61–79.
32. K. Furumoto, K.-i. Ogawara, S. Nagayama, Y. Takakura, M. Hashida, K. Higaki and T. Kimura, Important role of serum proteins associated on the surface of particles in their hepatic disposition, *J. Controlled Release*, 2002, **83**, 89–96.
33. S. Zanganeh, R. Spitler, M. Erfanzadeh, A. M. Alkilany and M. Mahmoudi, Protein corona: Opportunities and challenges, *Int. J. Biochem. Cell Biol.*, 2016, **75**, 143–147.

34. A. Salvati, A. S. Pitek, M. P. Monopoli, K. Prapainop, F. B. Bombelli, D. R. Hristov, P. M. Kelly, C. Åberg, E. Mahon and K. A. Dawson, Transferrin-functionalized nanoparticles lose their targeting capabilities when a biomolecule corona adsorbs on the surface, *Nat. Nanotechnol.*, 2013, **8**, 137.
35. V. Mirshafiee, M. Mahmoudi, K. Lou, J. Cheng and M. L. Kraft, Protein corona significantly reduces active targeting yield, *Chem. Commun.*, 2013, **49**, 2557–2559.
36. A. K. Murthy, R. J. Stover, W. G. Hardin, R. Schramm, G. D. Nie, S. Gourisankar, T. M. Truskett, K. V. Sokolov and K. P. Johnston, Charged gold nanoparticles with essentially zero serum protein adsorption in undiluted fetal bovine serum, *J. Am. Chem. Soc.*, 2013, **135**, 7799–7802.
37. Q. Dai, C. Walkey and W. C. Chan, Polyethylene glycol backfilling mitigates the negative impact of the protein corona on nanoparticle cell targeting, *Angew. Chem., Int. Ed.*, 2014, **53**, 5093–5096.
38. E. Blanco, H. Shen and M. Ferrari, Principles of nanoparticle design for overcoming biological barriers to drug delivery, *Nat. Biotechnol.*, 2015, **33**, 941.
39. A. D'Hollander, H. Jans, G. V. Velde, C. Verstraete, S. Massa, N. Devoogdt, T. Stakenborg, S. Muyldermans, L. Lagae and U. Himmelreich, Limiting the protein corona: a successful strategy for *in vivo* active targeting of anti-HER2 nanobody-functionalized nanostars, *Biomaterials*, 2017, **123**, 15–23.
40. V. Mirshafiee, R. Kim, S. Park, M. Mahmoudi and M. L. Kraft, Impact of protein pre-coating on the protein corona composition and nanoparticle cellular uptake, *Biomaterials*, 2016, **75**, 295–304.
41. E. Mahon, D. R. Hristov and K. A. Dawson, Stabilising fluorescent silica nanoparticles against dissolution effects for biological studies, *Chem. Commun.*, 2012, **48**, 7970–7972.
42. C. D. Walkey and W. C. Chan, Understanding and controlling the interaction of nanomaterials with proteins in a physiological environment, *Chem. Soc. Rev.*, 2012, **41**, 2780–2799.
43. J. C. Y. Kah, J. Chen, A. Zubieta and K. Hamad-Schifferli, Exploiting the protein corona around gold nanorods for loading and triggered release, *ACS Nano*, 2012, **6**, 6730–6740.
44. A. Cifuentes-Rius, H. de Puig, J. C. Y. Kah, S. Borros and K. Hamad-Schifferli, Optimizing the properties of the protein corona surrounding nanoparticles for tuning payload release, *ACS Nano*, 2013, **7**, 10066–10074.
45. S. Behzadi, V. Serpooshan, R. Sakhtianchi, B. Müller, K. Landfester, D. Crespy and M. Mahmoudi, Protein corona change the drug release profile of nanocarriers: the "overlooked" factor at the nanobio interface, *Colloids Surf., B*, 2014, **123**, 143–149.
46. A. Lesniak, F. Fenaroli, M. P. Monopoli, C. Åberg, K. A. Dawson and A. Salvati, Effects of the presence or absence of a protein corona on silica nanoparticle uptake and impact on cells, *ACS Nano*, 2012, **6**, 5845–5857.

47. S. Wan, P. M. Kelly, E. Mahon, H. Stöckmann, P. M. Rudd, F. Caruso, K. A. Dawson, Y. Yan and M. P. Monopoli, The "sweet" side of the protein corona: effects of glycosylation on nanoparticle–cell interactions, *ACS Nano*, 2015, **9**, 2157–2166.
48. X. Jiang, S. Weise, M. Hafner, C. Röcker, F. Zhang, W. J. Parak and G. U. Nienhaus, Quantitative analysis of the protein corona on FePt nanoparticles formed by transferrin binding, *J. R. Soc., Interface*, 2009, 7, S5–S13.
49. K. Saha, D. F. Moyano and V. M. Rotello, Protein coronas suppress the hemolytic activity of hydrophilic and hydrophobic nanoparticles, *Mater. Horiz.*, 2014, **1**, 102–105.
50. D. S. T. Martinez, A. J. Paula, L. C. Fonseca, L. A. V. Luna, C. P. Silveira, N. Durán and O. L. Alves, Monitoring the hemolytic effect of mesoporous silica nanoparticles after human blood protein corona formation, *Eur. J. Inorg. Chem.*, 2015, **2015**, 4595–4602.
51. O. V. Salata, Applications of nanoparticles in biology and medicine, *J. Nanobiotechnol.*, 2004, **2**, 3.
52. W. Hu, C. Peng, M. Lv, X. Li, Y. Zhang, N. Chen, C. Fan and Q. Huang, Protein corona-mediated mitigation of cytotoxicity of graphene oxide, *ACS Nano*, 2011, **5**, 3693–3700.
53. S. A. Alex, N. Chandrasekaran and A. Mukherjee, Impact of gold nanorod functionalization on biocorona formation and their biological implication, *J. Mol. Liq.*, 2017, **248**, 703–712.
54. C. Ge, J. Du, L. Zhao, L. Wang, Y. Liu, D. Li, Y. Yang, R. Zhou, Y. Zhao and Z. Chai, Binding of blood proteins to carbon nanotubes reduces cytotoxicity, *Proc. Natl. Acad. Sci. U. S. A.*, 2011, **108**, 16968–16973.
55. G. Maiorano, S. Sabella, B. Sorce, V. Brunetti, M. A. Malvindi, R. Cingolani and P. P. Pompa, Effects of cell culture media on the dynamic formation of protein–nanoparticle complexes and influence on the cellular response, *ACS Nano*, 2010, **4**, 7481–7491.
56. D. Docter, C. Bantz, D. Westmeier, H. J. Galla, Q. Wang, J. C. Kirkpatrick, P. Nielsen, M. Maskos and R. H. Stauber, The protein corona protects against size- and dose-dependent toxicity of amorphous silica nanoparticles, *Beilstein J. Nanotechnol.*, 2014, **5**, 1380.
57. M. Mahmoudi, M. P. Monopoli, M. Rezaei, I. Lynch, F. Bertoli, J. J. McManus and K. A. Dawson, The protein corona mediates the impact of nanomaterials and slows amyloid beta fibrillation, *ChemBioChem*, 2013, **14**, 568–572.
58. I. Y. Podolski, Z. Podlubnaya, E. Kosenko, E. Mugantseva, E. Makarova, L. Marsagishvili, M. Shpagina, Y. G. Kaminsky, G. Andrievsky and V. Klochkov, Effects of hydrated forms of C60 fullerene on amyloid β-peptide fibrillization *in vitro* and performance of the cognitive task, *J. Nanosci. Nanotechnol.*, 2007, **7**, 1479–1485.
59. S. Laurent, M. R. Ejtehadi, M. Rezaei, P. G. Kehoe and M. Mahmoudi, Interdisciplinary challenges and promising theranostic effects of nanoscience in Alzheimer's disease, *RSC Adv.*, 2012, **2**, 5008–5033.

60. K. Yokoyama and D. Welchons, The conjugation of amyloid beta protein on the gold colloidal nanoparticles' surfaces, *Nanotechnology*, 2007, **18**, 105101.
61. Y. D. Álvarez, J. V. Pellegrotti and F. D. Stefani, *Use of Nanoparticles in Neuroscience*, Springer, 2018, pp. 269–291.
62. S. Mirsadeghi, R. Dinarvand, M. H. Ghahremani, M. R. Hormozi-Nezhad, Z. Mahmoudi, M. J. Hajipour, F. Atyabi, M. Ghavami and M. Mahmoudi, Protein corona composition of gold nanoparticles/nanorods affects amyloid beta fibrillation process, *Nanoscale*, 2015, **7**, 5004–5013.
63. A. Lotfabadi, M. J. Hajipour, H. Derakhshankhah, A. Peirovi, S. Saffar, E. Shams, E. Fatemi, E. Barzegari, S. Sarvari, F. Moakedi, M. Ferdousi, F. Atyabi, A. A. Saboury and R. Dinarvand, Biomolecular Corona Dictates Aβ Fibrillation Process, *ACS Chem. Neurosci.*, 2018, 1725–1734.
64. H. Amiri, L. Bordonali, A. Lascialfari, S. Wan, M. P. Monopoli, I. Lynch, S. Laurent and M. Mahmoudi, Protein corona affects the relaxivity and MRI contrast efficiency of magnetic nanoparticles, *Nanoscale*, 2013, **5**, 8656–8665.
65. E. Mahon, A. Salvati, F. B. Bombelli, I. Lynch and K. A. Dawson, Designing the nanoparticle–biomolecule interface for "targeting and therapeutic delivery", *J. Controlled Release*, 2012, **161**, 164–174.
66. K. Hamad-Schifferli, How can we exploit the protein corona?, *Nanomedicine*, 2013, **8**, 1–3.
67. G. Caracciolo, O. C. Farokhzad and M. Mahmoudi, Biological identity of nanoparticles in vivo: clinical implications of the protein corona, *Trends Biotechnol.*, 2017, **35**, 257–264.
68. P. C. Ke, S. Lin, W. J. Parak, T. P. Davis and F. Caruso, A Decade of the Protein Corona, *ACS Nano*, 2017, **11**, 11773–11776.
69. U. Sakulkhu, L. Maurizi, M. Mahmoudi, M. Motazacker, M. Vries, A. Gramoun, M.-G. O. Beuzelin, J.-P. Vallée, F. Rezaee and H. Hofmann, *Ex situ* evaluation of the composition of protein corona of intravenously injected superparamagnetic nanoparticles in rats, *Nanoscale*, 2014, **6**, 11439–11450.
70. F. Chen, G. Wang, J. I. Griffin, B. Brenneman, N. K. Banda, V. M. Holers, D. S. Backos, L. Wu, S. M. Moghimi and D. Simberg, Complement proteins bind to nanoparticle protein corona and undergo dynamic exchange in vivo, *Nat. Nanotechnol.*, 2017, **12**, 387.
71. G. Stepien, M. Moros, M. Perez-Hernandez, M. Monge, L. Gutiérrez, R. M. Fratila, M. Las Heras, S. Menao Guillen, J. J. Puente Lanzarote, C. Solans, J. Pardo and J. M. de la Fuente, Effect of surface chemistry and associated protein corona on the long-term biodegradation of iron oxide nanoparticles *in vivo*, *ACS Appl. Mater. Interfaces*, 2018, 4548–4560.
72. N. Bertrand, P. Grenier, M. Mahmoudi, E. M. Lima, E. A. Appel, F. Dormont, J.-M. Lim, R. Karnik, R. Langer and O. C. Farokhzad, Mechanistic understanding of *in vivo* protein corona formation on polymeric nanoparticles and impact on pharmacokinetics, *Nat. Commun.*, 2017, **8**, 777.

73. R. García-Álvarez, M. Hadjidemetriou, A. Sánchez-Iglesias, L. M. Liz-Marzán and K. Kostarelos, *In vivo* formation of protein corona on gold nanoparticles. The effect of their size and shape, *Nanoscale*, 2018, 1256–1264.
74. L. K. Müller, J. Simon, C. Rosenauer, V. Mailänder, S. Morsbach and K. Landfester, The transferability from animal models to humans: challenges regarding aggregation and protein corona formation of nanoparticles, *Biomacromolecules*, 2018, **19**, 374–385.
75. C. Corbo, R. Molinaro, M. Tabatabaei, O. C. Farokhzad and M. Mahmoudi, Personalized protein corona on nanoparticles and its clinical implications, *Biomater. Sci.*, 2017, **5**, 378–387.
76. G. Caracciolo, D. Caputo, D. Pozzi, V. Colapicchioni and R. Coppola, Size and charge of nanoparticles following incubation with human plasma of healthy and pancreatic cancer patients, *Colloids Surf., B*, 2014, **123**, 673–678.
77. M. J. Hajipour, J. Raheb, O. Akhavan, S. Arjmand, O. Mashinchian, M. Rahman, M. Abdolahad, V. Serpooshan, S. Laurent and M. Mahmoudi, Personalized disease-specific protein corona influences the therapeutic impact of graphene oxide, *Nanoscale*, 2015, **7**, 8978–8994.

Subject Index